Environmental Sustainability and Economy

Environmental Sustainability and Economy

Edited by

Pardeep Singh

Pramit Verma

Daniela Perrotti

K.K. Srivastava

ELSEVIER

Elsevier
Radarweg 29, PO Box 211, 1000 AE Amsterdam, Netherlands
The Boulevard, Langford Lane, Kidlington, Oxford OX5 1GB, United Kingdom
50 Hampshire Street, 5th Floor, Cambridge, MA 02139, United States

Notices
Knowledge and best practice in this field are constantly changing. As new research and experience broaden our understanding, changes in research methods, professional practices, or medical treatment may become necessary.

Practitioners and researchers must always rely on their own experience and knowledge in evaluating and using any information, methods, compounds, or experiments described herein. In using such information or methods they should be mindful of their own safety and the safety of others, including parties for whom they have a professional responsibility.

To the fullest extent of the law, neither the Publisher nor the authors, contributors, or editors, assume any liability for any injury and/or damage to persons or property as a matter of products liability, negligence or otherwise, or from any use or operation of any methods, products, instructions, or ideas contained in the material herein.

Library of Congress Cataloging-in-Publication Data
A catalog record for this book is available from the Library of Congress

British Library Cataloguing-in-Publication Data
A catalogue record for this book is available from the British Library

ISBN: 978-0-12-822188-4

For information on all Elsevier publications visit our website at
https://www.elsevier.com/books-and-journals

Publisher: Candice Janco
Acquisitions Editor: Marisa LaFleur
Editorial Project Manager: Alice Grant
Production Project Manager: Sujatha Thirugnana Sambandam
Cover Designer: Greg Harris

Typeset by TNQ Technologies

Working together
to grow libraries in
developing countries

www.elsevier.com • www.bookaid.org

Contents

THEME 3 Circular economy and urban metabolism

CHAPTER 11 Earth, wood, and coffee: empirical evidence on value creation in the circular economy **197**

Stephan Kampelmann, Emmanuel Raufflet and Giulia Scialpi

CHAPTER 12 A story of resilience and local materials: sourcing bio-based materials in Norman wetlands, France **219**

Giulia Scialpi

CHAPTER 13   Consequential life cycle assessment to promote the recycling of metallurgic slag as new construction material 247
Andrea Di Maria and Karel Van Acker

THEME 4   Market and sustainability

CHAPTER 14   Material and energy services, human needs, and well-being 275
Kai Whiting, Luis Gabriel Carmona and Angeles Carrasco

CHAPTER 17   Challenges and opportunities at the crossroads of
Environmental Sustainability and Economy research 345
Daniela Perrotti, Pramit Verma, K.K. Srivastava and Pardeep Singh

Contributors

Alex O. Acheampong
Newcastle Business School, The University of Newcastle, Newcastle, NSW, Australia

Dan Banik
Centre for Development and the Environment, University of Oslo, Oslo, Norway

André C.S. Batalhão
School of Economics, Business Administration and Accounting at Ribeirão Preto, University of São Paulo, Ribeirão Preto, Brazil; Center for Environmental and Sustainability Research (CENSE), Nova Lisbon University, Caparica, Portugal; Global Organization Learning and Developing Network (GOLDEN) — Brazilian Chapter, University of São Paulo, São Paulo, Brazil

Adriana C.F. Caldana
School of Economics, Business Administration and Accounting at Ribeirão Preto, University of São Paulo, Ribeirão Preto, Brazil; Global Organization Learning and Developing Network (GOLDEN) — Brazilian Chapter, University of São Paulo, São Paulo, Brazil

Luis Gabriel Carmona
MARETEC—LARSyS, Instituto Superior Técnico, Universidade de Lisboa, Lisboa, Portugal; Faculty of Environmental Sciences, Universidad Piloto de Colombia, Bogotá, Colombia; Institute ForWARD (For Worldwide Alternative Research and Development), Bogotá, Colombia

Angeles Carrasco
Mining and Industrial Engineering School of Almadén, Universidad de Castilla—La Mancha, Almadén, Spain

Alexandre R. Choupina
School of Agronomy, Federal University of Jataí, Jataí, Brazil

Soumyendra Kishore Datta
Department of Economics, The University of Burdwan, Burdwan, West Bengal, India

Tanushree De
Department of Economics, Vivekananda Mahavidyalaya, Burdwan, West Bengal, India

Andrea Di Maria
Katholieke Universiteit Leuven (KU Leuven), Leuven, Belgium

Janet Dzator
Newcastle Business School, The University of Newcastle, Newcastle, NSW, Australia; Australia Africa Universities Network (AAUN) Partner, Newcastle, NSW, Australia

Michael Dzator
SAE, Central Queensland University, Mackay, QLD, Australia; Australia Africa Universities Network (AAUN) Partner, Newcastle, NSW, Australia

João H.P.P. Eustachio
School of Economics, Business Administration and Accounting at Ribeirão Preto, University of São Paulo, Ribeirão Preto, Brazil; Global Organization Learning and Developing Network (GOLDEN) — Brazilian Chapter, University of São Paulo, São Paulo, Brazil

Elsa Garavaglia
Department of Civil and Environmental Engineering, Politecnico di Milano, Milan, Italy

Stephan Kampelmann
Laboratory for Landscape, Urbanism, Infrastructure and Ecology, Faculty of Architecture La Cambre-Horta, Université Libre de Bruxelles, Bruxelles, Belgium

Deepika Kandpal
Department of Economics, University of Delhi, New Delhi, Delhi, India

Vaishali Kapoor
Deen Dayal Upadhyaya College, University of Delhi, New Delhi, Delhi, India

Negarsadat Madani
Local Environment Management & Analysis (LEMA), Department UEE, Université de Liège, Liège, Belgium

Medha Malviya
Independent Researcher in Environmental Economics and Economics of Climate Change

Matan Mayer
School of Architecture and Design, IE University, Segovia, Spain

Behzad Bamdad Mehrabani
Faculty of Architecture, Architectural Engineering and Urban Planning, Université Catholique de Louvain, Tournai, Belgium

Vishal Mishra
School of Biochemical Engineering, IIT (BHU), Varanasi, Uttar Pradesh, India

Daniela Perrotti
Research Institute for Landscape, Architecture and Built Environment, University of Louvain UCLouvain, Ottignies-Louvain-la-Neuve, Belgium

Emmanuel Raufflet
Département de Management, HEC Montréal, Montréal, QC, Canada

David A. Savage
Newcastle Business School, The University of Newcastle, Newcastle, NSW, Australia

Giulia Scialpi
Faculty of Architecture, Architectural Engineering and Urban Planning, University of Louvain UCLouvain, Ottignies-Louvain-la-Neuve, Belgium

Luca Sgambi
Faculty of Architecture, Architectural Engineering and Urban Planning, Université Catholique de Louvain, Tournai, Belgium

Shikha Singh
Integrative Ecology Laboratory (IEL), Institute of Environment & Sustainable Development (IESD), Banaras Hindu University, Varanasi, Uttar Pradesh, India

Jyoti Singh
School of Biochemical Engineering, IIT (BHU), Varanasi, Uttar Pradesh, India

Pardeep Singh
Department of Environmental Studies, PGDAV College, University of Delhi, New Delhi, India

Dhirendra Kumar Srivastava
Council of Science and Technology, Lucknow, Uttar Pradesh, India

K.K. Srivastava
PGDAV College, University of Delhi, New Delhi, India

Karel Van Acker
Katholieke Universiteit Leuven (KU Leuven), Leuven, Belgium; Center for Economics and Corporate Sustainability (CEDON), Katholieke Universiteit Leuven (KU) Leuven, Brussels, Belgium

Saumya Verma
Lady Shri Ram College, University of Delhi, New Delhi, Delhi, India

Pramit Verma
Integrative Ecology Laboratory (IEL), Institute of Environment & Sustainable Development (IESD), Banaras Hindu University Varanasi, U.P., India

Kai Whiting
Faculty of Architecture, Architectural Engineering and Urban Planning, Université Catholique de Louvain, Louvain-la-Neuve, Belgium

Priyanka Yadav
School of Biochemical Engineering, IIT (BHU), Varanasi, Uttar Pradesh, India

Abhinav Yadav
Integrative Ecology Laboratory (IEL), Institute of Environment & Sustainable Development (IESD), Banaras Hindu University, Varanasi, Uttar Pradesh, India

Environment and economy

The impact of transport infrastructure development on carbon emissions in OECD countries

Janet Dzator[1,3], Alex O. Acheampong[1], Michael Dzator[2,3]

[1]*Newcastle Business School, The University of Newcastle, Newcastle, NSW, Australia;* [2]*SAE, Central Queensland University, Mackay, QLD, Australia;* [3]*Australia Africa Universities Network (AAUN) Partner, Newcastle, NSW, Australia*

Chapter outline

1. Introduction

Research on transport infrastructure development and economic growth has been extensively investigated in the literature (Saidi and Hammami, 2017; Saidi et al., 2018). While several studies indicate that transport infrastructure contributes to economic growth, it could equally impact on the environment. Recently, research on transport infrastructure development and the environment has become a critical topic, both at the national and international level. The International Energy Agency (IEA) reports indicate that the transport sector in the Organisation for Economic Co-operation and Development (OECD) countries consume approximately 55% of the world's total energy in 2012 (Energy Information Administration, 2016). Additionally, IEA (2019) reports suggest that 24% of direct carbon (CO_2) emissions emanate from fuel consumption in the transport sector. The reports further revealed that trucks, cars, and buses, which are road vehicles, are responsible for approximately three-quarters

Environmental Sustainability and Economy. https://doi.org/10.1016/B978-0-12-822188-4.00006-3

of transport CO_2 emissions while the shipping and aviation sector's contribution to CO_2 emissions continues to soar. While transportation contributes substantially to global carbon emissions, empirical studies investigating the environment effect of the transport infrastructure development remain limited.

To inform sustainable environmental practice and policies to control climate change, this chapter seeks to probe the impact of transportation infrastructure development on the emissions of CO_2 for 26 OECD countries for the years between 1960 and 2018. This study concentrates on OECD countries because they play a significant role in the increase in global CO_2 emissions (Dzator and Acheampong, 2020; Marchal et al., 2011). Also, these countries are major actors in achieving the Paris Agreement on climate change and further play a major role in simulating global economic growth; therefore, concentrating on the OECD countries will enable policymakers to understand the role of transport infrastructures in CO_2 emissions. Our study adds to the literature in the following ways. To the best of the authors' knowledge, this is the first study to empirically probe the impact of transportation infrastructure development on CO_2 emissions in OECD countries. Furthermore, this research extends the prior literature by examining how transportation infrastructure development moderates the impact of GDP (economic growth) and energy use on CO_2 emissions. Also, this study utilized varieties of econometric estimation techniques such as random and fixed effect models (static models) and system-generalized method of moment (dynamic model) to check the consistency in our results. Finally, the findings that will emanate from this study will inform sustainable development and climate change policies in the OECD countries.

Our findings revealed that air transport infrastructure significantly increases carbon emissions while rail transport infrastructure does not directly influence CO_2 emissions. It was also revealed that the interaction between air transport infrastructure and GDP does not statistically influence CO_2 emissions while the interaction between rail transport infrastructure and GDP significantly induces higher CO_2 emissions. In addition, the interactions between air transport infrastructure and energy consumption, and rail transport infrastructure and energy consumption reduce carbon emissions. The remaining sections of the chapter are outlined as follows: Section 2 discusses the theoretical and empirical impact of transportation infrastructure development on CO_2 emissions. Data and methodology are discussed in Section 3, while the findings and discussions are presented in Section 4. Finally, conclusions and policy ramifications are discussed in Section 5.

2. Theoretical framework

Theoretically, Xie et al. (2017) propose three main mechanisms by which the development of transportation infrastructure influences CO_2 emissions. The authors indicate that the *population-scale effect*, *economic growth effect*, and *technological innovation effect* are the principal mechanisms through which transportation infrastructure development affects CO_2 emissions. First, the *population-scale effects* indicate that the development of transportation infrastructure could minimize the cost associated with travel, ensure accessibility, and improve the mobility of the population (Xie et al., 2017). These implications of transportation infrastructure development increase the concentration of manufacturing industries and the size of the population in the urban centers, where most of the infrastructure development occurs, and this in turn influences carbon emissions. The *economic growth effect* argues that transportation infrastructure development could fuel higher economic growth, which subsequently impacts on carbon emissions. Thus, the construction of transportation infrastructure improves interregional and international trade, gains from trade, and further widens market (Lakshmanan, 2011; Xie et al., 2017). These outcomes boost the general economic and sectorial growths (Lakshmanan, 2011; Tong et al., 2013), which subsequently affect carbon emissions.

On the *technological innovation effect*, Lakshmanan (2011) argues that transportation infrastructure development could ensure technological shifts, innovation, and commercialization of new knowledge. The technological effect of transportation infrastructure development has a critical outcome on the environment. From the endogenous growth theory, technological innovation ensures efficiency, but could also induce higher production, which will influence energy use and carbon emissions (Awaworyi Churchill et al., 2019; Dzator and Acheampong, 2020). In addition to the aforementioned mechanisms, transportation infrastructure development could impact on carbon emissions through *consumption effect*. It is evident in the literature that energy consumption, especially fossil energy use, represents the major sources of carbon emissions. It is estimated that transport energy demand will increase by an average rate of 1.4% from 104 quadrillion British thermal units (Btu) in 2012 to 144 quadrillion British thermal units (Btu) in 2040 (Energy Information Administration, 2016). The expected increase in transport energy demand could worsen carbon emissions.

In place of theoretical argument, a paucity of empirical studies exists in the literature. For instance, Xie et al. (2017) deploy the STIRPAT model to examine the impact of transportation infrastructure development on urban carbon emissions using panel data for 283 cities between 2003 and 2013. Using fixed effect model, the authors found that transportation infrastructure development is positively related to urban carbon emissions and intensity. The study further revealed that transportation infrastructure development increases urban carbon emissions through population-scale effect while it increases urban carbon emissions through economic growth and technological innovation effect. Saidi and Hammami (2017) further probe the interactions between freight transports, economic growth, and carbon emissions for 75 countries for the period 2000−2014. The findings from the system-generalized method of moment approach indicated that freight transport worsens carbon emissions at the global level, as well as lower income, middle income, and upper income countries.

Neves et al. (2017) utilized the ARDL approach to investigate the role of transport sector infrastructure and energy consumption on CO_2 emissions for 15 OECD countries for the period 1995−2014, and the study revealed that investment in rail infrastructure reduces energy consumption but increases carbon emissions. Contrarily, Georgatzi et al. (2020) studied the impact of transport activities on CO_2 emissions for 12 European countries for the period 1994−2014. The study indicated that transport infrastructure (road, rail, and inland water infrastructure) investment seems not to influence the emissions of CO_2. With the paucity of the empirical literature, this chapter extends the existing literature by probing the effect of transportation infrastructure development on CO_2 emissions for 26 OECD countries for the period between 1960 and 2018.

3. Methodology and data
3.1 Specification of empirical model
Our study utilized a panel data approach to examine the effect of transportation infrastructure development on the emissions of CO_2 in OECD countries. To estimate the effect of transportation

infrastructure development on CO_2 emissions, we follow Acheampong (2019) and Saidi and Hammami (2017) to specify the CO_2 emissions model as dependent on economic growth (GDP), economic growth squared (GDP^2), energy consumption (ENER), transportation infrastructure development (TR), and other control variables (X). Thus, the carbon emissions model is given in Eq. (1.1):

$$CO_2 = f\left(GDP, GDP^2, ENER, TR, X\right) \tag{1.1}$$

Therefore, the reduced form of the log-linear carbon emissions empirical equation to be estimated is specified in Eq. (1.2):

$$\ln CO_{2it} = \alpha_0 + \beta_1 \ln GDP_{it} + \beta_2 \ln GDP_{it}^2 + \beta_3 \ln ENER_{it} + \beta_4 \ln TR_{it} + \beta_5 \ln X_{it} + v_i + \varepsilon_{it} \tag{1.2}$$

Given that transport infrastructure development could influence CO_2 emissions through the consumption of energy and economic growth, Eq. (1.2) is augmented with the moderation term of transportation infrastructure development and GDP per capita ($\ln TR \times \ln GDP$) and the moderation term of transportation infrastructure development and energy consumption ($\ln TR \times \ln ENER$). Eq. (1.3) is thus used to estimate how transport infrastructure development moderates the impact of GDP and energy use on CO_2 emissions in the OECD countries.

$$\begin{aligned}
\ln CO_{2it} &= \alpha_0 + \beta_1 \ln GDP_{it} + \beta_2 \ln GDP_{it}^2 + \beta_3 \ln ENER_{it} + \beta_4 \ln TR_{it} + \delta_1 (\ln TR \times \ln GDP)_{it} \\
&\quad + \delta_2 (\ln TR \times \ln ENER)_{it} + \beta_5 \ln X_{it} + v_i + \varepsilon_{it}
\end{aligned} \tag{1.3}$$

where $i = 1 \ldots N$ and $t = 1960 \ldots 2018$; α_0 is the constant parameter; $\beta_1 \ldots \beta_5$ is the coefficient to be estimated; δ_1 and δ_2 capture the indirect effect of transportation infrastructure development; v_i and ε_{it} are the respective individual-effect and stochastic error terms. $\ln CO_{2it}$ is the CO_2 emissions of country i at time t; $\ln GDP$ and $\ln GDP^2$ are the respective GDP per capita and the GDP per capita squared; $\ln ENER$ is energy consumption; $\ln TR$ is transportation infrastructure development [comprises air and rail infrastructure development]; X is other covariates which are argued to exert some influence on CO_2 emissions. These covariates include population size, trade openness, urbanization, financial development, and foreign direct investment (see Acheampong, 2019; Dzator and Acheampong, 2020).

3.2 Econometric estimation strategy

The above equations were estimated with the random and fixed-effect estimators, which are capable of accounting for country-specific unobserved heterogeneity (Hsiao, 2014). One of the limitations of the random and fixed-effect models is their inability to account for the endogeneity of the independent variables which could affect the estimates. Therefore, the Blundell and Bond [hereafter, BB] (1998) System-GMM was utilized to overcome the endogeneity problem of the random and fixed-effect estimators. The BB (1998) system-GMM estimator controls for endogeneity by using the lagged difference and levels of the dependent variable as an instrument for level and first difference equations. Following the conventional approach, instruments validity was tested using the Sargan test while using the Arellano–Bond test to test for the first (AR1) and second (AR2) order autocorrelation. Therefore, to test the robustness of the findings from the random and fixed-effect models and present reliable findings, this study alternatively used the BB (1998) system-GMM with Windmeijer (2005) finite-sample correction to reestimate Eqs. (1.2) and (1.3).

3.3 Data

To probe the impact of transportation infrastructure development on the emissions of CO_2, this research uses a panel data approach for 26 OECD countries for the period 1960−2018. Appendix Table 1.A.1 presents the countries included in the study. Variable descriptions and descriptive statistics are shown in Table 1.1. From Table 1.1, the average CO_2 emission is 11.759%, and GDP per capita is 26.867%. Further, the average population size and urbanization are 16.685% and 4.244%, respectively. Also, the average energy consumption, trade openness, and foreign direct investment are 8.031%, 4.031%, and 0.118%, respectively. Financial development measured using credit to the private sector has an average of 4.191% while air and rail transport infrastructure has a mean of 6.091% and 8.933%, respectively. The World Bank (2019) World Development Indicators database served as the source of obtaining data for the study. The variables used in this study are expressed in their natural logarithm.

The extent of correlation among the variables is shown in Table 1.2. The correlation matrix suggests that CO_2 emissions correlate positively with GDP per capita, population size, financial development, energy consumption, urbanization, air and rail transport infrastructure, while they correlate negatively with trade openness and foreign direct investment. Also, multicollinearity among the independent variables is not an issue, since the correlation coefficient (not the level of significance) among the independent variables is relatively small.

Table 1.1 Descriptive statistics.

Variables	Symbols	Definitions	Mean	SD	Min	Max
Carbon emissions	lnco2kt	Carbon emissions (kt)	11.759	1.360	9.015	15.572
GDP	lngdpc	GDP (constant 2010 US$)	26.867	1.284	23.885	30.513
Population size	lnpop	Population, total	16.685	1.175	14.679	19.606
Energy consumption	lnener	Energy use (kg of oil equivalent per capita)	8.031	0.595	5.667	9.043
Trade openness	lntrade	Trade (% of GDP)	4.031	0.565	1.745	5.421
Urbanization	lnurbpopt	Urban population (% of total population)	4.244	0.209	3.322	4.585
FDI	lnfdi	Foreign direct investment, net inflows (% of GDP)	0.118	1.513	−7.234	4.461
Financial development	lncredit	Domestic credit to private sector (% of GDP)	4.191	0.737	1.748	5.399
Air transport infrastructure	lnair	Air transport, freight (million ton-km)	6.091	2.137	−2.303	10.669
Rail transport infrastructure	lnrail	Rail lines (total route-km)	8.933	1.242	4.060	12.225

Table 1.2 Correlation matrix.

	lnco2kt	lngdpc	lnpop	lnener	lntrade	lnurbpopt	lnfdi	lncredit	lnair	lnrail
lnco2kt	1									
lngdpc	0.904***	1								
lnpop	0.952***	0.874***	1							
lnener	0.273***	0.357***	0.0241	1						
lntrade	−0.720***	−0.685***	−0.717***	−0.0973	1					
lnurbpopt	0.322***	0.426***	0.207***	0.575***	−0.263***	1				
lnfdi	−0.354***	−0.354***	−0.404***	0.120***	0.569***	−0.0689	1			
lncredit	0.290***	0.515***	0.187***	0.480***	−0.195***	0.341***	−0.103	1		
lnair	0.721***	0.835***	0.668***	0.452***	−0.490***	0.526***	−0.207***	0.552***	1	
lnrail	0.653***	0.626***	0.673***	0.109***	−0.573***	0.0789	−0.199***	0.0406	0.399***	1

*P < .10, **P < .05, ***P < .01.

4. Empirical findings and discussions

This section presents and discusses the findings from the random and fixed-effect estimators. As presented in Table 1.3, Model [1]−[6] show the findings from random-effect estimator while Model [7]−[12] show the findings from the fixed-effect estimator. Considering the argument of this paper, we observe from Table 1.3 that air transport infrastructure significantly increases carbon emissions while rail transportation insignificantly reduces carbon emissions. Thus, rail transport infrastructure has not affected carbon emissions; however, the development of air transport infrastructure has contributed significantly to the increasing level of the emissions of CO_2 in the OECD countries. This evidence that air transport infrastructure worsens the emissions of CO_2 supports the earlier findings of Saidi and Hammami (2017) and Xie et al. (2017), which indicate that transportation infrastructure development increases carbon emissions. Our finding also conflicts with the result of Neves et al. (2017), which suggests that investment in rail infrastructure significantly worsens the emissions of CO_2 in OECD countries.

Further, the interaction effect results indicate that air transport infrastructure moderates the effect of economic growth to have no substantial impact of CO_2 emissions whereas the interaction term of rail transport infrastructure and economic growth significantly (only in the random effect model) worsens the emissions of CO_2 in the OECD countries. Thus, through economic growth, rail transport infrastructure can indirectly worsen the emissions of CO_2. Our evidence is similar to that of Xie et al. (2017), who reported that transportation infrastructure development moderates the impact of GDP per capita to reduce urban CO_2 emissions in China. In addition, the moderation term for air transport infrastructure and energy consumption significantly reduces carbon emissions while the moderation term of rail transport infrastructure and energy consumption insignificantly reduce carbon emissions. As OECD countries do not contribute to the global growth in transport energy use, this result implies that the development of transport infrastructure in OECD countries ensures efficiency in energy use, thereby reducing carbon emissions (Energy Information Administration, 2016).

We observe from the models of random-effect and fixed-effect that economic growth has a negative relationship with CO_2 emissions, while economic growth squared has a positive impact on CO_2 emissions. Inconsistent with the argument of Environmental Kuznets Curve (EKC) hypothesis, the respective effect of economic growth and economic growth squared on CO_2 emissions from these static econometric estimators suggests that the environmental effect of economic growth is rather U-shaped in OECD countries. Consistent with the pessimistic view, the results confirm that increasing population further worsens emissions of CO_2 in the OECD countries. The implication is that a consistent increase in population size in the OECD countries would increase the use of environmental and energy resources, thereby exerting pressure on increasing the emissions of CO_2. Also, energy consumption is found to induce higher CO_2 emissions significantly. The result is due to the intensive use of energy resources in the OECD countries to boost production and economic growth, thereby inducing higher carbon emissions. Previous empirical studies have confirmed that population size and energy consumption have been a major force behind carbon emissions in OECD countries (Dzator and Acheampong, 2020; Mensah et al., 2018; Ozcan et al., 2020).

Consistent with the technique effect, openness to trade is found to improve the emissions of CO_2 substantially. Thus, trade openness is associated with the transfer of environmentally efficient technologies into the OECD countries, thereby contributing to retarding the amount of CO_2 emissions.

Table 1.3 Random and fixed effects results.

Variables	Model 1	Model 2	Model 3	Model 4	Model 5	Model 6	Model 7	Model 8	Model 9	Model 10	Model 11	Model 12
	Random effect model						Fixed effect model					
GDP	-1.580^a	-1.545^b	-1.698	-1.493^a	-2.093^b	-1.514^b	-1.666^a	-2.107^b	-1.707	-2.058^b	-2.186^b	-2.061^b
	(0.881)	(0.730)	(1.255)	(0.788)	(0.848)	(0.743)	(0.950)	(0.880)	(1.316)	(0.950)	(0.911)	(0.898)
GDP squared	0.022	0.027^b	0.024	0.023	0.033^b	0.027^a	0.023	0.039^b	0.024	0.034^a	0.034^b	0.038^b
	(0.015)	(0.014)	(0.024)	(0.015)	(0.014)	(0.014)	(0.016)	(0.016)	(0.025)	(0.019)	(0.015)	(0.017)
Population size	1.400^c	1.018^c	1.401^c	1.036^c	1.363^c	1.012^c	1.534^c	0.743^c	1.533^c	0.783^c	1.467^c	0.737^c
	(0.096)	(0.077)	(0.103)	(0.078)	(0.104)	(0.077)	(0.200)	(0.202)	(0.211)	(0.208)	(0.210)	(0.205)
Energy consumption	1.083^c	1.269^c	1.086^c	1.283^c	1.361^c	1.356^c	1.098^c	1.271^c	1.099^c	1.288^c	1.369^c	1.362^c
	(0.073)	(0.086)	(0.081)	(0.079)	(0.146)	(0.113)	(0.079)	(0.093)	(0.085)	(0.084)	(0.166)	(0.121)
Trade openness	0.012	-0.086^a	0.011	-0.088^a	0.007	-0.089^a	-0.003	-0.110^b	-0.004	-0.112^b	-0.002	-0.113^b
	(0.048)	(0.052)	(0.044)	(0.052)	(0.048)	(0.052)	(0.055)	(0.048)	(0.050)	(0.048)	(0.055)	(0.048)
Urbanization	0.441	0.105	0.443	0.137	0.111	0.110	0.431	0.271	0.431	0.298	0.103	0.275
	(0.368)	(0.354)	(0.367)	(0.350)	(0.433)	(0.354)	(0.393)	(0.384)	(0.391)	(0.384)	(0.479)	(0.382)
FDI	-0.019	0.000	-0.019	-0.000	-0.019	0.000	-0.018	-0.000	-0.018	-0.001	-0.018	-0.000
	(0.016)	(0.002)	(0.016)	(0.002)	(0.016)	(0.002)	(0.016)	(0.002)	(0.016)	(0.002)	(0.016)	(0.002)
Financial development	0.006	-0.047^c	0.007	-0.049^c	0.007	-0.047^c	0.001	-0.043^c	0.001	-0.046^c	0.003	-0.043^c
	(0.031)	(0.017)	(0.030)	(0.017)	(0.030)	(0.017)	(0.033)	(0.014)	(0.033)	(0.014)	(0.033)	(0.014)
Air transport infrastructure	0.021^b		0.058		0.292^c		0.018^b		0.032		0.279^b	
	(0.009)		(0.397)		(0.106)		(0.008)		(0.389)		(0.123)	
Rail transport infrastructure		-0.000		-0.573^a		0.087		-0.000		-0.566		0.092
		(0.004)		(0.346)		(0.061)		(0.004)		(0.365)		(0.059)
Air transport infrastructure × GDP			-0.001						-0.001			
			(0.015)						(0.015)			

	(1)	(2)	(3)	(4)	(5)	(6)	(7)	(8)	(9)	(10)	(11)	(12)
Rail transport infrastructure × GDP				0.020[a] (0.012)						0.020 (0.013)		
Air transport infrastructure × energy consumption					−0.035[b] (0.014)						−0.034[b] (0.016)	
Rail transport infrastructure × energy consumption						−0.010 (0.007)						−0.011 (0.007)
Constant	4.132 (10.831)	6.274 (10.307)	5.545 (15.515)	7.749 (10.857)	10.191 (10.492)	5.161 (10.428)	3.309 (12.086)	17.267 (13.356)	3.817 (17.357)	18.410 (13.992)	9.985 (11.961)	15.928 (13.609)
Observations	640	378	640	378	640	378	640	378	640	378	640	378
r2_w	0.889	0.872	0.889	0.874	0.893	0.872	0.890	0.875	0.890	0.877	0.893	0.876
r2_o	0.969	0.937	0.969	0.937	0.971	0.937	0.968	0.880	0.968	0.885	0.970	0.880
r2_b	0.963	0.939	0.963	0.939	0.964	0.939	0.964	0.882	0.964	0.887	0.965	0.881
rho	0.902	0.980	0.908	0.980	0.908	0.981	0.941	0.994	0.941	0.993	0.931	0.994
rmse	0.081	0.039	0.081	0.039	0.080	0.039	0.079	0.037	0.079	0.037	0.078	0.037

Standard errors in parentheses.
[a] $P < .10$.
[b] $P < .05$.
[c] $P < .001$.

This finding is consistent with the evidence presented in the scholarly work of Acheampong (2018); Antweiler et al. (2001); Al-Mulali et al. (2015), and Shahbaz et al. (2013), which confirm that trade openness reduces carbon emissions. We further observe from Table 1.3 that foreign direct investment and urbanization play no substantial role in CO_2 emissions in the OECD countries. Thus, urbanization and the inflow of foreign investment into the OECD countries have not been major factors for influencing carbon emissions. From Table 1.3, it is further observed in most of the models that financial development substantially retards the emissions of CO_2 in OECD countries. With OECD developed financial sector, financial institutions can lend credits to firms at a lower cost which could enable these firms to invest in less carbon-intensive technologies and projects which leads to lowering of the emissions of CO_2 (Acheampong, 2019; Acheampong et al., 2020). Previous empirical studies have reported that financial development could contribute to carbon emissions mitigation (see Abbasi and Riaz, 2016; Shahbaz et al., 2013; Tamazian et al., 2009).

4.1 Robustness check

One of the major limitations of both random and fixed effect models is their inability to account for endogeneity. To avoid the inherent limitation of the random and fixed-effect models, we deploy the BB (1998) system-GMM to control for endogeneity in our model and check the consistency in the results. The system-GMM results are reported in Table 1.4. The system-GMM finding also shows that air transport infrastructure significantly increases carbon emissions while rail transport infrastructure has no substantial effect on the mitigation of CO_2 emissions. Similarly, the interaction term between air transport infrastructure and economic growth reveals an insignificant impact on the emissions of CO_2, whereas the interaction term of rail transport infrastructure and economic growth substantially generates higher CO_2 emissions. In addition, the moderation term of air transport infrastructure and energy consumption and the interaction term of rail transport infrastructure and energy consumption does not have any statistical impact on CO_2 emissions.

Unlike the random and fixed-effect models, the system-GMM results show that economic growth has a positive impact on CO_2 emissions while economic growth squared has a negative impact on CO_2 emissions. The evidence from the BB (1998) system-GMM confirms the argument of the Environmental Kuznets Curve. A similar finding is found in the study of Tamazian and Bhaskara Rao (2010). Thus, when we controlled endogeneity using the dynamic system-GMM approach, we found evidence for the EKC hypothesis while the static models (random and fixed effect models) negate its existence. This result supports Stern's (2004) critic of previous studies that ignore the possible endogeneity when testing the EKC hypothesis.

Consistent with the random and fixed effect results in Table 1.3, both population size and energy consumption significantly increase carbon emissions. Additionally, trade openness and foreign direct investment contribute significantly to the retardation of CO_2 emissions. Unlike the random and fixed effect results, the System-GMM results indicate that urbanization substantially improves the emissions of CO_2. In the scholarly work of Adams and Acheampong (2019) and Sadorsky (2014), these authors argue that the carbon emissions effect of urbanization is indeterminate, since the impact of urbanization depends on the econometric estimation technique. Similar to the results of Abbasi et al. (2016) and Shahbaz et al. (2013a,b), financial development significantly reduces the emissions of CO_2.

Table 1.4 System-GMM results.

Variables	Model 1	Model 2	Model 3	Model 4	Model 5	Model 6
Carbon emissions lagged	0.960[c]	0.964[c]	0.960[c]	0.954[c]	0.960[c]	0.963[c]
	(0.019)	(0.022)	(0.019)	(0.023)	(0.019)	(0.023)
GDP	0.063	0.109	0.109	0.371[c]	0.074	0.181[b]
	(0.057)	(0.084)	(0.122)	(0.133)	(0.058)	(0.090)
GDP squared	−0.002	−0.002	−0.003	−0.009[c]	−0.002[a]	−0.004[b]
	(0.001)	(0.002)	(0.002)	(0.003)	(0.001)	(0.002)
Population size	0.052[b]	0.056[b]	0.052[b]	0.074[c]	0.051[b]	0.059[b]
	(0.021)	(0.026)	(0.021)	(0.027)	(0.022)	(0.027)
Energy consumption	0.046[b]	0.065[c]	0.046[b]	0.077[c]	0.037	−0.004
	(0.021)	(0.022)	(0.021)	(0.022)	(0.031)	(0.049)
Trade openness	−0.023[c]	−0.024[b]	−0.024[c]	−0.024[b]	−0.022[c]	−0.022[b]
	(0.006)	(0.010)	(0.007)	(0.010)	(0.006)	(0.009)
Urbanization	−0.048[a]	−0.022	−0.048[a]	−0.021	−0.045	−0.024
	(0.027)	(0.025)	(0.027)	(0.026)	(0.028)	(0.025)
FDI	−0.001	−0.001	−0.001	−0.001	−0.001	−0.001
	(0.002)	(0.003)	(0.002)	(0.003)	(0.002)	(0.003)
Financial development	−0.013[c]	−0.014[b]	−0.013[c]	−0.018[c]	−0.013[c]	−0.017[b]
	(0.005)	(0.007)	(0.005)	(0.007)	(0.005)	(0.008)
Air transport infrastructure	0.013[c]		−0.009		0.002	
	(0.002)		(0.049)		(0.018)	
Rail transport infrastructure		−0.006		−0.267[c]		−0.075
		(0.005)		(0.094)		(0.056)
Air transport infrastructure × GDP			0.001			
			(0.002)			
Rail transport infrastructure × GDP				0.009[c]		
				(0.003)		
Air transport infrastructure × energy consumption					0.001	
					(0.002)	
Rail transport infrastructure × energy consumption						0.008
						(0.006)
Constant	−0.974	−2.015[a]	−1.509	−4.572[c]	−1.061	−2.431[b]
	(0.897)	(1.176)	(1.546)	(1.575)	(0.857)	(1.159)

Continued

Table 1.4 System-GMM results.—cont'd

Variables	Model 1	Model 2	Model 3	Model 4	Model 5	Model 6
Observations	640	378	640	378	640	378
Sargan	630.907	381.012	630.732	381.124	630.146	380.039
P(Sargan)	0.471	0.296	0.462	0.282	0.468	0.296
AR(1)	0.001	0.001	0.001	0.001	0.001	0.001
AR(2)	0.870	0.325	0.875	0.316	0.868	0.325

Heteroscedasticity robust standard errors in parentheses. Sargan-test refers to the overidentification test for the restrictions in system-GMM estimation. The AR (1) and AR (2) test are the Arellano—Bond test for the existence of the first and second-order autocorrelation in first differences.
[a]P <.10.
[b]P <.05.
[c]P <.01.

5. Conclusion and policy implications

This chapter probes the impact of infrastructure development on CO_2 emissions in 26 OECD countries for the time between 1960 and 2018 while accounting for economic growth, economic growth squared, energy consumption, population size, urbanization, trade openness, financial development, and foreign direct investment.

Our results indicated that the EKC hypothesis is only valid when endogeneity is controlled. The findings also show that population size and energy consumption significantly increase the emissions of CO_2 while trade openness, financial development, and foreign direct investment lower the emissions of CO_2. It was found that urbanization plays no substantial role in the emission of CO_2 when estimated with the static models (fixed and random effect models), but for dynamic system-GMM model, urbanization significantly lowers the emissions of CO_2. This supports the argument of Adams and Acheampong (2019) and Sadorsky (2014) that the environmental effect of urbanization is sensitive to the estimation technique.

Additionally, our study further indicated that air transport infrastructure significantly increases carbon emissions while rail transport infrastructure does not substantially impact on CO_2 emissions. It was also revealed that the interaction between air transport infrastructure and economic growth does not significantly lead to higher CO_2 emissions while the interaction between rail transport infrastructure and economic growth significantly lead to higher CO_2 emissions. In addition, the interactions between air transport infrastructure and energy consumption, and rail transport infrastructure and energy consumption reduce carbon emissions.

These findings have important ramifications for sustainable development and climate change policies in the OECD countries. Our results imply that omitting transport infrastructure, especially the road transport infrastructure from the OECD countries' carbon emissions models, could lead to an underestimation of actual carbon emissions and make strategies to curb carbon emissions mitigation unsustainable. It was also observed that rail and air transport infrastructure could reduce the emissions of CO_2 by promoting economic growth and ensuring energy efficiency. We, therefore, suggest that investment in sustainable transport infrastructures could promote economic growth and energy efficiency, thereby mitigating the emissions of CO_2. From the results, trade openness and foreign direct investment are imperative for lowering CO_2 emissions in the OECD countries. The role of trade openness and FDI in reducing carbon emissions could be attributed to the stringent environmental regulatory framework in the OECD that helps to decouple trade and FDI from carbon emissions. In the existing standard environmental regulations, economic liberalization policies will be consistent with strategies for controlling carbon emissions in the OECD countries. Finally, an effective transition toward renewable energy use and controlling population growth remain imperative for lowering the emission of CO_2.

Appendix

Table 1.A.1 Countries included in the study.

Australia, Austria, Belgium, Canada, Denmark, Finland, France, Germany, Greece, Hungary Ireland, Italy, Japan, Korea, Rep., Netherlands, New Zealand, Norway, Poland, Portugal, Slovak Republic, Spain, Sweden, Switzerland, Turkey, United Kingdom and the United States.

References

Abbasi, F., Riaz, K., 2016. CO_2 emissions and financial development in an emerging economy: an augmented VAR approach. Energy Pol. 90, 102−114. https://doi.org/10.1016/j.enpol.2015.12.017.

Acheampong, A.O., 2018. Economic growth, CO_2 emissions and energy consumption: what causes what and where? Energy Econ. 74, 677−692. https://doi.org/10.1016/j.eneco.2018.07.022.

Acheampong, A.O., 2019. Modelling for insight: does financial development improve environmental quality? Energy Econ. 83, 156−179. https://doi.org/10.1016/j.eneco.2019.06.025.

Acheampong, A.O., Amponsah, M., Boateng, E., 2020. Does financial development mitigate carbon emissions? Evidence from heterogeneous financial economies. Energy Econ. 88, 104768. https://doi.org/10.1016/j.eneco.2020.104768.

Adams, S., Acheampong, A.O., 2019. Reducing carbon emissions: the role of renewable energy and democracy. J. Clean. Prod. 240, 118245. https://doi.org/10.1016/j.jclepro.2019.118245.

Al-Mulali, U., Ozturk, I., Lean, H.H., 2015. The influence of economic growth, urbanisation, trade openness, financial development, and renewable energy on pollution in Europe. Nat. Hazards 79 (1), 621−644. https://doi.org/10.1007/s11069-015-1865-9.

Antweiler, W., Copeland, B.R., Taylor, M.S., 2001. Is free trade good for the environment? Am. Econ. Rev. 91 (4), 877−908.

Awaworyi Churchill, S., Inekwe, J., Smyth, R., Zhang, X., 2019. R&D intensity and carbon emissions in the G7: 1870−2014. Energy Econ. 80, 30−37. https://doi.org/10.1016/j.eneco.2018.12.020.

Blundell, R., Bond, S., 1998. Initial conditions and moment restrictions in dynamic panel data models. J. Econom. 87 (1), 115−143. https://doi.org/10.1016/S0304-4076(98)00009-8.

Dzator, J., Acheampong, A.O., 2020. The impact of energy innovation on carbon emissions mitigation: an empirical evidence from OECD countries. In: Hussain, C.M. (Ed.), Handbook of Environmental Materials Management. Springer International Publishing, Cham. https://doi.org/10.1007/978-3-319-58538-3_213-1.

Energy Information Administration (US), & Government Publications Office, 2016. In: International Energy Outlook 2016, with Projections to 2040. Government Printing Office.

Georgatzi, V.V., Stamboulis, Y., Vetsikas, A., 2020. Examining the determinants of CO_2 emissions caused by the transport sector: empirical evidence from 12 European countries. Econ. Anal. Pol. 65, 11–20. https://doi.org/10.1016/j.eap.2019.11.003.

Hsiao, C., 2014. Analysis of Panel Data. Cambridge university press.

International Energy Agency, 2019. Tracking Transport. IEA, Paris. https://www.iea.org/reports/tracking-transport-2019.

Lakshmanan, T.R., 2011. The broader economic consequences of transport infrastructure investments. J. Transport Geogr. 19 (1), 1–12. https://doi.org/10.1016/j.jtrangeo.2010.01.001.

Marchal, V., Dellink, R., Van Vuuren, D., Clapp, C., Chateau, J., Magné, B., Van Vliet, J., 2011. OECD Environmental Outlook to 2050, vol. 8. Organisation for Economic Co-operation and Development, pp. 397–413.

Mensah, C.N., Long, X., Boamah, K.B., Bediako, I.A., Dauda, L., Salman, M., 2018. The effect of innovation on CO_2 emissions of OCED countries from 1990 to 2014. Environ. Sci. Pollut. Control Ser. 25 (29), 29678–29698. https://doi.org/10.1007/s11356-018-2968-0.

Neves, S.A., Marques, A.C., Fuinhas, J.A., 2017. Is energy consumption in the transport sector hampering both economic growth and the reduction of CO_2 emissions? A disaggregated energy consumption analysis. Transport Pol. 59, 64–70. https://doi.org/10.1016/j.tranpol.2017.07.004.

Ozcan, B., Tzeremes, P.G., Tzeremes, N.G., 2020. Energy consumption, economic growth and environmental degradation in OECD countries. Econ. Modell. 84, 203–213.

Sadorsky, P., 2014. The effect of urbanisation on CO_2 emissions in emerging economies. Energy Econ. 41, 147–153. https://doi.org/10.1016/j.eneco.2013.11.007.

Saidi, S., Hammami, S., 2017. Modeling the causal linkages between transport, economic growth and environmental degradation for 75 countries. Transport. Res. Transport Environ. 53, 415–427. https://doi.org/10.1016/j.trd.2017.04.031.

Saidi, S., Shahbaz, M., Akhtar, P., 2018. The long-run relationships between transport energy consumption, transport infrastructure, and economic growth in MENA countries. Transport. Res. Pol. Pract. 111, 78–95. https://doi.org/10.1016/j.tra.2018.03.013.

Shahbaz, M., Kumar Tiwari, A., Nasir, M., 2013a. The effects of financial development, economic growth, coal consumption and trade openness on CO_2 emissions in South Africa. Energy Pol. 61, 1452–1459. https://doi.org/10.1016/j.enpol.2013.07.006.

Shahbaz, M., Solarin, S.A., Mahmood, H., Arouri, M., 2013b. Does financial development reduce CO_2 emissions in Malaysian economy? A time series analysis. Econ. Modell. 35, 145–152. https://doi.org/10.1016/j.econmod.2013.06.037.

Stern, D.I., 2004. The rise and fall of the environmental Kuznets Curve. World Dev. 32 (8), 1419–1439. https://doi.org/10.1016/j.worlddev.2004.03.004.

Tamazian, A., Bhaskara Rao, B., 2010. Do economic, financial and institutional developments matter for environmental degradation? Evidence from transitional economies. Energy Econ. 32 (1), 137–145. https://doi.org/10.1016/j.eneco.2009.04.004.

Tamazian, A., Chousa, J.P., Vadlamannati, K.C., 2009. Does higher economic and financial development lead to environmental degradation: evidence from BRIC countries. Energy Pol. 37 (1), 246–253. https://doi.org/10.1016/j.enpol.2008.08.025.

Tong, T., Yu, T.-H.E., Cho, S.-H., Jensen, K., De La Torre Ugarte, D., 2013. Evaluating the spatial spillover effects of transportation infrastructure on agricultural output across the United States. J. Transport Geogr. 30, 47–55. https://doi.org/10.1016/j.jtrangeo.2013.03.001.

Windmeijer, F., 2005. A finite sample correction for the variance of linear efficient two-step GMM estimators. J. Econ. 126 (1), 25–51.

World Bank, 2019. World Development Indicators. http://databank.worldbank.org/data/reports.aspx?source=world-development-indicators#.

Xie, R., Fang, J., Liu, C., 2017. The effects of transportation infrastructure on urban carbon emissions. Appl. Energy 196, 199–207. https://doi.org/10.1016/j.apenergy.2017.01.020.

Does transport infrastructure development contribute to carbon emissions? Evidence from developing countries

2

Michael Dzator[1,3], Alex O. Acheampong[2], Janet Dzator[2,3]

[1]*SAE, Central Queensland University, Mackay, QLD, Australia;* [2]*Newcastle Business School, The University of Newcastle, Newcastle, NSW, Australia;* [3]*Australia Africa Universities Network (AAUN) Partner, Newcastle, NSW, Australia*

Chapter outline

1. Introduction

Developing countries face huge gaps in infrastructure development. While infrastructure development, and especially transport infrastructure development, remains critical to sustaining economic development, policymakers in developing countries have enacted policies to close these gaps. Undoubtedly, transport infrastructure development could boost economic development (Marazzo et al., 2010; Saidi et al., 2018), but could also be detrimental to the environment in developing countries (Achour and Belloumi, 2016). For instance, the Energy Information Administration [EIA] (2016) report suggests that transport energy consumption in non-organisation for economic co-operation and development (OECD) countries dominates that of OECD countries. Statistically, it is projected that transport energy demand in non-OECD countries is expected to be 61% of the global transport energy demand (EIA, 2016). In other words, transportation energy consumption in the non-OECD countries is expected to nearly double from 47 quadrillion Btu in 2012 to 94 quadrillion Btu in 2040. These projections

Environmental Sustainability and Economy. https://doi.org/10.1016/B978-0-12-822188-4.00012-9

indicate that transport infrastructures could aggravate environmental problems, especially in carbon emissions in the developing world if appropriate measures are not put in place.

Several studies have investigated the influence of transport infrastructure development on economic growth (Esfahani and Ramırez, 2003; Saidi and Hammami, 2017; Saidi et al., 2018); however, there is a paucity of empirical studies on the role of transport infrastructure development on carbon emissions in developing countries. In theory, the role of transport infrastructure on the emissions of CO_2 is priori uncertain. For instance, transport infrastructure development could minimize travel costs, ensure accessibility and population mobility while also increasing the concentration of manufacturing industries in the urban areas (Xie et al., 2017). The ease of population mobility and the concentration of manufacturing industries could influence carbon emissions. Additionally, transportation infrastructure development is argued to increase economic growth and subsequently influence carbon emissions. Therefore, the construction of transportation infrastructure improves interregional and international trade, gains from trade, and further widens market (Lakshmanan, 2011; Xie et al., 2017). These outcomes boost the general economic and sectorial growths (Lakshmanan, 2011; Tong et al., 2013), which subsequently affect carbon emissions. Transport infrastructure development is further argued to ensure technological shifts, innovation, and commercialization of new knowledge, which have important implications for the environment (Lakshmanan, 2011; Xie et al., 2017). The technological effect of transportation infrastructure development has a critical outcome on the environment. From the endogenous growth theory, technological innovation ensures efficiency, but could also induce higher production, which will influence energy use and carbon emissions (Awaworyi Churchill, Inekwe et al., 2019; Dzator and Acheampong, 2020).

Despite the theoretical arguments, few studies have attempted to examine the environmental cost of transport infrastructure development. For instance, Xie et al. (2017) probed the impact of road infrastructure development on CO_2 emissions for 283 cities. The outcome of the study revealed that road infrastructure development is associated with increasing the intensity of CO_2 emissions in the cities. Indirectly, the study indicated that road infrastructure development decreases urban CO_2 emissions through population-scale effect while it increases urban CO_2 emissions through economic growth and technological innovation effect. In the same vein, Saidi and Hammami (2017) further probed the interactions between freight transports, GDP per capita, and CO_2 emissions for 75 countries. The study revealed that freight transport worsens CO_2 emissions for the full-sample, as well as countries at different stages of economic development. With the background and the scarcity of empirical studies on transportation infrastructure development on carbon emissions, this study aims to probe the impact of transportation infrastructure development on the emission of CO_2 in 113 developing countries for the years 1990–2018.

This chapter contributes to the literature in the following directions. First, contrary to the prior studies, our study not only examines the direct effect of transportation infrastructure on the emissions of CO_2 but also probes the transmission mechanism (indirect effect) by which the transportation infrastructure development impacts the emissions of CO_2. While our study involves a large sample of 113 developing countries, we further contribute to the literature by disaggregating this full sample into various income groups to avoid the assumption that the environmental cost of transportation infrastructure is homogenous in these countries. In addition, our study employed the generalized method of moment (IV-GMM) approach to address the inherent problems of endogeneity and omitted variable bias. Finally, this research is not just an empirical exercise, but its outcome will enable policymakers in developing countries to comprehend the futuristic trajectory of carbon emissions, which is critical for planning climate change policies. The rest of the sections are organized as follows: Section 2 discusses the methodology and data. Section 3 presents the empirical findings and discussions, while concluding remarks and policy recommendations are presented in Section 4.

2. Methodology and data
2.1 Empirical model and estimation approach

The principal aim of this chapter is to understand the role of transportation infrastructure development on carbon emissions in developing countries. To achieve the study objective, we used a panel data approach to probe the impact of transportation infrastructure development on CO_2 emissions in developing countries. Following Acheampong (2019) and Saidi and Hammami (2017), we specify that carbon emissions (CO_2) is a function of GDP per capita (GDP), GDP per capita squared (GDP^2), energy consumption ($ENER$), transportation infrastructure development (TR), and other control variables (X). Therefore, the reduced form of the log-linear carbon emissions empirical equation is specified in Eq. (2.1):

$$\ln CO_{2it} = \alpha_0 + \beta_1 \ln GDP_{it} + \beta_2 \ln GDP_{it}^2 + \beta_3 \ln ENER_{it} + \beta_4 \ln TR_{it} + \beta_5 \ln X_{it} + v_i + \varepsilon_{it} \quad (2.1)$$

To understand the transmission channels through which transportation infrastructure development impact on the emissions of CO_2, we augment Eq. (2.1) by incorporating the interaction between transportation infrastructure development and GDP per capita ($\ln TR \times \ln GDP$) and the interaction between transportation infrastructure development and energy consumption ($\ln TR \times \ln ENER$). Eq. (2.2) is, therefore, used to probe the indirect influence of transportation infrastructure development on CO_2 emissions in developing countries.

$$\ln CO_{2it} = \alpha_0 + \beta_1 \ln GDP_{it} + \beta_2 \ln GDP_{it}^2 + \beta_3 \ln ENER_{it} + \beta_4 \ln TR_{it}$$
$$+ \delta_1 (\ln TR \times \ln GDP)_{it} + \delta_2 (\ln TR \times \ln ENER)_{it} + \beta_5 \ln X_{it} + v_i + \varepsilon_{it} \quad (2.2)$$

where $i = 1...N$ and $t = 1990...2018$; α_0 is the constant parameter; $\beta_1...\beta_5$ is the coefficient to be estimated; δ_1 and δ_2 capture the indirect effect of transportation infrastructure development; v_i and ε_{it} are the respective individual effect the stochastic error terms. $\ln CO_{2it}$ is the emission of CO_2 of country i at time t; $\ln GDP$ and $\ln GDP^2$ are the respective GDP per capita and GDP per capita squared; $\ln ENER$ is energy consumption; $\ln TR$ is transportation infrastructure development [comprises air and rail infrastructure development]; X is other variables [population size, trade openness, urbanization, financial development, foreign direct investment], which the literature suggests influence CO_2 emissions.

Estimating the above equations in the presence of endogeneity, omitted variable bias and autocorrelation with Ordinary Least-Squares (OLS) could bias the findings. To avoid these limitations associated with OLS, the above equations are estimated using the generalized method of moment (IV-GMM) approach. According to Baum et al. (2002), the IV-GMM is capable of producing efficient estimates in the presence of unknown heteroscedasticity and also robust to autocorrelation. Following Acheampong et al. (2020), postestimation statistics such as the Kleibergen-Paap F-statistics and the Hansen J are used to test the validity of the instruments and fitness of the model.

2.2 Data

This study uses panel data for 113 developing countries[1] for the period 1990–2018. Table 2.1 summarizes the descriptive statistics for the variables. From Table 2.1, the mean of carbon emission is 8.972%, GDP per capita is 23.718%, population size is 16.630%, urbanization is 3.731%, energy consumption is 6.630%, trade openness is 4.180%, foreign direct investment is 0.603%, financial development is 3.001%. Air and rail transport infrastructure have a mean of 2.854% and 7.896%, respectively. Also, the standard deviation of carbon emission is 2.372, GDP per capita is 1.984, population size is 1.766, urbanization is 0.492, energy consumption is 0.786, trade openness is 0.640, foreign direct investment is 1.622, financial development is 0.940. Also, the standard deviation of air and rail transport infrastructure is 2.648 and 1.351, respectively. The data used in this paper were sourced from the World Bank (2019) World Development Indicators database.

To avoid presenting spurious results, the correlation among the independent variables are examined. The correlation matrix seen (Table 2.2) suggests that carbon emissions correlate positively with independent variables such as economic growth, population size, energy consumption, urbanization, financial development, and air and rail transportation infrastructure development, while they correlate negatively with trade openness and foreign direct investment. The highest correlation coefficient is 0.853, which is the correlation between population size and economic growth. In the statistics literature, having such high correlation coefficient is not problematic because having an observation above 100 or 200 can lead to correlation coefficient between 0.7 and 0.8 without affecting the regression results[2] (Acheampong et al., 2019; Drove, 2009). Besides, the correlation coefficient between the remaining covariates is relatively low, which indicates that multicollinearity is not an issue.

Table 2.1 Descriptive statistics.

Variables	Definitions	Mean	SD	Min	Max
lnco2kt	Carbon emissions (kt)	8.972	2.372	3.091	16.147
lngdpc	GDP (constant 2010 US$)	23.718	1.984	18.618	30.010
lnpop	Population, total	16.126	1.766	10.764	21.055
lnener	Energy use (kg of oil equivalent per capita)	6.630	0.786	4.170	8.688
lntrade	Trade (% of GDP)	4.180	0.640	−3.863	5.930
lnurbpopt	Urban population (% of total population)	3.731	0.492	1.689	4.520
lnfdi	Foreign direct investment, net inflows (% of GDP)	0.603	1.622	−12.509	4.638
lncredit	Domestic credit to private sector (% of GDP)	3.001	0.940	−0.910	7.850
lnair	Air transport, freight (million ton-km)	2.854	2.648	−9.350	10.137
lnrail	Rail lines (total route-km)	7.896	1.351	5.557	11.378

Note: All the variables are expressed in their natural logarithm.

[1]See the appendix for the countries.
[2]In estimating the models, when dropping either economic growth or population variable from the regression does not change the estimates, this shows that the high correlation between the two variables is not problematic.

Table 2.2 Correlation matrix.

	lnco2kt	lngdpc	lnpop	lnener	lntrade	lnurbpopt	lnfdi	lncredit	lnair	lnrail
lnco2kt	1									
lngdpc	0.934^c	1								
lnpop	0.769^c	0.853^c	1							
lnener	0.520^c	0.361^c	-0.0648	1						
lntrade	-0.238^c	-0.416^c	-0.557^c	0.287^c	1					
lnurbpopt	0.248^c	0.160^c	-0.281^c	0.718^c	0.257^c	1				
lnfdi	-0.110^b	-0.197^c	-0.270^c	0.211^c	0.418^c	0.132^c	1			
lncredit	0.506^c	0.464^c	0.304^c	0.259^c	0.204^c	0.147^c	0.0345	1		
lnair	0.646^c	0.704^c	0.671^c	0.0780^a	-0.279^c	0.00305	-0.185^c	0.459^c	1	
lnrail	0.783^c	0.751^c	0.720^c	0.388^c	-0.381^c	0.0157	-0.0856^a	0.249^c	0.471^c	1

[a]$P < .05.$
[b]$P < .01.$
[c]$P < .001.$

3. Results and discussions

From Table 2.3, the results suggest that GDP per capita lowers CO_2 emissions, while the GDP per capita squared increases CO_2 emissions. The theoretical implication is that GDP per capita has a U-shaped effect on CO_2 emissions, which conflicts with the argument that the impact of GDP per capita on CO_2 emissions is inverted U-shaped. This result conflicting with environmental Kuznets curve (EKC) argument is supported by the work of Acheampong (2019), Stern (2004), and Stern and Common (2001). Collaborating with the findings of Shi (2003), Zhu et al. (2012), and Yeh and Liao (2017), the evidence from Table 2.3 suggest that population size significantly increases CO_2 emissions in all developing countries. The results presented also indicate that CO_2 emissions worsen with increasing energy consumption, and the impact is statistically significant. This outcome indicates that increasing energy use, especially fossil energy in the developing countries, acts as a catalyst for worsening CO_2 emissions.

The evidence presented in Table 2.3 also suggests that trade openness does generate higher CO_2 emissions in developing countries. Thus, trade in developing countries results in the expansion of economic activities and energy consumption, thereby inducing higher CO_2 emissions. In the same vein, trade openness in developing countries is not encouraging the transfer of environmentally efficient technologies that could mitigate carbon emissions due to poor environmental regulation. This evidence collaborates with the results of Shahbaz et al. (2017), Adams and Acheampong (2019), Hakimi and Hamdi (2016), and Acheampong et al. (2019), who found that trade openness generates higher CO_2 emissions. Collaborating with the *pollution-halo hypothesis*, the results presented in Table 2.3 reveal that FDI is environmentally friendly by mitigating CO_2 emissions in developing countries. The theoretical and policy justification is that foreign direct investment facilitates the transfer and introduction of less carbon-intensive technologies and best environmental management practices to developing countries and thus reducing carbon emissions. Previous empirical research by Acheampong et al. (2019), Doytch and Uctum (2016), and Zhu et al. (2016) has indicated that FDI does not aggravate the emissions of CO_2. Contrarily, our result is inconsistent with the results of Hakimi and Hamdi (2016) and Shahbaz et al. (2018). In addition, evidence suggests that urbanization adds to increasing CO_2 emissions in developing countries. Thus, in support of the ecological modernization theory, urbanization worsens CO_2 emissions in the developing countries by increasing traffic congestions, overcrowding, and energy consumption (Acheampong, 2019; Poumanyvong and Kaneko, 2010). The results show that financial development significantly induces higher CO_2 emissions. This result implies that financial development in the developing countries mostly takes the form of credit to firms and households which enable them to access energy consumption appliances and machines, which contribute to higher carbon emissions (Acheampong, 2019; Khan, 2019).

The findings suggest that both air and rail transport infrastructure significantly contribute to higher CO_2 emissions. Thus, the development of transportation infrastructure in the developing countries reduce travel costs, ensure accessibility, and improve the mobility of the population and the concentration of manufacturing industries, which in turn aggravate carbon emissions (Xie et al., 2017). From the interaction effect results, the interaction between air transport infrastructure and economic growth play no role in CO_2 emissions while the interaction between rail transport infrastructure and GDP per capita significantly worsens the emissions of CO_2. The implication is that the development of rail infrastructure could boost economic growth, which subsequently generates higher CO_2 emissions.

Table 2.3 Results for the full sample.

	Model 1	Model 2	Model 3	Model 4	Model 5	Model 6
lngdpc	−0.212	−0.453	−0.423	0.329	−0.920^c	−0.586^a
	(0.196)	(0.321)	(0.377)	(0.651)	(0.227)	(0.327)
lngdpc2	0.010^b	0.018^c	0.014^a	−0.004	0.024^c	0.020^c
	(0.004)	(0.006)	(0.008)	(0.016)	(0.004)	(0.007)
lnpop	0.738^c	0.565^c	0.742^c	0.588^c	0.739^c	0.553^c
	(0.031)	(0.055)	(0.031)	(0.053)	(0.030)	(0.055)
lnener	0.956^c	0.832^c	0.956^c	0.864^c	1.214^c	1.208^c
	(0.029)	(0.052)	(0.029)	(0.055)	(0.048)	(0.140)
lntrade	0.100^c	0.259^c	0.100^c	0.245^c	0.071^b	0.285^c
	(0.027)	(0.062)	(0.027)	(0.061)	(0.030)	(0.063)
lnurbpopt	0.448^c	0.251^c	0.446^c	0.255^c	0.433^c	0.248^c
	(0.041)	(0.074)	(0.040)	(0.073)	(0.039)	(0.075)
lnfdi	−0.016	−0.051^c	−0.014	−0.048^c	−0.003	−0.056^c
	(0.011)	(0.017)	(0.011)	(0.017)	(0.011)	(0.017)
lncredit	0.168^c	0.237^c	0.172^c	0.225^c	0.174^c	0.234^c
	(0.018)	(0.030)	(0.020)	(0.031)	(0.018)	(0.030)
lnair	0.015^a		0.128		0.503^c	
	(0.008)		(0.140)		(0.064)	
lnrail		0.080^c		−0.795		0.420^c
		(0.023)		(0.514)		(0.114)
lnair×lngdpc			−0.005			
			(0.006)			
lnrail×lngdpc				0.034^a		
				(0.020)		
lnair×lnener					−0.071^c	
					(0.009)	
lnrail×lnener						−0.047^c
						(0.015)
Constant	−11.973^c	−7.690^a	−9.638^b	−14.184^b	−4.665^a	−8.639^b
	(2.378)	(4.190)	(4.307)	(6.565)	(2.712)	(4.126)
Observations	1352	808	1352	808	1352	808
r2	0.951	0.938	0.951	0.939	0.954	0.939
j	0.230	0.605	0.229	0.510	0.043	0.785
jp	0.631	0.437	0.632	0.475	0.835	0.376
widstat	620.184	456.886	169.236	134.874	501.331	457.744

Robust standard errors in parentheses. J is Hansen J-statistics, jp is the P-value of Hansen J-statistics. F-statistics is the F-statistics for weak instrument identification.
^{a}P < .10.
^{b}P < .05.
^{c}P < .01.

In addition, the interaction between air transport infrastructure development and energy consumption and the interaction between rail transport infrastructure and energy consumption significantly reduce carbon emissions. These results imply that transportation infrastructure development is ensuring technological shifts or incorporating more efficient technologies, which are improving energy efficiency, thereby reducing the emission of CO_2 in the developing countries (see Xie et al., 2017; Neves et al., 2017). Also, it is argued that road transport or private car usage has been the fundamental source of CO_2 emissions in developing countries. However, the development of air and rail infrastructure could reduce the demand for private car usage for long-distance travel, which, therefore, minimizes energy use and the emissions of CO_2 (Lin and Du, 2017; Neves et al., 2017).

3.1 Sensitivity analysis

It is contested that the effect of transport infrastructure development on CO_2 emissions depends on the extent of economic development of countries. To control for the heterogeneity, this section conducts sensitivity analysis by categorizing the full-sample into different income groupings. The results for low income, lower-middle income, and upper-middle income countries are presented in Tables 2.4–2.6, respectively. The results presented in Table 2.4 show that GDP per capita and GDP per capita squared do not affect CO_2 emissions in low income countries. This outcome reveals that for low income countries, GDP per capita is not a primary contributor to emissions of CO_2. In addition, it is observed from Table 2.5 that GDP per capita and GDP per capita squared have a respective significant negative and positive effect on CO_2 emissions in lower middle-income countries. Thus, in the lower-middle income countries, the relationship between GDP per capita and CO_2 emissions is a U-shaped relationship. Contrarily, the results presented in Table 2.6 reveal that GDP per capita and GDP per capita squared, respectively, have a statistically significant increase and reducing effect on CO_2 emissions in the upper-middle income countries. Thus, the argument of the EKC hypothesis that the relationship between GDP per capita and CO_2 emissions is an inverted U-shape holds for upper-middle income countries. These results suggest that the EKC hypothesis depends on the extent of economic development (see Acheampong, 2019).

The results indicate that population size reduces the emission of CO_2 in low income countries, whereas it results in higher CO_2 emissions in lower-middle income and upper-middle income countries. The implication is that household consumption tends to be low in the low income countries, thereby contributing less to carbon emissions; however, given the relatively high income in the lower-middle and upper-middle income countries, households' consumption tends to be unsustainable, thereby contributing higher to carbon emissions. Similar to the full sample results, energy consumption generates higher CO_2 emissions across all the income groups. However, the estimated elasticities of energy use on the emissions of CO_2 are larger in upper-middle income countries, followed by lower-middle income countries and low-income countries. Further evidence from Tables 2.4–2.6 reveal that trade openness fosters emissions of CO_2 in all the income groups.

Interestingly, the results suggest that foreign direct investment significantly generates higher emission of CO_2 in low income countries, whereas, for upper-middle income countries, it significantly reduces CO_2 emissions. On the other hand, the impact of FDI on the emissions of CO_2 in lower-middle income countries is negligible. Consistent with the literature argument (for instance, see Doytch and Uctum, 2016), our study validates that the *pollution haven hypothesis* exists in low income countries while the *pollution-halo hypothesis* exists in high income countries. The evidence presented in

Table 2.4 Results for low income countries.

	Model 1	Model 2	Model 3	Model 4	Model 5	Model 6
lngdpc	1.983	1.113	−1.359	1.208	1.363	0.514
	(1.942)	(3.650)	(2.019)	(3.394)	(2.026)	(3.498)
lngdpc2	−0.021	−0.010	0.050	−0.012	−0.007	0.004
	(0.042)	(0.082)	(0.043)	(0.076)	(0.044)	(0.078)
lnpop	−0.567^c	0.253	−0.542^c	0.260	−0.579^c	0.184
	(0.068)	(0.334)	(0.066)	(0.323)	(0.070)	(0.302)
lnener	0.608^c	1.189^c	0.576^c	1.194^c	0.481^c	1.752
	(0.105)	(0.288)	(0.104)	(0.285)	(0.116)	(3.035)
lntrade	0.232^c	0.286^c	0.218^c	0.285^c	0.224^c	0.296^c
	(0.072)	(0.068)	(0.063)	(0.067)	(0.070)	(0.077)
lnurbpopt	−0.332^c	−1.235^c	−0.228^b	−1.241^c	−0.334^c	−1.194^c
	(0.100)	(0.225)	(0.094)	(0.225)	(0.101)	(0.214)
lnfdi	0.041^b	−0.033	0.036^b	−0.033	0.036^b	−0.038
	(0.016)	(0.030)	(0.016)	(0.030)	(0.017)	(0.029)
lncredit	0.190^c	0.156^c	0.200^c	0.157^c	0.184^c	0.157^c
	(0.032)	(0.033)	(0.030)	(0.034)	(0.033)	(0.052)
lnair	−0.102^c		−1.511^c		−0.661	
	(0.025)		(0.492)		(0.424)	
lnrail		−0.522^b		−0.544		−0.015
		(0.230)		(1.103)		(2.360)
lnair×lngdpc			0.061^c			
			(0.021)			
lnrail×lngdpc				0.001		
				(0.050)		
lnair×lnener					0.093	
					(0.070)	
lnrail×lnener						−0.079
						(0.407)
Constant	−20.746	−17.347	17.951	−18.432	−12.809	−13.550
	(22.216)	(44.194)	(23.311)	(41.933)	(23.268)	(49.055)
Observations	112	62	112	62	112	62
r2	0.901	0.936	0.910	0.936	0.902	0.936
J	0.437	2.768	0.479	2.759	0.580	2.848
jp	0.804	0.251	0.787	0.252	0.748	0.241

Robust standard errors in parentheses. J is Hansen J-statistics, jp is the P-value of Hansen J-statistics. F-statistics is the F-statistics for weak instrument identification.
^{b}P $<.05.$
^{c}P $<.01.$

Table 2.5 Results for lower-middle income countries.

	Model 1	Model 2	Model 3	Model 4	Model 5	Model 6
lngdpc	−3.218^c	−2.670^c	−2.964^c	−4.047^c	−3.251^c	−2.672^c
	(0.282)	(0.498)	(0.788)	(0.903)	(0.289)	(0.529)
lngdpc2	0.061^c	0.051^c	0.055^c	0.087^c	0.062^c	0.051^c
	(0.005)	(0.010)	(0.016)	(0.021)	(0.006)	(0.011)
lnpop	1.230^c	1.033^c	1.220^c	1.060^c	1.208^c	1.034^c
	(0.054)	(0.097)	(0.061)	(0.114)	(0.061)	(0.098)
lnener	0.980^c	0.784^c	0.980^c	0.738^c	1.029^c	0.778^c
	(0.038)	(0.062)	(0.038)	(0.069)	(0.076)	(0.277)
lntrade	0.024	0.219^c	0.022	0.248^c	0.019	0.218^c
	(0.021)	(0.071)	(0.021)	(0.066)	(0.022)	(0.079)
lnurbpopt	1.001^c	0.731^c	0.997^c	0.757^c	0.980^c	0.732^c
	(0.077)	(0.146)	(0.078)	(0.169)	(0.082)	(0.149)
lnfdi	0.017	−0.029	0.016	−0.031	0.018	−0.029
	(0.014)	(0.021)	(0.015)	(0.021)	(0.015)	(0.022)
lncredit	0.266^c	0.381^c	0.263^c	0.390^c	0.267^c	0.381^c
	(0.026)	(0.035)	(0.030)	(0.036)	(0.026)	(0.038)
lnair	0.062^c		−0.077		0.169	
	(0.016)		(0.350)		(0.134)	
lnrail		0.220^c		1.641^c		0.215
		(0.033)		(0.630)		(0.211)
lnair×lngdpc			0.006			
			(0.014)			
lnrail×lngdpc				−0.056^b		
				(0.025)		
lnair×lnener					−0.017	
					(0.021)	
lnrail×lnener						0.001
						(0.032)
Constant	20.197^c	15.423^b	17.421^b	26.652^c	20.514^c	15.485^b
	(3.221)	(6.232)	(8.714)	(8.965)	(3.305)	(7.669)
Observations	553	365	553	365	553	365
r2	0.956	0.955	0.956	0.955	0.956	0.955
J	0.132	0.090	0.140	0.015	0.140	0.092
jp	0.716	0.765	0.709	0.903	0.708	0.761

Robust standard errors in parentheses. J is Hansen J-statistics, jp is the P-value of Hansen J-statistics. F-statistics is the F-statistics for weak instrument identification.
^{b}P < .05.
^{c}P < .01.

Table 2.6 Results for upper-middle income countries.						
	Model 1	**Model 2**	**Model 3**	**Model 4**	**Model 5**	**Model 6**
lngdpc	−0.131	0.779[c]	−0.914[c]	0.107	−0.207	0.671[b]
	(0.188)	(0.252)	(0.284)	(0.638)	(0.202)	(0.276)
lngdpc2	0.002	−0.017[c]	0.019[c]	0.001	0.003	−0.015[c]
	(0.004)	(0.005)	(0.006)	(0.016)	(0.004)	(0.006)
lnpop	1.071[c]	1.186[c]	1.097[c]	1.172[c]	1.069[c]	1.170[c]
	(0.025)	(0.036)	(0.025)	(0.037)	(0.024)	(0.040)
lnener	1.140[c]	1.093[c]	1.138[c]	1.068[c]	1.179[c]	1.450[c]
	(0.024)	(0.042)	(0.024)	(0.044)	(0.043)	(0.170)
lntrade	0.077[c]	0.049	0.089[c]	0.047	0.077[c]	0.039
	(0.028)	(0.046)	(0.027)	(0.048)	(0.028)	(0.047)
lnurbpopt	0.221[c]	0.267[c]	0.220[c]	0.282[c]	0.224[c]	0.272[c]
	(0.037)	(0.074)	(0.036)	(0.076)	(0.037)	(0.071)
lnfdi	−0.014[b]	0.019[a]	−0.008	0.017	−0.015[b]	0.016
	(0.007)	(0.011)	(0.007)	(0.011)	(0.007)	(0.010)
lncredit	0.064[c]	0.034[a]	0.084[c]	0.052[b]	0.064[c]	0.040[b]
	(0.014)	(0.018)	(0.016)	(0.022)	(0.014)	(0.019)
lnair	−0.003		0.394[c]		0.066	
	(0.006)		(0.090)		(0.054)	
lnrail		−0.004		0.682		0.340[b]
		(0.017)		(0.517)		(0.143)
lnair×lngdpc			−0.016[c]			
			(0.004)			
lnrail×lngdpc				−0.026		
				(0.020)		
lnair×lnener					−0.009	
					(0.007)	
lnrail×lnener						−0.045[b]
						(0.019)
Constant	−14.292[c]	−26.875[c]	−5.761[a]	−20.986[c]	−13.595[c]	−27.965[c]
	(2.418)	(3.355)	(3.347)	(6.291)	(2.527)	(3.188)
Observations	687	381	687	381	687	381
r2	0.984	0.983	0.984	0.983	0.984	0.983
j	0.108	0.119	0.061	0.123	0.169	0.127
jp	0.742	0.730	0.804	0.726	0.681	0.722

Robust standard errors in parentheses. J is Hansen J-statistics, jp is the P-value of Hansen J-statistics. F-statistics is the F-statistics for weak instrument identification.
[a]$P < .10$.
[b]$P < .05$.
[c]$P < .01$.

Table 2.4 reveals that urbanization impedes the emission of CO_2 in low income countries, whereas it induces higher CO_2 emissions in lower-middle and upper-middle income countries (see Tables 2.5 and 2.6). This depicts that the role of urbanization on the emissions of CO_2 is context-specific. Similar to the full sample results, financial development significantly induces higher CO_2 emissions in all the income groups.

It is observed from Table 2.4 that both air and rail transport infrastructure development significantly reduce emissions of CO_2 in low income countries. Conversely, both air and rail transport infrastructure significantly increase CO_2 emissions in lower-middle income countries (see Table 2.5) while their impact in the upper-middle income countries is insignificant (see Table 2.6). For the transmission channels, the interaction between air transport infrastructure and GDP per capita exerts a significant positive effect on CO_2 emissions while the moderation effect of rail transport infrastructure and GDP per capita plays no role in CO_2 emissions in low income countries (see Table 2.4).

On the other hand, the interaction term for air transport infrastructure and GDP per capita plays no role in CO_2 emissions while the interaction term of rail transport infrastructure and GDP per capita significantly reduce CO_2 emissions in lower-middle income countries (see Table 2.5). For upper-middle income countries (see Table 2.6), the interaction term for air transport infrastructure and GDP per capita significantly retards CO_2 emissions while the interaction term of rail transport infrastructure and GDP per capita is not sufficient to reduce CO_2 emissions. In addition, the moderation term of air transport infrastructure and energy consumption and rail transport infrastructure and energy consumption does not significantly impact CO_2 emissions in low income and lower-middle income countries (see Tables 2.4 and 2.5). For upper-middle income countries, the moderation term for air transport infrastructure and energy consumption has no influence on the emission of CO_2 while the moderation effect of rail transport infrastructure and energy consumption significantly reduces CO_2 emissions.

4. Conclusion and policy suggestions

This study contributes to policy and the debate on transportation infrastructure and climate change in developing countries. In lieu of this objective, our study employed the IV-GMM, which is capable of controlling endogeneity, to examine the impact air and rail transportation infrastructure development on CO_2 emissions for 113 developing countries between the years 1990 and 2018. Our study revealed that the EKC hypothesis does not exist in developing countries. In addition, factors such as population size, energy consumption, trade openness, urbanization, and financial development worsen carbon emissions while FDI mitigates the emission of CO_2 in the developing countries. Our study established that both air and rail transport infrastructure contribute directly to higher carbon emissions. Indirectly, it was observed that air transport infrastructure moderates GDP per capita to exert an insignificant effect on carbon emissions while rail transport infrastructure moderates GDP per capita to induce higher CO_2 emissions in developing countries. Also, air and rail transport infrastructure moderates the effect of energy consumption to reduce the emission of carbon in developing countries. It must be acknowledged that these results differ across low income, lower-middle income, and upper-middle income countries.

The policy implications of these findings are that for developing countries to curb carbon emissions, there should be regulation measures that encourage the adoption of environmentally efficient transport technologies. Additionally, industrial players, urban planners, and governments could coordinate and adopt sustainable transport approaches such as *park and ride*, *ride share*, or *car-pooling* to mitigate the contribution of the transport infrastructures to carbon emissions. Since developing countries are major importers of second-hand and old vehicles, which are environmentally inefficient, policymakers could incentivize the importation of zero-emissions vehicles and environmentally efficient nonoveraged vehicles by relatively lowering tariffs imposed on them. Additionally, policymakers in developing countries should focus on the digitization of the economy, which could reduce the demand for travel. Additionally, FDI is proven to mitigate carbon emissions in developing countries. Therefore, implementing policies that could provide an enabling environment for attracting FDI to developing countries is critical. However, it must also be noted that not all FDI is environmentally friendly; therefore, strengthening and enforcing existing environmental regulation and urging investments sectors that are environmentally efficient should be encouraged.

Additionally, adopting renewable energy and improving energy efficiency remains critical to mitigating carbon emissions in developing countries. While urbanization is associated with higher carbon emissions, urban planners should focus on building urban infrastructure facilities that are energy and environmentally efficient. With trade openness increasing CO_2 emissions, it is recommended that policymakers should consider trade liberalization when designing and implementing strategies to curb carbon emissions in developing countries. Furthermore, given that financial development induces higher carbon emissions, the financial sectors could help control carbon emissions by providing credits/loans at lower rates of interest to investors who are ready to embark on environmentally friendly projects.

Appendix

Table 2.A.1 Countries included in the study.
Low income countries: Afghanistan, Benin, Burkina Faso, Central African Republic, Chad, Congo, Dem. Rep., Ethiopia, Guinea, Guinea-Bissau, Haiti, Korea, Dem. People's Rep., Liberia, Madagascar, Malawi, Mali, Mozambique, Nepal, Niger, Rwanda, Sierra Leone, Somalia, Syrian Arab Republic, Tajikistan, Tanzania, Togo, Uganda, Yemen, Rep.
Lower-middle income countries: Angola, Bangladesh, Bhutan, Bolivia, Cabo Verde, Cambodia, Cameroon, Congo, Rep., Cote d'Ivoire, Djibouti, Egypt, Arab Rep., El Salvador, Eswatini, Ghana, Honduras, India, Indonesia, Kenya, Kiribati, Kyrgyz Republic, Lao PDR, Mauritania, Moldova, Mongolia, Morocco, Myanmar, Nicaragua, Nigeria, Pakistan, Papua New Guinea, Philippines, Senegal, Solomon Islands, Sudan, Tunisia, Ukraine, Uzbekistan, Vanuatu, Vietnam, Zambia, Zimbabwe
Upper-middle income countries: Algeria, Argentina, Armenia, Azerbaijan, Belarus, Bosnia and Herzegovina, Botswana, Brazil, Bulgaria, China, Colombia, Costa Rica, Cuba, Dominican Republic, Ecuador, Fiji, Gabon, Georgia, Guatemala, Guyana, Iran, Islamic Rep., Iraq, Jamaica, Jordan, Kazakhstan, Lebanon, Libya, Malaysia, Maldives, Marshall Islands, Mauritius, Mexico, Namibia, Paraguay, Peru, Romania, Russian Federation, Samoa, South Africa, Sri Lanka, Suriname, Thailand, Turkey, Turkmenistan, Venezuela, RB.

References

Acheampong, A.O., 2019. Modelling for insight: does financial development improve environmental quality? Energy Econ. 83, 156−179. https://doi.org/10.1016/j.eneco.2019.06.025.

Acheampong, A.O., Adams, S., Boateng, E., 2019. Do globalisation and renewable energy contribute to carbon emissions mitigation in sub-Saharan Africa? Sci. Total Environ. 677, 436−446. https://doi.org/10.1016/j.scitotenv.2019.04.353.

Acheampong, A.O., Amponsah, M., Boateng, E., 2020. Does financial development mitigate carbon emissions? Evidence from heterogeneous financial economies. Energy Econ. 88, 104768. https://doi.org/10.1016/j.eneco.2020.104768.

Achour, H., Belloumi, M., 2016. Investigating the causal relationship between transport infrastructure, transport energy consumption and economic growth in Tunisia. Renew. Sustain. Energy Rev. 56, 988−998. https://doi.org/10.1016/j.rser.2015.12.023.

Adams, S., Acheampong, A.O., 2019. Reducing carbon emissions: the role of renewable energy and democracy. J. Clean. Prod. 240, 118245. https://doi.org/10.1016/j.jclepro.2019.118245.

Administration, E.I., Office, G.P., 2016. International Energy Outlook 2016, with Projections to 2040. Government Printing Office.

Awaworyi Churchill, S., Inekwe, J., Smyth, R., Zhang, X., 2019. R&D intensity and carbon emissions in the G7: 1870−2014. Energy Econ. 80, 30−37. https://doi.org/10.1016/j.eneco.2018.12.020.

Baum, C.F., Schaer, M.E., Stillman, S., 2002. Instrumental Variables and GMM: Estimation and Testing. Working Paper. Boston College Economics.

Doytch, N., Uctum, M., 2016. Globalisation and the environmental impact of sectoral FDI. Econ. Syst. 40 (4), 582−594. https://doi.org/10.1016/j.ecosys.2016.02.005.

Dranove, D., 2009. Model Specification: Choosing the Right Variables for the Right Hand Side. Kellogg School of Management [On-line].

Dzator, J., Acheampong, A.O., 2020. The impact of energy innovation on carbon emissions mitigation: an empirical evidence from OECD countries. In: Hussain, C.M. (Ed.), Handbook of Environmental Materials Management. Springer International Publishing, Cham.

Esfahani, H.S., Ramırez, M.a.T., 2003. Institutions, infrastructure, and economic growth. J. Dev. Econ. 70 (2), 443−477.

Hakimi, A., Hamdi, H., 2016. Trade liberalisation, FDI inflows, environmental quality and economic growth: a comparative analysis between Tunisia and Morocco. Renew. Sustain. Energy Rev. 58, 1445−1456. https://doi.org/10.1016/j.rser.2015.12.280.

Khan, F., 2019. Crediting emissions. Nat. Energy 4 (8), 628. https://doi.org/10.1038/s41560-019-0453-8.

Lakshmanan, T.R., 2011. The broader economic consequences of transport infrastructure investments. J. Transport Geogr. 19 (1), 1−12. https://doi.org/10.1016/j.jtrangeo.2010.01.001.

Lin, B., Du, Z., 2017. Can urban rail transit curb automobile energy consumption? Energy Pol. 105, 120−127. https://doi.org/10.1016/j.enpol.2017.02.038.

Marazzo, M., Scherre, R., Fernandes, E., 2010. Air transport demand and economic growth in Brazil: a time series analysis. Transport. Res. E Logist. Transport. Rev. 46 (2), 261−269.

Neves, S.A., Marques, A.C., Fuinhas, J.A., 2017. Is energy consumption in the transport sector hampering both economic growth and the reduction of CO_2 emissions? A disaggregated energy consumption analysis. Transport Pol. 59, 64−70. https://doi.org/10.1016/j.tranpol.2017.07.004.

Poumanyvong, P., Kaneko, S., 2010. Does urbanisation lead to less energy use and lower CO_2 emissions? A cross-country analysis. Ecol. Econ. 70 (2), 434−444. https://doi.org/10.1016/j.ecolecon.2010.09.029.

Saidi, S., Hammami, S., 2017. Modeling the causal linkages between transport, economic growth and environmental degradation for 75 countries. Transport. Res. Transport Environ. 53, 415−427. https://doi.org/10.1016/j.trd.2017.04.031.

Saidi, S., Shahbaz, M., Akhtar, P., 2018. The long-run relationships between transport energy consumption, transport infrastructure, and economic growth in MENA countries. Transport. Res. Pol. Pract. 111, 78−95. https://doi.org/10.1016/j.tra.2018.03.013.

Shahbaz, M., Nasreen, S., Ahmed, K., Hammoudeh, S., 2017. Trade openness−carbon emissions nexus: the importance of turning points of trade openness for country panels. Energy Econ. 61, 221−232. https://doi.org/10.1016/j.eneco.2016.11.008.

Shahbaz, M., Nasir, M.A., Roubaud, D., 2018. Environmental degradation in France: the effects of FDI, financial development, and energy innovations. Energy Econ. 74, 843−857. https://doi.org/10.1016/j.eneco.2018.07.020.

Shi, A., 2003. The impact of population pressure on global carbon dioxide emissions, 1975−1996: evidence from pooled cross-country data. Ecol. Econ. 44 (1), 29−42. https://doi.org/10.1016/S0921-8009(02)00223-9.

Stern, D.I., 2004. The rise and fall of the environmental Kuznets Curve. World Dev. 32 (8), 1419−1439. https://doi.org/10.1016/j.worlddev.2004.03.004.

Stern, D.I., Common, M.S., 2001. Is there an environmental Kuznets Curve for sulfur? J. Environ. Econ. Manag. 41 (2), 162−178. https://doi.org/10.1006/jeem.2000.1132.

Tong, T., Yu, T.-H.E., Cho, S.-H., Jensen, K., De La Torre Ugarte, D., 2013. Evaluating the spatial spillover effects of transportation infrastructure on agricultural output across the United States. J. Transport Geogr. 30, 47−55. https://doi.org/10.1016/j.jtrangeo.2013.03.001.

World Bank, 2019. World Development Indicators. http://databank.worldbank.org/data/reports.aspx?source=world-development-indicators#.

Xie, R., Fang, J., Liu, C., 2017. The effects of transportation infrastructure on urban carbon emissions. Appl. Energy 196, 199−207. https://doi.org/10.1016/j.apenergy.2017.01.020.

Yeh, J.-C., Liao, C.-H., 2017. Impact of population and economic growth on carbon emissions in Taiwan using an analytic tool STIRPAT. Sustain. Environ. Res. 27 (1), 41−48. https://doi.org/10.1016/j.serj.2016.10.001.

Zhu, Q., Peng, X., 2012. The impacts of population change on carbon emissions in China during 1978−2008. Environ. Impact Assess. Rev. 36, 1−8. https://doi.org/10.1016/j.eiar.2012.03.003.

Zhu, H., Duan, L., Guo, Y., Yu, K., 2016. The effects of FDI, economic growth and energy consumption on carbon emissions in ASEAN-5: evidence from panel quantile regression. Econ. Modell. 58, 237−248. https://doi.org/10.1016/j.econmod.2016.05.003.

Interconnecting the environment with economic development of a nation

3

Shikha Singh, Abhinav Yadav

Integrative Ecology Laboratory (IEL), Institute of Environment & Sustainable Development (IESD), Banaras Hindu University, Varanasi, Uttar Pradesh, India

Chapter outline

1. Introduction

According to a report of *World Economic Outlook* (2020), economic growth is projected to rise by 3.3% in 2020 and 3.4% in 2021 in comparison to 2.9% in 2019. The increase in economic growth has resulted in worsening of environmental quality. As per International Energy Agency (2018) report, the level of production of goods and services doubled at a global scale from 1990 to 2015, which increased

45% of greenhouse gas (GHG) emissions and reached up to 50 gigatons (Gt) of carbon dioxide equivalent. It is well known that economic growth is crucial for giving greater opportunities to poor people, but it should be accomplished using more energy-efficient and environmentally sound models of growth. Moreover, continuous growth is necessary to develop more environmentally safe and adequate technologies to achieve sustainability goals. There is a strong link between the environment and sustainability, which is considered as a holistic approach for social well-being without any discrepancy. Now, one of the primary challenges is to assure that the developmental policies made for economic growth should be executed in a manner that works in harmony with environmental sustainability. The problems of environmental degradation and climate change have gained much attention over the past years, as they affect both developed and developing countries, though, the world has shifted toward a sustainable approach for meeting the demands of production and consumption. But still, the pace of this transition is not fast enough to reduce the extent of environmental problems.

However, in 1992, the relationship between environment and development was first discussed on a common platform at Rio in Brazil, which is known as "Earth Summit" or the United Nations Conference on Environment and Development (UNCED). This conference introduced the concept of "Sustainable Development," which created more awareness in both developed and developing countries. However, for developing countries, where poverty is a major issue, environmental problems are gaining more attention, which requires the acceleration of economic growth. From the past few decades, economic growth is the primary aim of policymakers, which is considered as the tool for sustainable development. A fast-growing economy is always in demand due to its favorable socio-economic outcomes; but at the same time, it also requires a healthy environment. In different studies, it has been established that the connection between economic development and the environment is intricate (Coondoo and Dinda, 2008; Grossman and Krueger, 1995; Lee and Lee, 2009; Lopez et al., 2014; Akbostanci et al., 2009). All human activities have a direct or indirect impact on the environment, which is formulated and implemented for economic growth to meet the needs of a country. The argument on the relationship of economic growth with the environment in the 1970s and for a significant part of the 1980s was mainly governed by the material balance paradigm. It strongly suggested that ceteris paribus, i.e., economic development contributes to ecological degradation and that if a financial framework is physically in a steady state it can be environmentally sustainable. The amount of resources used in production for the well-being of humans is restrained to a level that does not overutilize its natural assets and exhaust nature's sinks (Stagl, 1999; Smulders, 2000).

Understanding the role of the natural environment in the economic growth of a country is necessary to meet the goals of sustainable development. Environment consists of terrestrial and aquatic assets and also the atmosphere, altogether which are essential for the economy. For the production of goods and services, nature provides the material inputs and also acts as a sink in which the wastes arising from economic activities are dumped. Natural resources have been serving mankind for several decades, and still they are an important part of economic development. The environment functions as a life support system for mankind. Any variation in the social and economic fields may have a positive or negative effect on the earth's environment and vice-versa, and most of the time, the negative results are irreversible. It is necessary to address the environmental issues now, as we are facing the challenge of environmental degradation and climate change across the globe.

In today's world, one of the biggest challenges faced by several economies is accomplishing a harmony between reducing environmental degradation amid the requirement for economic development and environmental feasibility. However, for the protection of the environment, several policies have been adopted by countries around the world, for example, the reduction of GHG emissions and prevention of climate change. It is now a well-known fact that carbon dioxide emission and other GHGs are mainly responsible for global climate change and the greenhouse effect (Lee and Chang, 2009). In transportation, the CO_2 emissions contribute around 90% of all released GHGs (Nocera and Cavallaro, 2011). Therefore, CO_2 emissions have been a subject of considerable attention concerning the reduction of polluting gases. As per the IPCC report (2018), anthropogenic sources, such as emissions of GHG and land-use land-cover (LULC) changes, have resulted in approximately 1°C of global warming which is higher than preindustrial levels, i.e., from 1850 to 1900. Between 2030 and 2052, global warming will reach around 1.5°C; if in case it keeps on expanding at the current rate it will increase to, for example, about 2°C every decade. By 2100, global warming is supposed to reach 3–4°C above preindustrial levels with conceivably further warming in the future. It has been proposed that climate change will substantially decrease economic development in several developed and developing nations. Monetary and technological changes are fundamental to limit the worldwide temperature increase to 1.5–2.0°C above premodern levels. The major components of such transition include the rapid decoupling of financial development from energy consumption and emissions of CO_2 and the adoption of low and zero-carbon or carbon-negative techniques globally.

In the early 1990s, certain questions were a significant part of several important studies (Apergis and Payne, 2010a, 2010b; Chang et al., 2009; Cheng, 1999). Will the constant economic growth prove to be more detrimental to the environment and earth or will higher incomes help in the recovery of the degraded environment? These studies proposed an inverted U-shaped relationship between per capita income and environmental deterioration related to pollutants such as SO_2, NOx, etc., and emission of carbon which was commonly known as Environment Kuznets Curve (EKC). The hypothesis of EKC received enormous attention in research and formulation of policies. This relationship for the first time stated the fact that economic growth could benefit the environment in several ways. EKC hypothesis describes an inverted U-shaped relationship, which means that as the income level increases the different indicators of environmental degradation initially decline and subsequently fall when income crosses some threshold limit.

Despite decades of research, there is no general acceptance of how development and environment are connected and the factors determining this relationship. This is mostly because of the absence of appropriate data and observational proof and a short period of time to allow robust connections and reliable projections. There is still significant scientific inaccuracy in characterizing the discussion, for example, on the utilitarian relationship between certain air and water contaminants and economic development, between climate change and economic development, and between natural resource utilization and economic development. Some of the researches in the literature describe "growth optimism" concerning the effect of growth and environmental quality and accomplishing sustainability of the environment, while others discover evidence for "growth pessimism" i.e., economic development is causing harm to the earth, in short, and/or long term. Thus, the objective of this chapter is to explore the relationship between the natural environment and economic growth and vice-versa. The current work gives a general survey of the existing literature on EKC, which deals with the "energy-environment growth" nexus for both specific and multiple-country studies. Studies from different countries show that the connection between economic growth and pollution emissions differs significantly. These studies may be useful to draft policy recommendations for conservation of energy, emission reduction, and better efficiency in economic growth. For certain pollutants and certain countries, there is proof for EKC; however, its existence locally does not imply that it is a predefined, predictive, or robust relationship on a larger scale and across time, and it is not meant for all forms of environmental constraints. Therefore, it is preferable to perform a separate investigation for each country to determine the major factors influencing certain indicators of the environment precisely.

Therefore, the goal of this chapter is mainly to discuss the role of nature in building a country's economy and also the connection between growth and environment, which can be best described through EKC. So, based on this, the chapter consists of seven sections including introduction. The next section includes the role of natural resources and the environment in economic growth. The third section discusses the relationship between the environment and growth. In Section 4 which is the main part of the chapter, the nature of the interaction between growth and environment is explained through EKC. This includes a selective review of literature on the relevance of EKC focusing on different studies conducted in various countries. It also provides theoretical and empirical evidences for EKC. Section 5 includes a short discussion on the effect of growth on the environment in the form of pollution. It also highlights the connection between CO_2 emissions and economic growth. Section 6 emphasizes on regulations and policy recommendations. The last section is the conclusion of the chapter.

2. Role of natural resources and the environment in economic growth

The natural environment assumes two vital roles in economic development. First, the environment supplies resources, which work as raw material for the manufacturing of goods and services. Manufacturing of goods relies upon (a) advancements of fundamental services (b) movements of removed natural goods (organic and geographical, inexhaustible and exhaustible), and (c) ecosystem services provided by natural frameworks. The Millennium ecosystem services (2005) has divided the services provided by nature into four broad categories: (1) *Provisioning services*: products like fresh water, air, food, wood, medicines, biochemical, etc. (2) *Regulating services:* regulation of natural processes, climate, water purification, erosion, disease control, air quality, etc. (3) *Supporting services:* soil formulation, photosynthesis, nutrient and water cycling, etc. (4) *Cultural services*: tourism, recreation, spiritual enrichment, intellectual development, etc. (Fig. 3.1).

The amount of wastes discharged to the earth's atmosphere depends upon the volume of material yield and on the development of services originated from another type of presumption by society, which is implemented to cope up with the undesirable effects of goods produced inside the economic structure. In literature, such kinds of services are outlined with the shorthand "by-products management capital" (Toman, 2003). The term "by-products management" refers both to pollution control (decrease in unwanted harmful results in comparison with wanted goods) and end-of-pipe solutions that reduce the harm caused by physical emissions. The inputs can be direct or indirect. Second, the natural environment functions as a sink to the pollution which arises from economic productions and utilizations. Models incorporate harmful air, water, and solid contaminations, which are dispersed in the environment, which is additionally a storehouse for solid and hazardous waste.

At the point when the elements of the natural environment are truly impeded, economic development decelerates or can be negative. This is the situation when access to natural resources goes down quickly, for example, aquatic resources, woods, and minerals are being drained, or when nature's ability to assimilate or disperse waste and toxins is surpassed and when ecological quality is decreased. When the environmental quality is degraded, it has detrimental impacts on economic profitability,

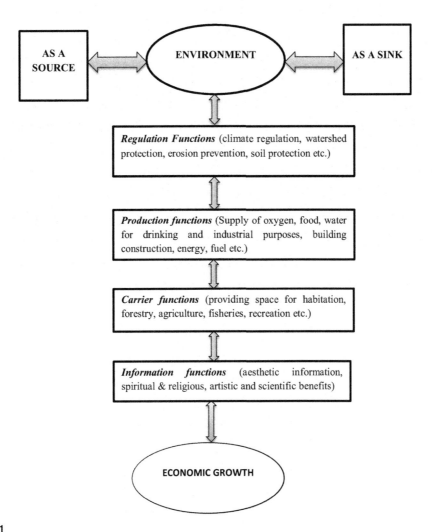

FIGURE 3.1

Environmental functions supporting economic growth.

Source: Heywood, V.H., Watson, R.T., 1995. Global Biodiversity Assessment, vol. 1140. Cambridge University Press, Cambridge.

leading to a decline in the efficiency of diverse environmental services and some natural resources. It also has an immediate adverse impact on having spread out the different pathways through which natural assets and the environment are related to economic development and human well-being. Development might also be constrained because of policy measures which need huge investments for the reduction of pollution, which have lower economic profitability and returns in contrast with alternative costs.

From the past few decades, the environment is under immense pressure due to overexploitation by humans to meet their needs by being dependent on air, and water resources. Along with this, the environmental degradation is also caused by deforestation, species loss, climatic change, etc. Unrestricted utilization of natural assets has led to their quantitative and qualitative deterioration. For example, overfishing caused a reduction in catch per effort and subsequent changes in several aquatic systems disturbing the balance in ecological species. Overexploitation of groundwater has led to the deprivation of safe drinking water in several cities. Irrational exploitation of groundwater in coastal regions has led to seawater intrusion and salinization. For example, saltwater intrusion and salinization in coastal Bangladesh, coastal farms, and forests in southeast United States and California are caused by sea-level rise, storms and tides, droughts, and water resources management practices. Extensive deforestation has caused the loss of biodiversity and vegetation, which results in degradation and depletion of the soil. Several environmental problems such as eutrophication of water resources, global warming, climate change, and depletion of ozone have been caused due to industrialization.

This section of the chapter aims to explore the role of natural assets of the environment in the economic growth and well-being of humans. It also discusses how these assets are being overexploited in meeting the demands of present generations. Presently, these resources are utilized in such a manner that nothing will be left for our future generations. Ecosystems have a broad range of effects on both quality and quantity of development and vice-versa. The natural environment is essential for economic activity and growth, as it provides the resources we require to produce goods and services, and uptakes and treats the undesired byproducts in the form of pollution and waste. The environment manages threats to economic and social activity, as it helps to control flood risks, regulate the local climate, and retain the supply of clean water and other resources. This serves as the basis for economic activity and social welfare, and so balancing and restoring the natural assets is an important aspect in sustaining growth for the long term. Environmental performance and economic growth should run parallel to achieve sustainable development goals.

3. The environment and economic growth linkage

During the 1980s, there was a significant milestone in understanding the relation of economic development with the environment which identified the connection that existed between them. The environmental issues were incorporated in the planning process with an emphasis on a sustainable approach in the developmental process (World Commission on Environment and Development, 1987; Pearce and Warford, 1993). Grossman and Krueger (1991) supported inverted U-shaped linkage of economic growth (estimated by a rise in income per capita) with certain environmental quality indicators in their revolutionary effort of studying the probable impact of the North American Free Trade Agreement (NAFTA). The relationship is termed as Environmental "Kuznets curve" (EKC).

The growth-environment nexus has drawn the attention of researchers and academicians for discussions in different countries for a long time, and there is remarkable literature available on this relationship. Panayotou (1993), Grossman and Krueger (1993), and Selden and Song (1994) concluded that the positive or negative connection between economic development and the quality of the environment cannot be constant along developmental trajectories of a nation. It can change from positive to negative at an income level where people of the country need and support a strong economic base and a healthier natural environment. The environmental quality may produce positive or negative external factors that can stimulate economic growth by affecting human health. The relationship between energy vectors, economic development, and the quality of the environment was a matter of contradiction among the policymakers. This implies that understanding the dynamic linkage is essential to know the current policies on energy and environment, and it is one of the basic frameworks for designing robust economic guidelines with substantial goals.

The relationship of economic growth with the environment is, and may always remain, a subject of controversy. Most of the countries have achieved economic development without considering the environmental consequences. They are now facing several environmental problems like air and water pollution, pesticides in the food, ultraviolet rays penetrating the ozone layer, emission of greenhouse gases causing global warming, and so on. Some of the complexities in meeting the challenges of economic growth are the occurrence of new pollution problems, failure in dealing with rising global temperatures, and the ever-increasing population. However, due to advancements in the field of technology, great progress has been made in providing sanitation facilities; improved air quality in major cities, and constant progression in human conditions. Economic growth degrades the environment through increasing pollution, while the degraded environment in turn limits the probability of further economic growth. Exhaustion of resources and waste production get accelerated as agricultural production and industrialization increase. On the other hand, at higher developmental levels, the organizational transformation towards information-based services, effective technological approach, and the necessity for improving the quality of environment results in stabilization with a gradual reduction in environmental deterioration (Panayotou, 1993).

Nature is a sink for the undesirable byproduct of commercial activities that have normally not been given much attention. Nature disseminates harmful air, water, and solid pollutants as a sink, which is the storehouse for millions of tons of garbage and hazardous chemicals (Fig. 3.2). Once the environment's capacity to absorb waste crosses a certain limit, environmental quality is affected and may limit the process of economic growth. This may be because degraded environmental quality demands cleaner technologies or mitigation efforts. It further lowers the investment returns, or maybe the damage caused to the ecosystem is such that it is beyond restoration. It finally establishes a new less productive and stable state.

Expandable and sustainable natural resources contribute to the generation of numerous products and services. If the structure of yield and the strategies for production were permanent, at that point harm to the earth would be inseparably connected to the size of the worldwide monetary movement. Yet, considerable proof proposes that improvement leads to a rise in an additional change in what an economy produces (Syrquin, 1989). On a primary level, the efforts promoting the change in the composition and strategies of production might be adequately strong to more than balance the adverse impacts of increased economic activities on the environment.

However, the linkage of economic growth with the environment is often explained with the help of the EKC. This postulate establishes a relationship that is not linear but is relevant for different nations. The primary indicators utilized to depict the variations in conditions of the environment have been designed and used in numerous countries. A higher rate of economic development is a major and long-term goal of both government and people, mainly in developing nations. The accelerated growth of the economy is closely linked to an increase in the manufacture and intake of commodities and services; as a result, this leads to an increase in the maximized goods of the people and consumption of income for each individual. Though evidences are supporting the existence of the EKC relationship for some countries, still it cannot be used for all types of environmental damage and across all the countries and income levels. Therefore, in the next section, we will discuss the EKC.

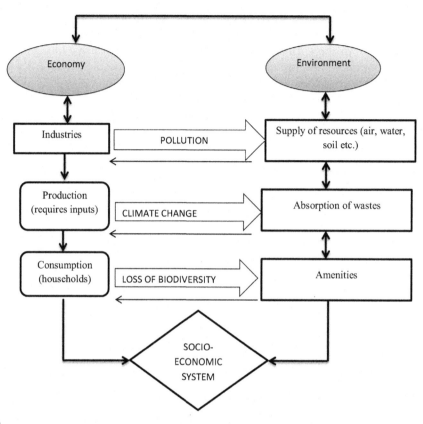

FIGURE 3.2

The link between economic growth and environment.

4. EKC hypothesis: explaining the relation between environment and economic growth

It is a well-known fact that the pollution and economy are closely related to each other since the history of mankind. However, the connection between environmental damage and economic improvement is unpredictable and complex. This linkage depends upon several factors like the economic size, the organization of the industry, the origin of the innovation, the need for better quality of environment, etc. All of these aspects are related to each other. The inconsistent link between the quality of income and per capita income can be identified practically and graphically utilizing the advanced tools of economic studies and can also be illustrated by the EKC. This theory was based on the proposition that at different levels of income growth, the distribution of income is not equal. Despite that, as the level of the economy expands, the distribution of income inclines to become even.

The EKC was introduced in 1955 for the first time by Kuznets to analyze the per capita income and environmental quality nexus. This assumes that during initial phases of economic development, the

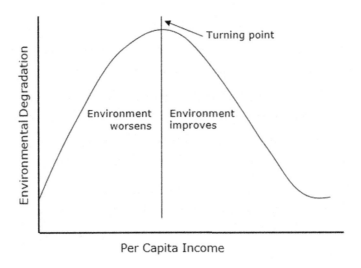

FIGURE 3.3

Environment Kuznets Curve.

damage to the environment goes up or increases at a higher rate. But, after a certain threshold limit of economic growth, the movement tends to become inverted at a higher degree of economic development (Usenata, 2018). Grossman and Krueger (1991) were the first to apply the EKC, which resulted in an inverse linkage of GDP per capita with an indicator of the quality of the environment (Fig. 3.3). In the early phase of industrialization, pollution in the EKC increases at a faster rate because individuals are more fascinated toward employments and income in comparison to healthy air and clean water. Most of the population is inefficient to even think about paying for reduction of pollution, and also environmental policies are correspondingly weak. With increase in income, the balance gets shifted. The leading industries tend to become cleaner, people start respecting the earth, and regulatory organizations gradually become more operational. Along the curve, pollution decreases in the medium income range but later on falls toward preindustrial levels in higher income range. At relatively lower degrees of per capita GDP, emissions increase with economic progression. At the final stage, when the economic growth increases along with the expansion of agronomy and extraction of different resources, the rate of resource exhaustion starts to transcend the rate of resource restoration, generating more hazardous wastes. A shift toward information-based industries increased environmental concerns, and implementation of environmental guidelines, improved techniques, and greater environmental costs causes progressive reduction of environmental deterioration at higher levels of development. As income exceeds the EKC turning point (Fig. 3.3), it is expected that progress toward the improvement of environmental quality begins (Arrow et al., 1995). Thus, the EKC indicates that economic development can be utilized to address the environmental issues as economic development is probably compatible with environmental recovery (Kijima et al., 2010).

The turning point at which the deviation or environmental pollution begins to decline in per capita income of the countries is a matter of argument among economists. This uncertainty brings up numerous issues like: Does pollution obey the " i.e., first increasing and subsequent decrease with income increment? What is the income level at which the reversal happens? Do all types of pollution have the same tendency of first rising and then falling? Is pollution a decrease in advanced economies due to insignificant changes or due to legislation? Since the middle of 1990s, several researchers attempted empirical surveys on EKC for various countries applying different environmental indicators, e.g., SO_2, CO_2, CO, SPM, and so on (Dinda, 2004; He, 2007; Kijima et al., 2010; Iwata et al., 2010; Shafik and Bandyopadhyay, 1992; Shapiro and Walker, 2018). However, a few studies have discovered limited evidence on EKC (e.g., Dinda, 2001; Harbaugh et al., 2002; Iwata et al., 2011; Soytas et al., 2007). For example, it was shown that the EKC theory for CO_2 discharges does not hold for non-organisation for economic co-operation and development (OECD) nations (Iwata et al., 2011). The proof from the previous works proposes that differences across the countries witnessed socially, financially, politically influence the environment and different phases of financial growth in several nations. Hence, the conclusion drawn from an individual country utilizing certain ecological indicators could not be inferred for other countries or probably for different kinds of contaminants. In a nutshell, the EKC is particularly a country-specific or indicator-specific phenomenon as the outcomes fluctuate according to the countries and the quality of the environment (Dinda, 2004; Nasir and Rehman, 2011).

4.1 Factors affecting EKC hypothesis

Certain factors that affect EKC. Each factor is described individually in the subsections below.

4.1.1 Income elasticity

With an increase in income, individuals accomplish a better quality of living and care more for the environment. The urge for a healthier environment creates basic changes in the economy that diminishes the degradation of the environment. The most widely recognized clarification for the state of an EKC is the idea that when a nation accomplishes an adequately elevated requirement of living, individuals append expanding an incentive to natural conveniences (Selden and Song, 1994; Baldwin, 1995). When individuals reach a certain level of income, readiness to pay for the betterment of the environment quality prevails to a higher degree (Roca, 2003).

4.1.2 Scale, technological, and composition effects

The environmental quality is affected by financial development through scale, technological, and composition effects (Grossman and Krueger, 1991). More natural resources are being harnessed as the increasing output needs more raw materials as inputs to fulfill the requirements. This generates higher GHG emissions and a large amount of wastes as byproducts. As a result, economic growth exhibits a scale effect that adversely affects the environment. However, the composition effect has positive effects on economic growth. With the increment in income, the framework of the economy also changes and gradually shifts toward cleaner technologies that produce less pollution. As the economy moves from rural to urban, farming to industrial, environmental deterioration will also increase. Nonetheless, it starts falling with a change from the power sector to administrative and information technologies.

4.1.2.1 International trade

International trade can be the reason for environmental degradation, as it prompts increment in the size of the economy that accelerates pollution. It is one of the most significant components that can clarify EKC. Although, in the past years, several economic experts proposed that trading is not the main controller of environmental degradation (Jones and Rodolfo, 1995; Lee and Roland-Holst, 1997), but still it can be said that uncontrolled trade has inverse impacts on the environment, in the form of increasing pollution. The quality of the environment is degraded as a result of the scale effect as rising of the trade volume enhances the size of the economy, leading to pollution. Trade can also recover environment as income lifts and environmental regulations are followed strictly that leads to reduction in pollution levels.

4.1.3 Market mechanism
It has been proposed in a hypothesis (World Bank, 1992; Unruh and Moomaw, 1998) that the presence of an internal "self-administrative market system" for the environmental assets that are being the subject of trade in markets may control ecological degradation from increasing. Economic growth may enhance the market system with the aim that a growing economy may continuously move toward nonmarket to market energy assets which are less polluting (Kadekodi and Agarwal, 1999).

4.1.4 Regulation
4.1.4.1 Formal regulation
As the economy of a country grows, it develops its social organizations to establish strong environmental guidelines (Dasgupta et al., 2001). Pollution continues to increase unless the environmental regulations are not strictly reinforced (Hettige et al., 2000). Developing nations are shifting more toward market-driven guidelines (Panayotou, 1999; Dasgupta et al., 2002). Environmental standards can be better implemented if the regulators have data about polluters, damages, local ecological quality, emission reductions, etc. In the case of less developed countries or developing countries the environmental regulatory institutions are weak. Regular monitoring of the sources of pollution and the implementation of strict guidelines can reduce pollution in such countries.

4.1.4.2 Informal regulation
Societies utilize different measures to control pollution from local manufacturing industries in the form of "informal regulation" at the places where formal regulations are weak or absent. Agarwal et al. (1982) reported a circumstance in India where a paper plant after facing complaints from the local communities installed pollution controlling equipment and paid the communities as compensation for the damage caused. Nongovernmental organizations (NGOs) and social sectors may follow "informal regulation" (Dasgupta et al., 2002; Afsah et al., 1996). Informal regulation must be followed at the local level.

4.2 Survey of literature on the EKC hypothesis validity
Since the past decades, several investigations have been conducted on the environment-growth nexus and have been subject of interest for scientists. The EKC proposes that the linkage of the environment with economic development is an inverted U-shaped curve (Ang, 2007; Saboori et al., 2012; Omri, 2013). This means that there is an increase in economic loss with yield until it reaches a threshold point at which it starts declining. The graph first explained the income and inequality nexus as described by Kuznets (1955). Later, this example has motivated several researchers for further long-term investigations and recommended this hypothesis (Grossman and Krueger, 1991; Stern, 2004; Dinda, 2004). Dinda and Coondoo (2006) and Managi and Jena (2008) have also emphasized on this hypothesis. In Table 3.1, we have given examples of some studies conducted for both specific and

Table 3.1 Examples from the existing empirical studies on EKC (includes both country-specific and multiple-country studies).

S. No.	Study	Periods	Country	Evidence for EKC
1.	Roca et al. (2001)	1973−96	Spain	*No evidence*
2.	Haisheng et al. (2005)	1990−2002	China	*Support the EKC hypothesis*
3.	Du et al. (2012)	1995−2009	China	*No evidence for EKC hypothesis*
4.	Ang (2007)	1960−2000	France	*Support the EKC hypothesis*
5.	Halicioglu (2009)	1960−2005	Turkey	*Support the EKC hypothesis.*
6.	Ghosh (2010)	1971−2006	India	*Support the EKC hypothesis*
7.	Iwata et al. (2010)	1960−2003	France	*Support the EKC hypothesis*
8.	Ozturk and Acaravci (2010)	1968−2005	Turkey	*No evidence*
9.	Pao and Tsai (2011)	1990−2007	Russia	*No evidence*
10.	Fosten et al. (2012)	1850−2002	United Kingdom	*Support the EKC hypothesis*
11.	Jayanthakumaran and Liu (2012)	1990−2007	China	*Support the EKC hypothesis*
12.	Ozturk and Acaravci (2013)	1960−2007	Turkey	*Support the EKC hypothesis*
13.	Shahbaz et al. (2013).	1971−2011	Malaysia	*Support the EKC hypothesis*
14.	Tiwari et al. (2013)	1966−2009	India	*Support the EKC hypothesis*
15.	Yang and Zhao (2014)	1970−2008	India	*Support the EKC hypothesis*
16.	Grossman and Krueger (1991)	1977, 1982, 1988	52 Cities of 32 countries	*Support the EKC hypothesis.*
17.	Shafik and Bandyopadhyay (1992)	1972−88	47 Cities in 31 countries	*Support the EKC hypothesis*
18.	Panayotou (1997)	1982−94	30 Developed and developing countries	*Support the EKC hypothesis*
19.	Torras and Boyce (1998)	1977−91	18−52 Cities in 19−42 countries	*No evidence*
20.	Dinda et al. (2000)	1979−82, 1983−86, and 1987−90	39 Cities in 26 countries	*Support the EKC hypothesis*
21.	Orubu and Omotor (2011)	1990−2002	47 African countries	*Support the EKC hypothesis*

Table 3.1 Examples from the existing empirical studies on EKC (includes both country-specific and multiple-country studies).—cont'd

S. No.	Study	Periods	Country	Evidence for EKC
22.	Iwata et al. (2012)	1960−2003	11 OECD countries	*EKC does not hold for most countries*
23.	Liao and Cao (2013)	1971−2009	132 Countries	*Support the EKC hypothesis*
24.	Babu and Datta (2013)	1980−2008	Developing countries	*No evidence*
25.	Wang et al. (2013)	2005−2011	150 Nations	*No evidence*
26.	Apergis and Payne (2010a, 2010b)	1980−2010	7 Central American countries	*Support the EKC hypothesis.*
27.	Cho et al. (2014)	1971−2000	22 OECD countries	*Support the EKC hypothesis*
28.	Cowan et al. (2014)	1990−2010	The BRICS countries	*Support the EKC hypothesis*
29.	Osabuohien et al. (2014)	1995−2010	50 African countries	*Support the EKC hypothesis*
30.	Shafiei and Salim (2014)	1980−2011	29 OECD countries	*Support the EKC hypothesis*

Source: Tiba, S., Omri, A., 2017. Literature survey on the relationships between energy, environment and economic growth. Renew. Sustain. Energy Rev. 69, 1129−1146.

multiple countries, and the validity of the EKC hypothesis is also given. Generally, the impact of economic development on emissions of CO_2 for several countries was investigated for 87 (high, middle, and low income) countries within the framework of the EKC hypothesis by the cross-correlation approach (You and Lv, 2018; Shahbaz and Sinha, 2019). This investigation supported the inverted U-shaped EKC hypothesis for the relationship between income and environmental degradation, specifically for 16 of the upper-middle income nations. Only 6% of the studied countries show the U-shaped relationship between globalization and environmental degradation.

4.2.1 Environment-growth nexus studies for specific countries

From the period 1960 to 1999, the nexus on economic development-CO_2 emissions were examined for Australia (Friedl and Getzner, 2003). The observed evidences revealed an N-shaped curve. Results of Martinez-Zarzoso and Begochea-Morancho (2004) showed a quadratic relationship between income and environment, which gives evidence in support of the EKC hypothesis. For Turkey, Akbostanci et al. (2009) investigated the EKC assumption over the periods 1968−2003 and 1992−2001 (using the cointegration techniques for time series and provincial data), which established a monotonical nexus for times series interpretation. The relationship between CO_2 emissions and income was examined by He and Richard (2010) over the period 1948−2004 for Canada and found a small evidence supporting EKC. Likewise, for Tunisia, Fodha and Zaghdoud (2010) examined the relationship for the period from 1961 to 2004. It was concluded that CO_2 is correlated with per capita output, but in the case of $CO_{2,}$ a uniformly increasing relationship linked to economic growth was observed. For Malaysia, Lau et al. (2014) applied the Bounds tests and Granger causality method and tested this relationship including FDI and trade for the period 1970−2008. They concluded that for short and long term, the inverted U-shaped linkage exists between economic growth and environment.

4.2.2 Environment-growth nexus studies for multiple countries

EKC hypothesis was investigated by Jaunky (2011) for 36 high income nations for the period 1980−2005, the results revealed a unidirectional linkage between income and the environment for both short and long term. Likewise, a study was conducted by De Bruyn et al. (1998) using a time series model for the countries the Netherlands, West Germany, the UK, and the United States for the connection between income and pollutant emissions. It was concluded that income has a substantial positive influence on the environment. Moreover, the absence of a critical causal relationship between income and emission of pollutants was observed by Richmond and Kaufmann (2006). For BRIC nations, Pao and Tsai (2010) examined the nexus for growth and environment over the period 1971−2005, which resulted in a unidirectional cause-and-effect relationship. In a study, Lopez et al. (2014) examined the connection between income and environment during the period 1980−2010 for Ecuador, and no evidence for the EKC hypothesis was found.

4.2.3 Theoretical evidence for EKC

Grossman and Krueger (1991, 1995) in their work concluded that in the initial phases of development when the economy is mainly governed by agriculture and related activities, the level of pollution is relatively low. But later when the economy is converted to huge industries, pollution increases. However, when the economy shifts to modern technologies and services, pollution again decreases. Antle and Heidebrink (1995) argue that at low levels of income, economic development leads to deterioration of the environment but as the income increases the desire for improvement of the environment also tends to increase. This follows a path of development that includes both economic growth and enhancement of environmental quality causing replacement of old and unsafe technologies with a better one which results in improvement of the environment.

The theoretical evidences have experienced rapid and remarkable modifications by the researchers who have contributed to the existing documents on EKC. For example, Banerjee and Newman (1991) investigated the theory of insurance with the neoclassical theory of economic development. Their results were following the Kuznets theory. Andreoni and Levinson (2001) used a static-partial equilibrium model and concluded that as the income increases more capital investment is required to prevent the pollution. Thus, EKC rises as more income is being returned to pollution-control technologies. Similarly, Ekins (1997) utilized different effects (scale, composition, and technique) in a single equation structure from which he concluded that if technique and composition effect exceeds the scale effect, the EKC can probably rise. McConnell (1997) examined a model that considers overlapping generations suggesting that the production does not lead to pollution; rather it is the consumption that causes pollution. The static models have often faced objections as they are not able to attribute policy-making and planning in the long term. This results in false and indefinite projections and conclusions about the accurate interlinkage between pollution and output.

There are certain studies by Jones et al. (1995), Seldon and Song (1995), and Ansuategi and Perrings (2000), which describe the dynamic models of EKC for different economies. Stokey (1998) used an infinite horizon two-country growth model where a significant level of technological effects came into effect. This is the critical point for EKC. Some researchers support EKC (Grossman and

Krueger, 1991; Seldon and Song, 1994). While some of the studies like those by Gershuny and Weber (2009) and Saboori and Soleymani (2011) did not find any evidence for EKC, the EKC provides information about the interpretation of the linkage between economic development and its effect on the pollution of the environment.

4.2.4 Empirical evidences for EKC

There a several investigations in the 1990s which have determined the presence of an inverted U-shaped relationship between pollutants and income. These studies establish a robust relationship between growth and the environment (Larson et al., 2012). In general, the EKC hypothesis had resulted in the opinion that if countries become economically efficient they can control the degradation of the environment. Thus, we can say that economic development is favorable for the environment. In most of the empirical examinations, certain pollutants (like NOx, SO_2, SPM, CO_2, CO, etc.) have been considered for the study. Grossman and Krueger's model was further investigated by Harbaugh et al. (2002) by utilizing upgraded information for some of the pollutants used in the previous studies. But their results did not find any evidence in support of EKC. In certain studies, EKC exists for SO_2 (Shafik and Bandyopadhyay, 1992; Panayatou 1993; Cole et al., 1997; Stern and Common, 2001). In one of the studies conducted by Taguchi (2013), it was concluded that SO_2 emissions obey the inverted U-shaped relationship. But in the case of CO_2, it increases with a rise in income. Similarly, Rothman (1998) used several ecological indicators and observed that CO_2 emission does not decrease with per capita income rise. Nonetheless, it was found in another study by Coondoo and Dinda (2002) that in the case of developing nations there is no significant cause-and-effect relationship, while in developed nations there exists a causal relationship between emission and income. The most recent investigations on EKC consist of studies by Galleotti et al. (2006), Lipford and Yandle (2010), Akpan and Chuku (2011), Taguchi (2013), Osabuhein et al. (2013), and several others. The main focus of these studies was to enhance the technique and investigations of the Kuznets Curve. However, any specific conclusion cannot be drawn from these studies. But still, some findings reveal that there is an empirical connection between environmental quality and environment, and also environmental degradation decreases with an increase in income, although it is not clear that a definite turning point is true for all pollutants. Lastly, there are several other factors including income growth that can result in the recovery of the environment.

5. Relating pollution and economy

Presently, environmental pollution is one of the major challenges faced by each nation. The available environmental resources are draining quickly, making it insufficient for the future generations. An enormous population, especially poor people, is suffering from this stress. From one viewpoint, each country is attempting to increase economic development to fulfill the expectations for everyday needs; at the same time, ecological issues are getting complex because of the unnecessary utilization of resources. The social and economic well-being of humanity is connected to the environment in which they live. Any modification in the social and economic sectors will affect the earth's environment and vice-versa. In several instances, negative outcomes are irreversible.

Economy and pollution appear to have been intrinsically connected since the past. However, the connection between environmental damage and economic development is intricate. When considering a communal shift for achieving the goals of a circular economy, it is necessary to consider an increasingly coordinated structure for investigating the empirical proof that links pollution and economic growth. The implications of this relationship should also be focused on human welfare and the accomplishment of the objectives of sustainability. This section primarily builds up the major linkage of pollution with economic growth by examining in brief the studies already available in the previous literary works.

In natural and social sciences, the connection between environment and economic improvement has been broadly discussed across different aspects. The connections between pollution and economic development are complicated along with a few potential input circles that are dependent on determining factors and outcomes of economic development, the flexibility of ecosystems, and a definitive dependence of the economy on nature. The accomplishment of the sustainability goals is a possibility for amending and regulating the arguments on pollution and financial growth. The modernized environmental progression, which began in industrialized nations during the 1960s, blamed economic growth for the pollution of the environment. The investigations of Stern (2004), Dinda (2004), Luzzati and Orsini (2009), Halicioglu (2009), Acaravci and Ozturk (2010), and Al-Mulali et al. (2015) give a broad literature analysis which examined the EKC theory and economic development-pollution nexus.

Although the world is aware about the environmental pollution and climate change, still the world is going to observe a remarkable decline in worldwide emission of pollutants, for example, GHGs, SPM, ozone-forming gases, and different natural sources of pollution. The environmental scholars and nongovernmental agencies have strengthened attention to global warming and climate change in recent decades. Despite that, evidence has continued the declaration that the determined worldwide risk to the earth and atmosphere remains a serious problem for 21st century population (Alola, 2019a, 2019b; Alola et al., 2019; Bekun et al., 2019; Bekun et al., 2019). The emission of carbon dioxide (CO_2) that accounts for the major contribution in GHG discharges is recorded by the International Energy Agency (IEA, 2019) and is reported to have expanded by 1.7% in 2018. As reported by IEA, the rise in emissions of CO_2 is mainly due to expansion in the worldwide economic activity and higher energy consumptions.

5.1 CO_2 emission and economic growth

Carbon dioxide emissions are the major contributors to GHG emissions resulting in environmental deterioration (Stern, 2004). So, it is necessary to continuously monitor the effect of emissions of carbon on the environment. It can also result in the destruction of the subsequent generations, and its connection with factors like globalization, energy consumption, and economic growth has been analyzed in different studies. The connection between CO_2 emissions, globalization, energy utilization, and financial growth has been a subject of focus for the formulation of policies and scientific investigations (Shahbaz et al., 2017, 2018a, 2018b; Akadiri and Akadiri, 2019a, 2019b). A general observation in all the studies is that the globalization, financial growth, and energy utilization remarkably contribute to higher carbon releases. Carbon dioxide (CO_2) emission is the main topic in different ongoing debates on sustainable development and ecological protection. It is a well-known fact that energy is a major cause of emissions of CO_2, but still it is a necessity for both production and utilization in economic development. Thus, the connection between CO_2 emissions and economic development has a huge complexity for the environmental and economic regulations. A significant part of the CO_2 emissions originate from fuel consumption, which is essential for the transportation facilities and industrial sectors that are closely related to economic development. Hence understanding the indivisible connection between CO_2 emissions and economic development is important for implementing economic and environmental policies.

A modern monetary framework cannot exist without an effective ecological system as a source of natural resources, from one perspective, and as a sink of waste materials, on the other (Stanojević et al., 2013). Authors, for example, Borhan et al. (2012), list the following as key components of ecological degradation: "industrialization, transport, population, poverty, soil degradation, traffic, misuse of open access resources because of ill-defined property rights, etc." The connection between the emission of CO_2 and financial development is determined using the Auto-Regressive Distributed Lag (ARDL) limits for studying 19 countries from Europe over the time of 1960–2005 (Acaravci and Ozturk 2010). Similarly, ARDL was used for Turkey over the time of 1968–2005 (Ozturk and Acaravci, 2010). For India Granger approach was used (Tiwari, 2011). Hossain (2012), utilized time series information for Japan over the period 1960–2009. Likewise, for Iran, a new time series method was used during the time of 1967–2007 (Lotfalipour, 2010). Such investigations are helpful to formulate the techniques from the perspective of energy conservation, emission mitigation, and economic performance. Further, Liao et al. (2013) inspected the recorded connection of economic growth and CO_2 emission based on three perspectives: sources of information, model determinations, and evaluation strategies. They utilized the CO_2 emissions and economic growth panel data index of 132 nations within the period of 1971–2009 with an adaptable econometric model.

Liu (2005) discovered observational proof for CO_2 outflows in OECD nations for the EKC hypothesis. A few investigations have discovered little proof for EKC (e.g., Dinda, 2001; Harbaugh et al., 2002; Iwata et al., 2011; Soytas et al., 2007). The heterogeneous observational proof rising out of the literature proposes that variations in social, financial, governmental, and biological and physical factors across the countries can influence nature and different phases of financial advancement in various nations. The discoveries got from one nation utilizing explicit ecological indicators cannot be summed up for different nations, as well as for different types of pollution.

6. Policy implications

The environmental problems are highly complex due to tradeoffs between environment and economic development. Economic growth requires more utilization of natural assets to meet the demand for social welfare, which creates tremendous pressure on the environment. The needs of future generations are also compromised to fulfill the needs of present generations. So, the natural resources should be used rationally. A sustainable approach toward economic growth is also necessary so that natural resources do not face overexploitation. However, it is quite difficult to maintain a balance between the environment and growth. But, to make the economic development sustainable on a global scale, it is important to reinforce environmental strategies; so manufactures will depend less on pollution resulting output (e.g., nonrenewable energy sources) and more on new technologies, skills, and knowledge (i.e., human capital). The poor communities are the ones that are most affected by the adverse impacts of pollution and environmental deterioration. The low income section of the society cannot afford to drink safe water, breathe clean air, and have proper sanitation facilities. So, it is recommended that policies should focus on the poor communities because a strict policy can strengthen and support the strategies of economic development. The involvement of governments is preferable rather than private markets. These private markets do not promote pollution reductions whether it may be air pollution or disposal of wastes.

Developed countries can help in contributing toward the betterment of environmental quality. They can reduce the emissions by designing clean methodologies and changing their unsustainable demand pattern. Such countries should not develop and shift their hazardous chemicals and industries to less developed countries. Strict regulations need to be implemented to stop such kind of activities. Countries must design and execute strict legal policies to stop the disposal of dangerous chemicals and hazardous wastes from developed nations. Transportation of such chemicals should be prohibited. Public awareness is one of the most important steps toward a reduction in pollution through economic activities. People should be educated and trained about the consequences of overutilizing natural resources and the effects of pollution on the environment. Regular monitoring of sources of pollution is also very important. However, eminent advances against pollution have additionally been made in developing nations. Presently, the environmental authorities in developing nations are shifting from customary order-and-control arrangements toward market-based guidelines. Pollution charges have been demonstrated practically in developing nations, with operational practice in China (Wang and Wheeler, 1996), Colombia, Malaysia, and the Philippines. Guidelines can emphatically impact marketability and efficiency through the channel of advancement. Strict regulations produce more prominent advancement as compliance requires more exertion toward artistic products and new process solutions (Porter and van der Linde, 1995). It subsequently becomes vital to effectively survey the stringency of environmental policies, as strict policies can positively affect the economy in general. Sustainable economic development strategies rely upon the quality and the control of inexhaustible and nonfeasible regular assets and on the status of the environment. Efficient ecological policy is probably going to require the utilization of numerous instruments (market-based, direct regulation, development of infrastructure and technology, provision of information, product labeling policies, etc.), each handling a diverse part of the issue while preventing duplication and pointless legislative burdens. Valuing ecological inputs effectively deals with the sustainable supply and utilization of natural assets. A uniform and comprehensible environmental policy gives a more prominent conviction about the estimation of investments and empowers long-term business interest in innovation and development. Thus, a strict environmental policy can control the effects of economic growth and reduce environmental risks. It can also enhance the resilience of the economy toward the environmental risks.

6.1 Application of EKC in policy-making

The EKC can be used in designing policies that reduce the effects of pollution on the environment. Given below are some examples from the literature that have utilized the EKC hypothesis for the implication of policies.

Case study 1: Purcel (2020) conducted a survey of literature on pollution-growth nexus from theoretical and empirical viewpoints using the EKC. In this study, firstly, the author provides an empirical and theoretical review of the linkage between environmental degradation and economic development with the help of the Kuznets curve. The results of the study indicate that there is a long-term relationship between income and pollution. This confirms that EKC is a phenomenon that is to be observed for a longer period. Moreover, the impacts of environmental policies are observable after years from execution. The developing nations can enhance their quality of the environment by not repeating the same mistakes done by developed nations and must shift toward a pollution pattern that is not harmful to the environment for a lower pay level.

Hence, based on these results some policy implications have been suggested. Some developing countries have been able to keep low degrees of pollution together with monetary development and reached at the EKC limit for a lower level of income as compared to developed countries (Yao et al., 2019). Thus, these nations could be treated as a positive model and inspected more in detail to understand the elements that add to initiating a downward bend curve in pollution. For a greater efficiency of the environment, the developing nations should be involved in adopting greener technologies both at the regional and global level. Better information on the household, financial, social, and political structures, along with a consistent modulation of ecological guidelines and policies, may improve the quality of the environment.

Case study 2: Al-Mulali and Ozturk (2016) tested the impact of energy prices on pollution (CO_2 emission) and examined the EKC hypothesis for 27 advanced nations for the period 1990–2012. The panel nonstationary techniques were used for the investigation. The outcomes of the study show that urban growth, GDP, and consumption of nonrenewable energy accelerate the emission of CO_2. It also confirms the existence of EKC as there is an inverted U-shaped linkage between GDP and pollution. The authors suggested different policy implications based on the results of the study. The countries must use environmentally safer and more effective sources of renewable energy (hydro, solar, geothermal, wind, etc.) and higher grades of coal to reduce the consumption of energy and pollution. To reduce energy consumption (which further helps in the reduction of pollution), the taxes on energy prices should be increased. To accomplish progressively reasonable energy costs, governments should increase the taxes to incorporate engine fuels as well as to consolidate other fossil fuel products. The rate of taxes needs to be adjusted in such a manner that it does not cause any damage to the environment.

Case study 3: Apergis and Ozturk (2015) examined the EKC theory for 14 Asian countries in the period 1990–2011 by using the Generalized Method of Moments (GMM) methodology (Arellano and Bond, 1991) in a multivariate framework. The major goal was to focus on what way both income and policies of these countries affect the linkage between income and environment. The results supported the EKC hypothesis, i.e., the degradation of the environment was found to be increasing with income during initial phases. But later, it decreased with per capita income after reaching a certain limit. The authors have proposed some policies according to the empirical results of the study. The organizational limitations have a very important role to play as systems that encourage technology diffusion, sharing information on energy performance, reduction of emissions, and strengthening the institutional capacity for Asian countries. Additionally, the concerned government offices at different levels should be promoted to share successful administrative experiences. The policymakers in Asian nations should implement energy policies that diminish the carbon power of energy radiated per unit of energy used to achieve the long-term sustainability goals. However, appropriate institutional operations have a very important role in the decrease of greenhouse gas outflows on a global scale.

7. Conclusion

This chapter aims to explain the relationship between economic growth and the environment at a very basic and hypothetical level. It gives a general survey of literature about the existing evidences on environment-growth nexus. The connection between economic development and the environment is controlled by various drivers and accomplishing continued development needs decoupling financial development from its ecological effects, not only regionally but also on a global scale. It is obvious from the available literature that besides the fundamental connections between economic growth and environment, an intricate, multifaceted, and dynamic connection between monetary development, environment, and climate change also exists. It has both biological and anthropogenic (financial, political, and social) input impacts, which are connected. This chapter analyzes different investigations that have found evidence for the EKC theory. The EKC investigations conclude a specific recorded connection between income and ecological quality. Thus, some indicators of air quality, especially pollutants, confirm the verification of EKC. But still, there is no understanding in the available literary works about the pay level at which natural deterioration starts to recover.

In this chapter, we look at the exploration in general and locate that unequivocal proof for an EKC relationship is extremely insufficient, that there are significant indicators which show a uniform expanding relationship, that even there might be an EKC relationship. The majority of the total population is still on the segment of the curve that is increasing. The economic development based on this relationship would bring about further ecological damage and the survey of literature concerning environmental quality even in the rich and developed nations shows that it is still degrading. Thus, we conclude from the perspective of the sustainability of the environment that the connection between income and environment is still a complicated one, and stringent environmental regulation and policies are mandatory if future economic growth is to be consistent with sustainable development.

Economic development seems, by all accounts, to be an influential method for improving ecological quality in developing nations. Since, financial development is beneficial for the environment, at that point approaches that encourage development, for example, deregulation of trade, and financial rebuilding and value reform should likewise be useful for nature. Thus there is a general recommendation that the environment needs no specific consideration, either regarding regional environmental strategies or global pressure and assistance. The natural assets can be best targeted toward accomplishing rapid financial development by shifting rapidly from the environmentally adverse phase of development to the environmentally favorable range of the Kuznets curve.

With respect to the evidence found in the literature about the linkage of economic development and quality of the environment, a few inferences can be drawn and a few targets need development for research in the future. Primarily, we need cost-effective models, which appropriately address the physical and natural assumption of economical movement with significant criticism between the economic system and the environment. Furthermore, distinguishing evidence of the prevailing investigation defining the EKC should be given more attention in research. Thirdly, the estimation of basic models, rather than weak structure models, might be expected to recognize the real mechanism. Further, the investigation of degradation can give more insight into which collection of interpretations is predominant, for example, innovative advancement and organizational changes. More efforts should be made to focus on time series investigations, which can establish a good notion of the improvement of pollution with explicit periods of progressions in each country. Politically, strong policy measures are important to promote economic sustainability. Finally, while adopting new and advanced technologies, societies should be careful about the potential ill-defined pollutants and hazardous wastes.

References

Acaravci, A., Ozturk, I., 2010. On the relationship between energy consumption, CO2 emissions and economic growth in Europe. Energy 35 (12), 5412–5420.

Afsah, S., Laplante, B., Wheeler, D., 1996. Controlling Industrial Pollution: A New Paradigm. World Bank, Policy Research Department working paper 1672.

Agarwal, A., Chopra, R., Sharma, K., 1982. The State of India's Environment 1982. Centre for Science and Environment, New Delhi, India.

Akadiri, S.S., Akadiri, A.C., 2019a. Interaction between CO2 emissions, energy consumption and economic growth in the middle east: panel causality evidence. Int. J. Energy Technol. Pol. (in press).

Akadiri, S.S., Akadiri, A.C., 2019b. The role of natural gas consumption in economic growth. Strat. Plann. Energy Environ. In press.

Akbostancı, E., Türüt-Aşık, S., Tunç, G.İ., 2009. The relationship between income and environment in Turkey: is there an environmental Kuznets curve? Energy Pol. 37 (3), 861–867.

Akpan, U.F., Chuku, A., 2011. Economic Growth and Environmental Degradation in Nigeria: Beyond the Environmental Kuznets Curve.

Al-Mulali, U., Ozturk, I., 2016. The investigation of environmental Kuznets curve hypothesis in the advanced economies: the role of energy prices. Renew. Sustain. Energy Rev. 54, 1622–1631.

Al-Mulali, U., Saboori, B., Ozturk, I., 2015. Investigating the environmental Kuznets curve hypothesis in Vietnam. Energy Pol. 76, 123–131.

Alola, A.A., 2019a. Carbon emissions and the trilemma of trade policy, migration policy and health care in the US. Carbon Manag. 10 (2), 209–218.

Alola, A.A., 2019b. The trilemma of trade, monetary and immigration policies in the United States: accounting for environmental sustainability. Sci. Total Environ. 658, 260–267.

Alola, A.A., Alola, U.V., Saint Akadiri, S., 2019a. Renewable energy consumption in Coastline Mediterranean Countries: impact of environmental degradation and housing policy. Environ. Sci. Pollut. Control Ser. 1–13.

Andreoni, J., Levinson, A., 2001. The simple analytics of the environmental Kuznets curve. J. Publ. Econ. 80 (2), 269–286.

Ang, J.B., 2007. CO2 emissions, energy consumption, and output in France. Energy Pol. 35 (10), 4772–4778.

Ansuategi, A., Perrings, C., 2000. Transboundary externalities in the environmental transition hypothesis. Environ. Resour. Econ. 17 (4), 353–373.

Antle, J.M., Heidebrink, G., 1995. Environment and development: theory and international evidence. Econ. Dev. Cult. Change 43 (3), 603–625.

Apergis, N., Ozturk, I., 2015. Testing environmental Kuznets curve hypothesis in Asian countries. Ecol. Indicat. 52, 16–22.

Apergis, N., Payne, J.E., 2010a. A panel study of nuclear energy consumption and economic growth. Energy Econ. 32 (3), 545–549.

Apergis, N., Payne, J.E., 2010b. Renewable energy consumption and economic growth: evidence from a panel of OECD countries. Energy Pol. 38 (1), 656–660.

Arellano, M., Bond, S., 1991. Some tests of specification for panel data: Monte Carlo evidence and an application to employment equations. Rev. Econ. Stud. 58 (2), 277–297.

Arrow, K., Bolin, B., Costanza, R., Dasgupta, P., Folke, C., Holling, C.S., Pimentel, D., 1995. Economic growth, carrying capacity, and the environment. Ecol. Econ. 15 (2), 91–95.

Babu, S., Datta, S.K., 2013. The relevance of environmental Kuznets curve (EKC) in a framework of broad-based environmental degradation and modified measure of growth–a pooled data analysis. Int. J. Sustain. Dev. World Ecol. 20 (4), 309–316.

Baldwin, R., 1995. Does sustainability require growth? In: Goldin, I., Winters, L.A. (Eds.), The Economics of Sustainable Development. Cambridge Univ. Press, Cambridge, UK, pp. 19–47.

Banerjee, A.V., Newman, A.F., 1991. Risk-bearing and the theory of income distribution. Rev. Econ. Stud. 58 (2), 211−235.

Bekun, F.V., Alola, A.A., Sarkodie, S.A., 2019a. Toward a sustainable environment: nexus between CO2 emissions, resource rent, renewable and non-renewable energy in 16-EU countries. Sci. Total Environ. 657, 1023−1029.

Bekun, F.V., Emir, F., Sarkodie, S.A., 2019b. Another look at the relationship between energy consumption, carbon dioxide emissions, and economic growth in South Africa. Sci. Total Environ. 655, 759−765.

Borhan, H., Ahmed, E.M., Hitam, M., 2012. The impact of CO2 on economic growth in ASEAN 8. Proc. Soc. Behav. Sci. 35, 389−397.

Chang, T.H., Huang, C.M., Lee, M.C., 2009. Threshold effect of the economic growth rate on the renewable energy development from a change in energy price: evidence from OECD countries. Energy Pol. 37, 5796−5802.

Cheng, B.S., 1999. Causality between energy consumption and economic growth in India: an application of cointegration and error-correction modeling. Indian Econ. Rev. 39−49.

Cho, C.H., Chu, Y.P., Yang, H.Y., 2014. An environment Kuznets curve for GHG emissions: a panel cointegration analysis. Energy Sources B Energy Econ. Plann. 9 (2), 120−129.

Cole, M.A., Rayner, A.J., Bates, J.M., 1997. The environmental Kuznets curve: an empirical analysis. Environ. Dev. Econ. 2 (4), 401−416.

Coondoo, D., Dinda, S., 2002. Causality between income and emission: a country group-specific econometric analysis. Ecol. Econ. 40 (3), 351−367.

Coondoo, D., Dinda, S., 2008. Carbon dioxide emission and income: a temporal analysis of cross-country distributional patterns. Ecol. Econ. 65 (2), 375−385.

Cowan, W.N., Chang, T., Inglesi-Lotz, R., Gupta, R., 2014. The nexus of electricity consumption, economic growth and CO2 emissions in the BRICS countries. Energy Pol. 66, 359−368.

Dasgupta, S., Mody, A., Roy, S., Wheeler, D., 2001. Environmental regulation and development: a cross-country empirical analysis. Oxf. Dev. Stud. 29 (2), 173−187.

Dasgupta, S., Laplante, B., Wang, H., Wheeler, D., 2002. Confronting the environmental Kuznets curve. J. Econ. Perspect. 16 (1), 147−168.

De Bruyn, S.M., van den Bergh, J.C., Opschoor, J.B., 1998. Economic growth and emissions: reconsidering the empirical basis of environmental Kuznets curves. Ecol. Econ. 25 (2), 161−175.

Dinda, S., 2001. A Note on Global EKC in Case of CO2 Emission. Economic Research Unit, Indian Statistical Institute, Kolkata. Mimeo.

Dinda, S., 2004. Environmental Kuznets curve hypothesis: a survey. Ecol. Econ. 49 (4), 431−455.

Dinda, S., Coondoo, D., 2006. Income and emission: a panel data-based cointegration analysis. Ecol. Econ. 57 (2), 167−181.

Dinda, S., Coondoo, D., Pal, M., 2000. Air quality and economic growth: an empirical study. Ecol. Econ. 34 (3), 409−423.

Du, L., Wei, C., Cai, S., 2012. Economic development and carbon dioxide emissions in China: provincial panel data analysis. China Econ. Rev. 23 (2), 371−384.

Ekins, P., 1997. The Kuznets curve for the environment and economic growth: examining the evidence. Environ. Plann. 29 (5), 805−830.

Energy, I.G., 2018. CO2 Status Report 2017. International Energy Agency.

Fodha, M., Zaghdoud, O., 2010. Economic growth and pollutant emissions in Tunisia: an empirical analysis of the environmental Kuznets curve. Energy Pol. 38 (2), 1150−1156.

Fosten, J., Morley, B., Taylor, T., 2012. Dynamic misspecification in the environmental Kuznets curve: evidence from CO2 and SO2 emissions in the United Kingdom. Ecol. Econ. 76, 25−33.

Friedl, B., Getzner, M., 2003. Determinants of CO2 emissions in a small open economy. Ecol. Econ. 45 (1), 133–148.

Galeotti, M., Lanza, A., Pauli, F., 2006. Reassessing the environmental Kuznets curve for CO2 emissions: a robustness exercise. Ecol. Econ. 57 (1), 152–163.

Gershuny, P., Weber, W.L., 2009. A Law and Economics Approach to Climate Change Regulation in California Using the Environmental Kuznets' Curve. Available at: SSRN 1613266.

Ghosh, S., 2010. Examining carbon emissions economic growth nexus for India: a multivariate cointegration approach. Energy Pol. 38 (6), 3008–3014.

Grossman, G., Krueger, A., 1991. Environmental Impacts of a North American Free Trade Agreement', WP-3914. National Bureau of Economic Research, Cambridge, MA.

Grossman, G., Krueger, A.B., 1993. Environmental Impacts of a North American Free Trade Agreement, the US-Mexico Free Trade Agreement. MIT Press, Cambridge, MA.

Grossman, G.M., Krueger, A.B., 1995. Economic growth and the environment. Q. J. Econ. 110 (2), 353–377.

Haisheng, Y., Jia, J., Yongzhang, Z., Shugong, W., 2005. The impact on environmental Kuznets curve by trade and foreign direct investment in China. Chin. J. Populat. Resourc. Environ. 3 (2), 14–19.

Halicioglu, F., 2009. An econometric study of CO2 emissions, energy consumption, income and foreign trade in Turkey. Energy Pol. 37 (3), 1156–1164.

Harbaugh, W.T., Levinson, A., Wilson, D.M., 2002. Reexamining the empirical evidence for an environmental Kuznets curve. Rev. Econ. Stat. 84 (3), 541–551.

He, J., 2007. Is the Environmental Kuznets Curve Valid for Developing Countries? A Survey. Working Paper 07-03. University de Sherbrooke.

He, J., Richard, P., 2010. Environmental Kuznets curve for CO2 in Canada. Ecol. Econ. 69 (5), 1083–1093.

Hettige, H., Dasgupta, S., Wheeler, D., 2000. What improves environmental compliance? Evidence from Mexican industry. J. Environ. Econ. Manag. 39 (1), 39–66.

Heywood, V.H., Watson, R.T., 1995. Global Biodiversity Assessment, vol. 1140. Cambridge University Press, Cambridge.

Hossain, S., 2012. An Econometric Analysis for CO2 Emissions, Energy Consumption, Economic Growth, Foreign Trade and Urbanization of Japan.

https://www.iea.org/geco/emissions/. Retrieved July 13, 2019.

https://www.ipcc.ch/2018/10/08/summary-for-policymakers-of-ipcc-special-report-on-global-warming-of-1-5c-approved-by-governments/.

Iwata, H., Okada, K., Samreth, S., 2010. Empirical study on the environmental Kuznets curve for CO2 in France: the role of nuclear energy. Energy Pol. 38, 4057–4063.

Iwata, H., Okada, K., Samreth, S., 2011. A note on the environmental Kuznets curve for CO2: a pooled mean group approach. Appl. Energy 88, 1986–1996.

Iwata, H., Okada, K., Samreth, S., 2012. Empirical study on the determinants of CO2 emissions: evidence from OECD countries. Appl. Econ. 44 (27), 3513–3519.

Jaunky, V.C., 2011. The CO2 emissions-income nexus: evidence from rich countries. Energy Pol. 39 (3), 1228–1240.

Jayanthakumaran, K., Liu, Y., 2012. Openness and the environmental Kuznets curve: evidence from China. Econ. Modell. 29 (3), 566–576.

Jones, L.E., Manuelli, R.E., 1995. A Positive Model of Growth and Pollution Controls (No. w5205). National Bureau of Economic Research.

Jones, L.E., Rodolfo, E.M., 1995. A Positive Model of Growth and Pollution Controls. NBER Working Paper 5205.

Kadekodi, G., Agarwal, S., 1999. Why an Inverted U-Shaped Environmental Kuznets Curve May Not Exist? Institute of Economic Growth, Delhi. Mimeo.

Kijima, M., Nishide, K., Ohyama, A., 2010. Economic models for the environmental Kuznets curve: a survey. J. Econ. Dynam. Contr. 34 (7), 1187−1201.

Kuznets, S., 1955. Economic growth and income inequality. Am. Econ. Rev. 49, 1−28.

Larson, D.F., Dinar, A., Blankespoor, B., 2012. Aligning Climate Change Mitigation and Agricultural Policies in Eastern Europe and Central Asia. The World Bank.

Lau, L.S., Choong, C.K., Eng, Y.K., 2014. Investigation of the environmental Kuznets curve for carbon emissions in Malaysia: do foreign direct investment and trade matter? Energy Pol. 68, 490−497.

Lee, C.C., Chang, C.P., 2009. Stochastic convergence of per capita carbon dioxide emissions and multiple structural breaks in OECD countries. Econ. Modell. 26 (6), 1375−1381. https://doi.org/10.1016/j.econmod.2009.07.003.

Lee, C.C., Lee, J.D., 2009. Income and CO2 emissions: evidence from panel unit root and cointegration tests. Energy Pol. 37 (2), 413−423.

Lee, H., Roland-Holst, D., 1997. The environment and welfare implications of trade and tax policy. J. Dev. Econ. 52 (1), 65−82.

Liao, H., Cao, H.S., 2013. How does carbon dioxide emission change with the economic development? Statistical experiences from 132 countries. Global Environ. Change 23 (5), 1073−1082.

Lipford, J.W., Yandle, B., 2010. Environmental Kuznets curves, carbon emissions, and public choice. Environ. Dev. Econ. 15 (4), 417−438.

Liu, X., 2005. Explaining the relationship between CO2 emissions and national income—the role of energy consumption. Econ. Lett. 87 (3), 325−328.

López-Menéndez, A.J., Pérez, R., Moreno, B., 2014. Environmental costs and renewable energy: Re-visiting the environmental Kuznets curve. J. Environ. Manag. 145, 368−373.

Lotfalipour, M.R., Falahi, M.A., Ashena, M., 2010. Economic growth, CO2 emissions, and fossil fuels consumption in Iran. Energy 35 (12), 5115−5120.

Luzzati, T., Orsini, M., 2009. Investigating the energy-environmental Kuznets curve. Energy 34 (3), 291−300.

Managi, S., Jena, P.R., 2008. Environmental productivity and Kuznets curve in India. Ecol. Econ. 65 (2), 432−440.

Martinez-Zarzoso, I., Begochea-Morancho, A., 2004. Pooled mean group estimation of an environmental Kuznets curve for CO$_2$. Econ. Lett. 82, 121−126.

McConnell, K.E., 1997. Income and the demand for environmental quality. Environ. Dev. Econ. 2 (4), 383−399.

MEA, 2005. Ecosystems and Human Well-Being: Wetlands and Water.

Nasir, M., Rehman, F., 2011. Environmental Kuznets curve for carbon emissions in Pakistan: an empirical investigation. Energy Pol. 39, 1857−1864.

Nocera, S., Cavallaro, F., 2011. Policy effectiveness for containing CO2 emissions in transportation. Proc. Soc. Behav. Sci. 20, 703−713.

Omri, A., 2013. CO2 emissions, energy consumption and economic growth nexus in MENA countries: evidence from simultaneous equations models. Energy Econ. 40, 657−664.

Orubu, C.O., Omotor, D.G., 2011. Environmental quality and economic growth: searching for environmental Kuznets curves for air and water pollutants in Africa. Energy Pol. 39 (7), 4178−4188.

Osabuohien, E.S., Efobi, U.R., Gitau, C.M., 2013. External intrusion, internal tragedy: environmental pollution and multinational corporations in Sub-Saharan Africa. In: Principles and Strategies to Balance Ethical, Social and Environmental Concerns with Corporate Requirements. Emerald Group Publishing Limited.

Osabuohien, E.S., Efobi, U.R., Gitau, C.M.W., 2014. Beyond the environmental Kuznets curve in Africa: evidence from panel cointegration. J. Environ. Pol. Plann. 16 (4), 517−538.

Ozturk, I., Acaravci, A., 2010. CO2 emissions, energy consumption and economic growth in Turkey. Renew. Sustain. Energy Rev. 14 (9), 3220−3225.

Ozturk, I., Acaravci, A., 2013. The long-run and causal analysis of energy, growth, openness and financial development on carbon emissions in Turkey. Energy Econ. 36, 262−267.

Panayotou, T., 1993. Empirical Tests and Policy Analysis of Environmental Degradation at Different Stages of Economic Development (No. 992927783402676). International Labour Organization.

Panayotou, T., 1997. Demystifying the environmental Kuznets curve: turning a black box into a policy tool. Environ. Dev. Econ. 2 (4), 465−484.

Panayotou, T., 1999. The economics of environments in transition. Environ. Dev. Econ. 4 (4), 401−412.

Pao, H.T., Tsai, C.M., 2010. CO2 emissions, energy consumption and economic growth in BRIC countries. Energy Pol. 38 (12), 7850−7860.

Pao, H.T., Tsai, C.M., 2011. Multivariate granger causality between CO_2 emissions, energy consumption, FDI (foreign direct investment) and GDP (gross domestic product): evidence from a panel of BRIC (Brazil, Russian Federation, India, and China) countries. Energy 36 (1), 685−693.

Pearce, D.W., Warford, J.J., 1993. World without End: Economics, Environment, and Sustainable Development. Oxford University Press.

Porter, M.E., Van der Linde, C., 1995. Toward a new conception of the environment-competitiveness relationship. J. Econ. Perspect. 9 (4), 97−118.

Purcel, A.A., 2020. New insights into the environmental Kuznets curve hypothesis in developing and transition economies: a literature survey. Environ. Econ. Pol. Stud. 1−47.

Richmond, A.K., Kaufmann, R.K., 2006. Energy prices and turning points: the relationship between income and energy use/carbon emissions. Energy J. 27 (4).

Roca, J., 2003. Do individual preferences explain environmental Kuznets curve? Ecol. Econ. 45 (1), 3−10.

Roca, J., Padilla, E., Farré, M., Galletto, V., 2001. Economic growth and atmospheric pollution in Spain: discussing the environmental Kuznets curve hypothesis. Ecol. Econ. 39 (1), 85−99.

Rothman, D.S., 1998. Environmental Kuznets curves—real progress or passing the buck?: a case for consumption-based approaches. Ecol. Econ. 25 (2), 177−194.

Saboori, B., Soleymani, A., 2011. CO2 emissions, economic growth and energy consumption in Iran: a cointegration approach. Int. J. Environ. Sci. 2 (1), 44−53.

Saboori, B., Sulaiman, J., Mohd, S., 2012. Economic growth and CO2 emissions in Malaysia: a cointegration analysis of the environmental Kuznets curve. Energy Pol. 51, 184−191.

Selden, T.M., Song, D., 1994. Environmental quality and development: is there a Kuznets curve for air pollution emissions? J. Environ. Econ. Manag. 27 (2), 147−162.

Selden, T.M., Song, D., 1995. Neoclassical growth, the J curve for abatement, and the inverted U curve for pollution. J. Environ. Econ. Manag. 29 (2), 162−168.

Shafiei, S., Salim, R.A., 2014. Non-renewable and renewable energy consumption and CO2 emissions in OECD countries: a comparative analysis. Energy Pol. 66, 547−556.

Shafik, N., Bandyopadhyay, S., 1992. Economic Growth and Environmental Quality: Time-Series and Cross-Country Evidence, vol. 904. World Bank Publications.

Shahbaz, M., Sinha, A., 2019. Environmental Kuznets curve for CO_2 emissions: a literature survey. J. Econ. Stud.

Shahbaz, M., Khan, S., Tahir, M.I., 2013. The dynamic links between energy consumption, economic growth, financial development and trade in China: fresh evidence from multivariate framework analysis. Energy Econ. 40, 8−21.

Shahbaz, M., Shahzad, S.J.H., Mahalik, M.K., Sadorsky, P., 2017. How strong is the causal relationship between globalization and energy consumption in developed economies? A country-specific time-series and panel analysis. Appl. Econ. 1−16.

Shahbaz, M., Lahiani, A., Abosedra, S., Hammoudeh, S., 2018a. The role of globalization in energy consumption: a quantile cointegrating regression approach. Energy Econ. 71, 161−170.

Shahbaz, M., Shahzad, S.J.H., Mahalik, M.K., 2018b. Is globalization detrimental to CO2 emissions in Japan? New threshold analysis. Environ. Model. Assess. 23 (5), 557−568.

Shapiro, J.S., Walker, R., 2018. Why is pollution from US manufacturing declining? The roles of environmental regulation, productivity, and trade. Am. Econ. Rev. 108 (12), 3814−3854.

Smulders, S., Bretschger, L., 2000. Explaining Environmental Kuznets Curves: How Pollution Induces Policy and New Technologies. Centre for Economic Research, pp. 2000–2095.

Soytas, U., Sari, R., Ewing, T., 2007. Energy consumption, income, and carbon emissions in the United States. Ecol. Econ. 62, 482–489.

Stagl, S., 1999. Delinking Economic Growth from Environmental Degradation? A Literature Survey on the Environmental Kuznets Curve Hypothesis.

Stanojević, D., Mitić, P., Rakić, S., 2013. Strategija bankarskog sektora u kontekstu održivog razvoja-perspektive Srbije. Poslovna ekonomija: Časopis za Poslovnu Ekonomiju, Preduzetništvo i Finasije 437–446.

Stern, D.I., 2004. The rise and fall of the environmental Kuznets curve. World Dev. 32 (8), 1419–1439.

Stern, D.I., Common, M.S., 2001. Is there an environmental Kuznets curve for sulfur? J. Environ. Econ. Manag. 41 (2), 162–178.

Stokey, N.L., 1998. Are there limits to growth? Int. Econ. Rev. 1–31. Survey on the environmental Kuznets curve hypothesis.

Syrquin, M., Chenery, H.B., 1989. Patterns of Development, 1950 to 1983 (No. 41). World Bank, Washington.

Taguchi, H., 2013. The environmental Kuznets curve in Asia: the case of sulphur and carbon emissions. Asia Pac. Dev. J. 19 (2), 77–92.

Tiba, S., Omri, A., 2017. Literature survey on the relationships between energy, environment and economic growth. Renew. Sustain. Energy Rev. 69, 1129–1146.

Tiwari, A.K., 2011. Energy consumption, CO2 emissions and economic growth: evidence from India. J. Int. Bus. Econ. 12 (1), 85–122.

Tiwari, A.K., Shahbaz, M., Hye, Q.M.A., 2013. The environmental Kuznets curve and the role of coal consumption in India: cointegration and causality analysis in an open economy. Renew. Sustain. Energy Rev. 18, 519–527.

Toman, M.T., Jemelkova, B., 2003. Energy and economic development: an assessment of the state of knowledge. Energy J. 24 (4).

Torras, M., Boyce, J.K., 1998. Income, inequality, and pollution: a reassessment of the environmental Kuznets curve. Ecol. Econ. 25 (2), 147–160.

Unruh, G.C., Moomaw, W.R., 1998. An alternative analysis of apparent EKC-type transitions. Ecol. Econ. 25, 221–229.

Usenata, N., 2018. Environmental Kuznets Curve (EKC): A Review of Theoretical and Empirical Literature.

Wang, H., Wheeler, D., Wang, H., 1996. Pricing Industrial Pollution in China: An Economic Analysis of the Levy System. The World Bank.

Wang, Y., Kang, L., Wu, X., Xiao, Y., 2013. Estimating the environmental Kuznets curve for ecological footprint at the global level: a spatial econometric approach. Ecol. Indicat. 34, 15–21.

WCED, S.W.S., 1987. World commission on environment and development. Our Common Future 17, 1–91.

https://www.imf.org/en/Publications/WEO/Issues/2020/01/20/weo-update-january2020.

World Bank, 1992. World Development Report. Oxford Univ. Press, New York.

Yang, Z., Zhao, Y., 2014. Energy consumption, carbon emissions, and economic growth in India: evidence from directed acyclic graphs. Econ. Modell. 38, 533–540.

Yao, S., Zhang, S., Zhang, X., 2019. Renewable energy, carbon emission and economic growth: a revised environmental Kuznets Curve perspective. J. Clean. Prod. 235, 1338–1352.

You, W., Lv, Z., 2018. Spillover effects of economic globalization on CO2 emissions: a spatial panel approach. Energy Econ. 73, 248–257.

The role of economic institutions in electricity consumption, economic growth, and CO_2 emissions linkages: evidence from sub-Saharan Africa

Alex O. Acheampong[1], Janet Dzator[1,2], David A. Savage[1]

[1]*Newcastle Business School, University of Newcastle, Newcastle, NSW, Australia;* [2]*Australia Africa Universities Network (AAUN) Partner, Newcastle, NSW, Australia*

Chapter outline

1. Introduction

This chapter examines the link between economic institutions, economic growth, carbon emissions, and electricity consumption in sub-Saharan Africa (SSA). SSA contributes less to global carbon

Environmental Sustainability and Economy. https://doi.org/10.1016/B978-0-12-822188-4.00002-6

emissions; however, the continent suffers the huge cost of climate change (Acheampong, 2019; Adams and Klobodu, 2018). Recent studies have estimated that climate change could cost SSA approximately 20−30 billion per year (Mekonnen, 2014). As the agriculture sector employs the majority of SSA labor force, Asafu-Adjaye (2014) and Simbanegavi and Arndt (2014) estimated that climate change could decrease the region agriculture productivity and total economic output by 28% and 8%, respectively. Carbon dioxide emissions have been the major driver of climate change. While research has proven that energy consumption and economic growth have been the primary contributors to carbon emissions, it is required that policies and strategies for controlling carbon emissions should focus on improving energy efficiency and decoupling economic growth from emissions. However, reducing energy consumption implies limiting economic growth since energy is crucial for sustaining higher economic growth (Acheampong, 2018; Asafu-Adjaye, 2000). It is stated, "these conflicting arguments make economic, energy and environmental conservations policies at odds with one another" (Acheampong, 2018, p. 677). Because of these arguments, researchers have embarked on two major groups of empirical works.

The first category of the scholarly studies, which is pioneered by the seminal paper of Kraft and Kraft (1978), has examined the nexus between economic growth and energy consumption. However, the findings from these studies remain ambiguous and controversial (Ang, 2007, 2008; Antonakakis et al., 2017). For instance, the findings from some studies suggest that economic growth is caused by energy consumption (Dergiades et al., 2013; Wandji, 2013), while others also contest that it is economic growth that causes energy consumption (Esso, 2010; Kraft and Kraft, 1978; Lee and Chang, 2007). Another group of empirical studies suggests that energy consumption and economic growth are interdependent (Akinlo, 2008; Dagher and Yacoubian, 2012; Eggoh et al., 2011; Esso, 2010; Lean et al., 2010c; Mishra et al., 2009), whereas the final group of the empirical findings suggests that energy consumption and economic growth are independent of each other (Akarca et al., 1980; Akinlo, 2008; Lee, 2006).

The second set of empirical studies aims at testing the Environmental Kuznets Curve (EKC) hypothesis. The EKC hypothesis suggests that there is an inverted U-shaped relationship between economic growth and carbon emissions. Thus, at the initial stage of economic growth, environmental pollution increases but there is an improvement in environmental quality after a certain threshold of economic growth (Grossman and Krueger, 1995). While some scholars have supported the EKC hypothesis (see Ahmad et al., 2017; Apergis et al., 2015; Apergis and Payne, 2009, 2010), others have debunked it (Acheampong, 2019; Stern, 2004; Stern and Common, 2001). One of the major limitations in the previous EKC studies is their failure to consider or investigate the causality relationship between economic growth and carbon emissions (Lean and Smyth, 2010a; Soytas et al., 2007). While the policy goal of most economies is to achieve energy efficiency and economic growth while reducing carbon emissions, recent studies have utilized an integrated approach to study the linkages between energy consumption, carbon emissions, and economic growth (see Acheampong, 2018; Arouri et al., 2012; Farhani and Ozturk, 2015; Lean and Smyth, 2010a; Soytas and Sari, 2009; Soytas et al., 2007). This strand of empirical research helps to address the limitation in the EKC literature by considering the causality between carbon emissions and economic growth. Additionally, using a unified framework to investigate the linkages between energy consumption, carbon emissions, and economic growth prevents model misspecification and further helps in proper policy recommendations (Acheampong, 2018; Ang, 2007, 2008; Soytas et al., 2007, 2009).

Undoubtedly, the effectiveness of environmental, economic, and energy policies depends on the efficacy of institutional quality (Abid, 2017; Acemoglu et al., 2001; Acemoglu and Robinson, 2010; Bhattacharya et al., 2017; Fosu, 2012); however, this strand of empirical studies failed to consider economic institutions when studying the energy-carbon emissions-economic growth link (see Table 4.1). The question is, why consider economic institutions when discussing the dynamic link between economic growth, electricity consumption, and carbon emissions? Institutions are defined as the "rules of the game" of society that constrain and shape human interaction (North, 1989, 1990, 1992). In other words, institutions define the incentives and penalties, structure and shape social behavior and collective actions, and thus are a precondition for sustainable development (Alonso and Garcimartín, 2013). In this regard, Acemoglu et al. (2005) argue that economic institutions are the

Table 4.1 Summary of empirical studies on economic growth, energy consumption, and carbon emissions links.

Author (s)	Study period	Countries	Methodology	Empirical results
Lean and Smyth (2010a)	1980−2006	5 ASEAN countries	DOLS, GRC	ELEC $\rightarrow$ RGDP; CO_2 $\rightarrow$ RGDP CO_2 $\rightarrow$ ELEC
Cowan et al. (2014)	1990−2010	BRICS	GRC	ELEC $\leftrightarrow$ RGDP in Russia RGDP $\rightarrow$ ELEC in South Africa ELEC $\neq$ RGDP in Brazil, India and China CO_2 $\leftrightarrow$ RGDP in Russia CO_2 $\rightarrow$ RGDP in South Africa RGDP $\rightarrow CO_2$ in Brazil CO_2 $\neq$ RGDP in India and China ELEC $\rightarrow$ CO_2 in India ELEC $\neq CO_2$ in Brazil, Russia, China and South Africa
Ang (2008)	1971−99	Malaysia	VECM	**SR** RGDP $\rightarrow$ENER **LR** ENER $\leftrightarrow$ RGDP; CO_2 $\rightarrow$ RGDP
Ang (2007)	1960−2000	France	ARDL, VECM	**SR** ENER $\rightarrow$ RGDP **LR** RGDP $\rightarrow CO_2$; ENER $\rightarrow CO_2$
Jahangir Alam, Ara Begum, Buysse, and Van Huylenbroeck (2012)	1972−2006	Bangladesh	ARDL, VECM	**SR** ENER $\rightarrow$ RGDP; ELEC $\neq$ RGDP; ENER $\rightarrow CO_2$; CO_2 $\rightarrow$ RGDP **LR** ELEC $\leftrightarrow$ RGDP; ENER $\leftrightarrow$ CO_2; CO_2 $\rightarrow$ RGDP
Acheampong (2018)	1990−2016	116 countries	GMM-PVAR	ENER $\rightarrow$ RGDP; RGDP $\leftrightarrow$ CO_2; ENER $\leftrightarrow$ CO_2

Continued

Table 4.1 Summary of empirical studies on economic growth, energy consumption, and carbon emissions links.—cont'd

Author (s)	Study period	Countries	Methodology	Empirical results
Mirza and Kanwal (2017)	1971–2009	Pakistan	ARDL, VECM	RGDP $\leftrightarrow$ CO_2; ENER $\leftrightarrow$ CO_2; ENER $\leftrightarrow$ RGDP
Soytas and Sari (2009)	1960–2000	Turkey	VAR	CO_2 $\rightarrow$ ENER; RGDP $\neq$ CO_2; RGDP $\neq$ ENER
Soytas et al. (2007)		USA	VAR	RGDP $\neq$ ENER; CO_2 $\rightarrow$ ENER; RGDP $\neq$ CO_2
Zhang and Cheng (2009)	1960–2007	China	VAR, VECM	RGDP $\rightarrow$ ENER; ENER $\rightarrow$ CO_2; RGDP $\neq$ CO_2
Salahuddin and Gow (2014)	1980–2012	GCC	PMG, SUR, GRC	ENER $\leftrightarrow$ CO_2; RGDP $\rightarrow$ ENER; RGDP $\neq$ CO_2
Halicioglu (2009)	1960–2005	Turkey	ARDL	**SR** ENER $\leftrightarrow$ CO_2; CO_2 $\leftrightarrow$ RGDP; ENER $\leftrightarrow$ RGDP **LR** ENER $\rightarrow$ CO_2; CO_2 $\leftrightarrow$ RGDP; ENER $\rightarrow$ RGDP
Al-mulali, Lee, Hakim Mohammed, and Sheau-Ting (2013)	1980–2008	LAC	CCR	ENER $\leftrightarrow$ CO_2; CO_2 $\leftrightarrow$ RGDP; ENER $\leftrightarrow$ RGDP
Omri (2013)	1990–2011	MENA	System-GMM	ENER $\leftrightarrow$ RGDP; ENER $\rightarrow$ CO_2; CO_2 $\leftrightarrow$ RGDP
Apergis and Payne (2010)	1992–2004	CIS	FMOLS, GRC	**SR** ENER $\rightarrow$ CO_2; RGDP $\rightarrow$ CO_2; ENER $\leftrightarrow$ RGDP **LR** CO_2 $\leftrightarrow$ ENER
Apergis and Payne (2009)	1971–2004	CA	FMOLS, GRC	**SR** ENER $\rightarrow$ CO_2; RGDP $\rightarrow$ CO_2; ENER $\leftrightarrow$ RGDP **LR** CO_2 $\leftrightarrow$ ENER

$\neq$ *No causality;* $\rightarrow$ *unidirectional causality;* $\leftrightarrow$ *bidirectional causality;* ARDL, *Autoregressive distributed lag;* CA, *Central America;* CCR, *Conical cointegration regression;* CIS, *Commonwealth Independent States;* CO_2, *Carbon emissions;* DOLS, *Dynamic ordinary least squares;* ELEC, *Electricity;* ENER, *Total energy consumption;* FMOLS, *Fully modified ordinary least square;* GCC, *Gulf Cooperation Council countries;* GMM-PVAR, *Generalized method of moments—panel vector autoregression;* GRC, *Granger causality;* LAC, *Latin America and Caribbean;* LR, *Long-run;* MENA, *Middle East and North Africa countries;* PMG, *Pooled mean group;* RGDP, *Economic growth;* SR, *Short-run;* SUR, *Seemingly unrelated regression;* VAR, *Vector autoregression;* VECM, *Vector error correction model.*

fundamental cause of economic development, since they structure property right, and without property right individuals will not have the motivation for human or physical capital investment as well as adopting more efficient technologies, determine the efficiency of resource allocation, and determine transaction cost. Consistently, Bhattacharya et al. (2017) opine that economic institutions that support the promotion of property right protection also support voluntary exchange and help governments to design and implement sustainable energy and environmental policies without which these policies will not bear the desirable outcome.

The argument by these scholars is an indication that economic institutions that ensure property right protection and lower transaction cost are very critical for achieving higher economic growth, efficiency in electricity consumption, and the mitigation of carbon emissions. Therefore, incorporating economic institutions when studying the links between economic growth, electricity consumption, and carbon emissions is very crucial to prevent omission bias. In this regard, the current study extends the literature by investigating the dynamic linkages between economic institutions, economic growth, electricity consumption, and carbon emissions for a panel of 45 sub-Saharan Africa (SSA) countries over the period 1960–2017. SSA is an interesting case for this study due to its energy efficiency, security, and accessibility problems. Additionally, as indicated earlier, SSA remains the most vulnerable region to climate change, and if pragmatic policy measures are not taken its economic development will worsen. Furthermore, despite the richness of energy resources in SSA, out of the total population of 915 million, only 290 million people have access to electricity. On the other hand, the total number of people without access to electricity has increased by 100 million since 2000, which is very astronomical (International Energy Agency, 2014). Finally, economic institutions and economic growth have also been anemic in SSA. Therefore, the results that will emanate from this study would help guide policymakers designing economic, energy, and environmental policies, as well as institutional reforms.

This study contributes to the literature in three main directions: (i) this study incorporates economic institutions in economic growth, electricity consumption, and carbon emissions nexus. (ii) This study also employs an econometric approach that could capture the full dynamic interactions between economic institutions, economic growth, electricity consumption, and CO_2 emissions. Thus, this study utilizes GMM-PVAR, which is an advanced econometric technique, developed by Love and Zicchino (2006), to examine the interactions among these variables. (iii) Given that the actual causality between economic institutions, economic growth, electricity consumption, and CO_2 could be different across regions within SSA because of different institutional and economics arrangements. Therefore, this study further disaggregates the full sample into western, southern, and central eastern Africa countries. The remaining sections are organized as follows. Methodology and data are presented in Section 2. Empirical findings are presented in Section 3, followed by policy analysis in Section 4. Conclusion remarks are presented in Section 5.

2. Methodology and data
2.1 Empirical model specification

This study examines the dynamic links between economic institutions, electricity consumption, carbon emissions, and economic growth. To achieve this objective, the current study adopts the generalized method of moment panel vector autoregression (GMM-PVAR)[1] empirical methodology of Acheampong (2018), Abrigo and Love (2015), and Love and Zicchino (2006). Thus, the empirical

---

[1]The GMM-PVAR was implemented in STATA using the code developed by Love et al. (2006).

model used to estimate the nexus between economic institutions (*e_overall*), electricity consumption (*lnelect*), economic growth (*lngdpc*), and carbon emission (*lnco$_2$gdp*) is given in Eq. (4.1) below:

$$Z_{it} = \mu_i + \Phi(I)Z_{it-1} + v_i + \theta_t + \varepsilon_{it} \tag{4.1}$$

where $i = 1\ldots\ldots45$ and $t = 1960 \ldots\ldots 2017$, Z_{it} is the vectors of endogenous variables, which are $e\_overall_{it}$, $lnrgdpc_{it}$, $lnco_2gdp_{it}$, $lnelect_{it}$. The lag operator of the endogenous variables is given as $\Phi(I)$, v_i and θ_t are the individual and time-specific effect, respectively, while the stochastic error term is given as ε_{it}. Eq. (4.1) could have been estimated using Ordinary Least Squares (OLS) and fixed effect. However, the inclusion of the lagged of the dependent variable within the system equation would correlate with the panel fixed effect resulting in bias and inconsistent estimates of the parameters. In this regard, Baltagi (2008) argues that using the lagged instruments and first differencing are the best strategies to eliminate the individual fixed effect and produce consistent estimates. Following this argument, this study adopted the Blundell and Bond [hereafter, BB] (1998) system generalized method of moment (GMM) to estimate the above equation. Unlike the Arellano and Bond (1991) GMM approach, the generalized method of moment of BB (1998) utilizes the difference of lag of the dependent variable as instruments for equations in levels and also incorporates the lagged levels of the dependent variable as instruments for equations in first differences (see Acheampong 2018; Čeh Časni et al., 2016). Therefore, to eliminate the time fixed and country-specific effect from Eq. (4.1), we estimate Eq. (4.2), using the first difference (Δ) approach as specified in Eq. (4.2).

$$\Delta Z_{it} = \Delta\mu_i + \Phi(I)\Delta Z_{t-1} + \Delta v_i + \Delta\theta_t + \Delta\varepsilon_{it} \tag{4.2}$$

We further follow the standard approach of Hamilton (1994) and Lutkepohl (2005) to estimate the impulse response functions (IRF). According to Love et al. (2006), The IRF describes the reaction of one variable to the structural innovation in another variable within a system while all other shocks remain constant. According to Abrigo and Love (2015), the IRF should not be interpreted as causality since the errors are contemporaneously correlated. In our model, we orthogonalized the errors such that we could isolate the shocks to one of the vector autoregression errors. Additionally, Sims (1980) opines that variables in the vector autoregression model should have recursive causal ordering based on their degree of exogeneity. In this regard, we assume that a current shock to economic institutions affects economic growth, electricity consumption, and carbon emissions contemporaneously while economic institutions are affected by the lags of economic growth, electricity consumption, and carbon emissions. Thus, current economic institutions are influenced by previous economic growth, electricity consumption, and environmental pollution (see Acheampong, 2018). Thus, in the empirical model, economic institutions were ordered first, followed by economic growth, electricity consumption, and carbon emissions. In analyzing the IRF, one thousand (1000) Monte Carlo replication or simulations were conducted to generate the standard errors.

The GMM-PVAR is advantageous in our study, since it builds an endogenous system and treats all the variables in an unrestricted way (Love et al., 2006). This feature helps this study to account for the dynamic endogenous interactions between economic institutions, economic growth, electricity consumption, and carbon emissions. The GMM-PVAR also provides a better understanding of the source of heterogeneities by accounting for cross-sectional dynamic heterogeneities. The GMM-PVAR further helps to capture the time variations in the coefficient and the variance of the shocks. Compared to the traditional VAR model, the GMM-PVAR is robust, and consistent results are obtained by increasing the estimation observations (Acheampong, 2018).

2.2 Data

The study used annual data for 45 SSA countries over the period 1960–2017. The countries for the study can be found in Appendix Table 4.1. Real GDP per capita was used as the proxy for economic growth. Carbon emission (kg) per 2010 US dollars as a percentage of GDP was used as the proxy for CO_2 emissions. Electric power consumption (kWh per capita) was used as the measure for electricity consumption. We obtained these data from the World Bank Development Indicators (2018). Additionally, this study used the Fraser Institute World Economic Freedom index as the measure for economic institutions. The World Economic Freedom Index is a composite index that comprises the legal system and property right, freedom to trade internationally, regulations size of government and sound money. The World Economic Freedom index was used a proxy for economic institutions, since it is highly correlated with other measures of economic institutions such as the risk of appropriation index and corruption perception index (Góes, 2016).

2.3 Stationarity check

Before estimating the empirical models, we conducted four types of panel unit root tests the Levin-Lin-Chu (LLC) unit root test (Levin et al., 2002), Im-Pesaran-Shin (IPS) unit root test (Im et al., 2003), Augmented Dickey-Fuller (ADF) unit root test (Dickey et al., 1979), and the Philips-Perron (PP) unit root test (Phillips et al., 1988). As presented in Table 4.2, except the economic growth series, most of the unit tests show that the remaining variables are stationary at levels. However, at the first difference, all the unit root tests show that all the variables are stationary.

3. Results and discussion
3.1 GMM-PVAR causality estimates

The results shown in Table 4.3 present the GMM-PVAR causality among the variables: From Table 4.3, the empirical findings suggest that economic growth improves economic institutions by 0.783% while economic institutions are found to exert an insignificant effect on economic growth. The implications are that economic institutions do not contribute to the economic growth of the SSA countries; however, an improvement in economic growth contributes to the improvement in the efficiency of economic institutions. This evidence supports Sachs (2003) argument that economic institutions are not the primary cause of economic growth. Consistently, Baliamoune-Lutz and Boko's (2012) study reveals that institutions have a negligible effect SSA economic development. Additionally, our results further are consistent with the results of Alonso and Garcimartín (2013) and Baliamoune-Lutz et al. (2012), which indicate that economic growth is a major determinant of economic institutions.

Our empirical evidence suggests that economic institutions significantly reduce carbon emissions while it is found to have no causal effect on electricity consumption. The estimate shows carbon emissions will decrease by 0.054% when there is an improvement in economic institutions. Thus, promoting economic institutions will enhance property right protection, which in turn attracts investment in the development of environmentally friendly technologies. The development of environmentally friendly technologies will contribute significantly to carbon emissions mitigation. Our results confirm the empirical findings of Tamazian and Bhaskara (2010) and Bhattacharya et al. (2017) that economic institutions contribute to the reduction in carbon emissions. The results further reveal that electricity consumption significantly increases carbon emissions, such that 1% increase in

Table 4.2 Panel unit root test.

	Levels				First difference			
	LLC	IPS	ADF	PP	LLC	IPS	ADF	PP
lngdpc	1.118	9.278	67.530	79.327	−32.764***	−32.981***	1053.01***	1065.87***
lnelect	−2.865***	1.240	62.555*	60.909*	−25.434***	−25.955***	548.985***	597.257***
lnco$_2$gdp	−12.664***	−11.953***	225.145***	253.489***	−50.420***	−48.153***	1595.34***	1700.56***
E_overall	−2.625***	−1.527*	109.328***	126.426***	−31.094***	−28.240***	743.164***	782.614***

Probabilities for Fisher tests are computed using an asymptotic Chi-square distribution. All other tests assume asymptotic normality. Levin-Lin-Chu t (LLC), Im-Pesaran-Shin W-stat (IPS), Augmented Dickey-Fuller-Fisher Chi-square (ADF), Philips-Perron-Fisher Chi-square (PP). The LLC unit root test assumes a common unit root process while the IPS, ADF, and PP unit root tests assume individual unit root process. *P <.1, **P <.05, ***P <.01.*

Table 4.3 GMM-PVAR causality main results for the full sample.

	Dependent variables			
	Independent variables			
	de_ovrall$_t$	*dlngdpc$_t$*	*dlnelect$_t$*	*dlnco$_2$gdp$_t$*
de_ovrall$_{t-1}$	0.002	−0.001	−0.007	−0.054∗∗
	(0.017)	(0.006)	(0.013)	(0.025)
dlngdpc$_{t-1}$	0.783∗∗∗	0.191∗∗∗	0.199	−0.491∗
	(0.254)	(0.069)	(0.127)	(0.258)
dlnelect$_{t-1}$	0.077	−0.001	−0.123	0.141∗∗
	(0.083)	(0.016)	(0.080)	(0.066)
dlnco$_2$gdp$_{t-1}$	0.064	0.011	0.037	−0.105
	(0.046)	(0.009)	(0.028)	(0.066)

Standard error presented in parenthesis are heteroscedasticity-robust standard errors. ∗P < .1, ∗∗P < .05, ∗∗∗P < .01.

electricity consumption is associated with 0.141% increase in carbon emissions. Thus, consistently depending on electricity, which is generated using fossil energy, will increase carbon emissions. Furthermore, electricity consumption and carbon emissions exert no effect on economic growth. Similarly, the empirical results suggest that economic growth has a negligible impact on electricity consumption but reduces carbon emissions significantly. Additionally, economic institutions and carbon emissions do not affect electricity consumption. From Panel A of Table 4.4, economic institutions are not influenced by electricity consumption and carbon emissions.

3.2 Impulse response function analysis

We report the findings from impulse response function (IRF) and the 95% confidence interval band that was generated using 1000 Monte Carlo simulations. The first columns of Figs. 4.1−4.4 show the response of electricity consumption, economic growth, and economic institutions to a shock to carbon

Table 4.4 GMM-PVAR causality main results.

	Panel A: West Africa countries			
	de_ovrall$_t$	*dlngdpc$_t$*	*dlnelect$_t$*	*dlnco$_2$gdp$_t$*
de_ovrall$_{t-1}$	−0.017 (0.027)	0.004 (0.007)	0.008 (0.032)	0.008 (0.025)
dlngdpc$_{t-1}$	0.859∗∗ (0.379)	0.278∗∗∗ (0.100)	0.534∗∗ (0.218)	−0.935∗∗∗ (0.305)
dlnelect$_{t-1}$	0.109∗ (0.059)	−0.025 (0.018)	−0.120 (0.074)	0.065 (0.053)
dlnco$_2$gdp$_{t-1}$	0.040 (0.083)	0.023 (0.015)	−0.037 (0.046)	−0.126∗ (0.067)
	Panel B: Southern Africa countries			
	de_ovrall$_t$	*dlngdpc$_t$*	*dlnelect$_t$*	*dlnco$_2$gdp$_t$*
de_ovrall$_{t-1}$	0.018 (0.020)	0.017∗ (0.009)	−0.011 (0.008)	−0.060∗ (0.033)
dlngdpc$_{t-1}$	0.724∗ (0.415)	0.459∗∗∗ (0.092)	0.371∗∗∗ (0.093)	−0.685∗∗ (0.294)

Continued

Table 4.4 GMM-PVAR causality main results.—cont'd

Panel B: Southern Africa countries			
	de_ovrall$_t$	*dlngdpc$_t$*	*dlnelect$_t$*	*dlnco$_2$gdp$_t$*
dlnelect$_{t-1}$	0.242 (0.172)	0.010 (0.039)	0.176** (0.075)	0.127 (0.098)
dlnco$_2$gdp$_{t-1}$	0.174* (0.091)	0.015 (0.020)	0.020 (0.032)	−0.170*** (0.062)

Panel C: Central and east Africa countries			
	de_ovrall$_t$	*dlngdpc$_t$*	*dlnelect$_t$*	*dlnco$_2$gdp$_t$*
de_ovrall$_{t-1}$	−0.028 (0.037)	−0.014* (0.008)	−0.006 (0.021)	−0.142*** (0.043)
dlngdpc$_{t-1}$	0.701** (0.304)	0.464*** (0.087)	0.249* (0.143)	−0.573 (0.482)
dlnelect$_{t-1}$	−0.055 (0.106)	−0.009 (0.020)	−0.221*** (0.073)	0.115 (0.123)
dlnco$_2$gdp$_{t-1}$	0.051 (0.032)	0.003 (0.010)	0.024 (0.019)	−0.126** (0.051)

Standard error presented in parenthesis are heteroscedasticity-robust standard errors. $*P < .1$, $**P < .05$, $***P < .01$.

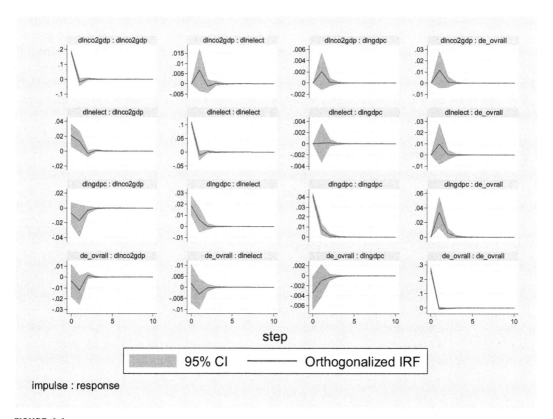

FIGURE 4.1

IRF for Full sample.

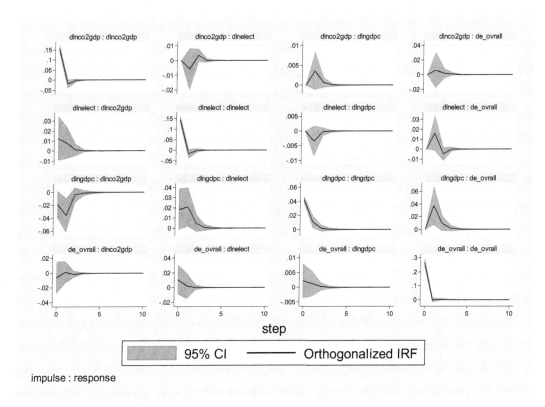

FIGURE 4.2

IRF for west Africa countries.

emissions while the second columns show the response of carbon emissions, economic growth, and economic institutions to a shock to electricity consumption. Similarly, the third columns of Figs. 4.1–4.4 show the response of carbon emissions, electricity consumption, and economic institutions to a shock to economic growth. Finally, the fourth columns of Figs. 4.1–4.4 depict the response of economic growth, carbon emissions, and electricity consumption to a shock to economic institutions.

Fig. 4.1 reveals that economic growth, electricity consumption, and carbon emissions react negatively to a shock to economic institutions. This result is consistent with Acemoglu et al.'s (2001) observation that the SSA region has retarded economic growth, since it has a relatively weak economic institution that increases the risk of expropriation and rent-seeking. Additionally, economic institutions are inimical to economic growth, since SSA governments implement undesirable economic policies that do not benefit the general population which increase the risk of political instability thereby retarding economic growth (Fosu et al., 2006). The role of economic institutions in reducing electricity consumption and carbon emissions could be via economic growth. Thus, a shock to economic institutions reduces economic growth, which implies less production and hence lower carbon emissions.

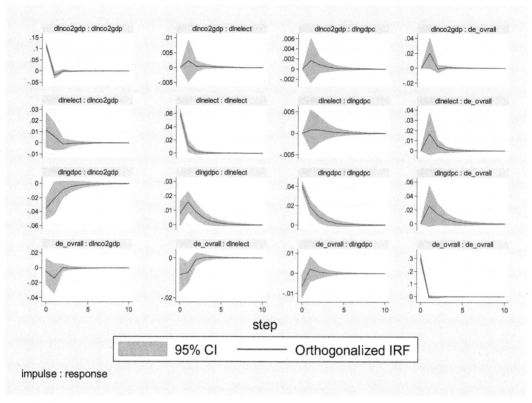

FIGURE 4.3

IRF for southern Africa countries.

The reducing effect of economic institutions on electricity consumption implies that improving domestic economic institutions could increase efficiency in electricity consumption through technological effect. These results are inconsistent with the observation by Abid (2016) that institutions expand economic growth thereby increasing CO_2 emissions in SSA.

The evidence further reveals that a positive innovation to economic growth increases economic institutions and electricity consumption while reducing carbon emissions. Thus, embarking on liberalization policies and investing both human and physical capital to boost economic growth could enhance the efficacy of the region's economic institutions while contributing to carbon emissions mitigation. Additionally, a shock to electricity consumption increases carbon emissions and economic institutions while it exerts an insignificant impact on economic growth. The IRF findings further suggest that a shock to carbon emissions increases economic growth, economic institutions, and electricity consumption. The positive effect of carbon emissions on electricity consumption and economic growth has been found in previous studies (Acheampong, 2018; Lean and Smyth, 2010a; Soytas et al., 2007).

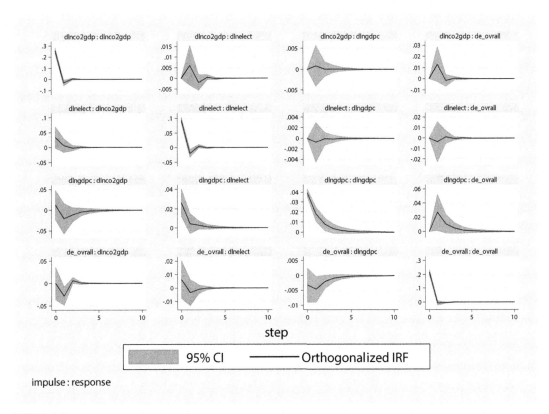

FIGURE 4.4

IRF for central and eastern Africa countries.

3.3 Further analysis

Because of different institutional and economics arrangement among the regions within SSA, we conduct further analysis about how the link between economic institutions, economic growth, electricity consumption, and carbon emissions differ across regions within sub-Saharan Africa.

3.3.1 Regional analysis

The results shown in Table 4.4 present the GMM-PVAR causality results for the subsamples: Panel A of Table 4.4 gives the estimates for the West Africa countries, while Panel B and C show the estimates for the southern and central-eastern Africa countries, respectively. From Table 4.4, the results indicate that economic institutions improve economic growth in southern Africa countries while it reduces economic growth in central eastern African countries. For the west Africa countries, economic institutions exert an insignificant positive effect on economic growth. Thus, the role of economic institutions in ensuring property right protection and reduction in transaction cost has been responsible for the higher economic growth in southern Africa countries. However, extractive institutions, which increase the risk of expropriation and transaction, have been limiting economic growth in central eastern African countries.

Also, electricity consumption and carbon emissions have an insignificant effect on economic growth in all regions. Thus, electricity consumption has not been a major contributor to economic growth in these regions. This result further adds to the argument of classical economic theory that energy is not a major input in the production function. Therefore, the insignificant effect of electricity consumption on economic growth could be attributed to energy poverty in these regions (see Karekezi, 2002).

Economic institutions and carbon emissions have an insignificant effect on electricity consumption in all regions. Thus, economic institutions and carbon emissions are not a major determinant of electricity consumption in the regions. On the other hand, the results indicate that economic growth increases electricity consumption in all the regions. Thus, 1% increase in economic growth is associated with 0.534%, 0.371%, and 0.249% increase in electricity consumption in west Africa, southern Africa, and central-eastern Africa countries, respectively. Thus, increasing economic production to achieve higher economic growth results in higher demand for electricity in all the regions. The results suggest that economic growth improves economic institutions in all regions. Electricity consumption significantly impacts economic institutions in west Africa countries while its impacts in southern, central, and eastern African countries are insignificant. The results also suggest that carbon emissions significantly improve economic institutions in southern Africa countries while their impact in west, central, and eastern African countries are negligible.

Furthermore, the results suggest that economic institutions reduce carbon emissions in southern and central eastern African countries while their impact is negligible in west Africa countries. This observation result indicates that economic institutions enable governments to design and implement sustainable environmental policies, which have been contributing to carbon emission reduction (See, Bhattacharya et al., 2017). We observe that economic growth contributes to carbon emissions mitigation in west and southern Africa countries while its impact on carbon emissions in central and eastern Africa countries is negligible. Thus, increasing economic growth is associated with an improvement in environmental quality in southern and central eastern African countries. These results support Beckerman (1992) and Meadow et al.'s (1992) argument that economic growth is critical for improving environmental quality. It is observed that electricity consumption has an insignificant positive effect on carbon emissions in all the regions. Thus, electricity consumption is not a major contributor to carbon emissions in the regions since energy consumption in entire sub-Saharan Africa is relatively less (Acheampong, 2018; Karekezi, 2002).

3.3.2 IRF analysis for the regions

The IRF results for western, southern, and central eastern Africa countries are presented in Figs. 4.2–4.4, respectively. The results indicate that a standard deviation shock to economic institutions is associated with increased economic growth in all the regions. Additionally, a shock to economic institutions increases electricity consumption in west Africa and central eastern Africa countries while it reduces electricity consumption in southern Africa countries. Also, carbon emissions react negatively to a shock to economic institutions in all the regions.

Electricity consumption and economic institutions react positively to a shock to economic growth in all the regions. Also, a positive innovation to economic growth reduces carbon emissions. A shock to electricity consumption reduces economic growth in west Africa countries while it increases economic growth in southern and central eastern Africa countries. A shock to electricity consumption also increases carbon emissions in all the regions. Lastly, a shock to carbon emissions increases economic growth in all the regions. A shock to carbon emissions increases electricity consumption in southern and central eastern Africa countries while it reduces electricity consumption in west Africa countries.

3.4 Granger causality results

To complement the GMM-PVAR estimates, we further conducted the Granger causality between the variables under consideration (see Table 4.5). The results from the Granger causality suggest economic growth unidirectionally causes economic institutions for the full sample and west Africa countries. Additionally, bidirectional causality exists between economic growth and economic institutions in southern and central eastern Africa countries and confirms the bidirectional causality between economic growth and economic institutions found by Góes (2016).

Table 4.5 Results from Granger causality analysis.

	de_ovrall_t	$dlngdpc_t$	$dlnelect_t$	$dlnco_2gdp_t$
Panel A: Full sample				
de_ovrall_{t-1}		0.036	0.341	4.588**
		(0.850)	(0.559)	(0.032)
$dlngdpc_{t-1}$	9.548***		2.478	3.625*
	(0.002)		(0.115)	(0.057)
$dlnelect_{t-1}$	0.862	0.005		4.579***
	(0.353)	(0.946)		(0.032)
$dlnco_2gdp_{t-1}$	1.940	1.588	1.699	
	(0.164)	(0.208)	(0.192)	
Panel B: West Africa countries				
	de_ovrall_t	$dlngdpc_t$	$dlnelect_t$	$dlnco_2gdp_t$
de_ovrall_{t-1}		0.353	0.064	0.109
		(0.553)	(0.801)	(0.741)
$dlngdpc_{t-1}$	5.140**		5.996**	9.378***
	(0.023)		(0.014)	(0.002)
$dlnelect_{t-1}$	3.352*	2.065		1.503
	(0.067)	(0.151)		(0.220)
$dlnco_2gdp_{t-1}$	0.238	2.293	0.653	
	(0.626)	(0.130)	(0.419)	
Panel C: Southern Africa countries				
	de_ovrall_t	$dlngdpc_t$	$dlnelect_t$	$dlnco_2gdp_t$
de_ovrall_{t-1}		3.595*	2.035	3.411*
		(0.058)	(0.154)	(0.065)
$dlngdpc_{t-1}$	3.041*		16.021***	5.418**
	(0.081)		(0.000)	(0.020)

Continued

Table 4.5 Results from Granger causality analysis.—cont'd

	Panel C: Southern Africa countries			
	de_ovrall$_t$	*dlngdpc$_t$*	*dlnelect$_t$*	*dlnco$_2$gdp$_t$*
dlnelect$_{t-1}$	1.986	0.072		1.664
	(0.159)	(0.788)		(0.197)
dlnco$_2$gdp$_{t-1}$	3.672*	0.598	0.373	
	(0.055)	(0.439)	(0.541)	
	Panel D: Central and Eastern Africa countries			
	de_ovrall$_t$	*dlngdpc$_t$*	*dlnelect$_t$*	*dlnco$_2$gdp$_t$*
de_ovrall$_{t-1}$		3.161*	0.084	10.883***
		(0.075)	(0.772)	(0.001)
dlngdpc$_{t-1}$	5.330**		3.004*	1.415
	(0.021)		(0.083)	(0.234)
dlnelect$_{t-1}$	0.267	0.194		0.869
	(0.606)	(0.660)		(0.351)
dlnco$_2$gdp$_{t-1}$	2.611	0.112	1.675	
	(0.106)	(0.738)	(0.196)	

*Probability values are presented in parenthesis. *P <.1, **P <.05, ***P <.01.*

From Table 4.5, our findings further reveal that no causality exists between economic growth and electricity consumption for the full sample and central eastern Africa countries, and this result is consistent with the studies that support neutrality hypothesis (Akarca and Long, 1980; Akinlo, 2008; Cowan et al., 2014; Jahangir Alam et al., 2012; Lee, 2006). Contrarily, economic growth unidirectionally causes electricity consumption in west and southern Africa countries and this result is consistent with studies that support the conservation hypothesis (Cong et al., 2011; Esso, 2010; Kraft et al., 1978; Lean et al., 2010b).

Furthermore, the Granger causality result indicates that economic growth unidirectionally causes carbon emissions for the full sample, west and southern Africa countries. This result adds to findings of previous studies which indicate that economic growth unidirectionally causes carbon emissions (Ang, 2007; Apergis et al., 2010). Contrarily, no causal relationship exists between economic growth and carbon emissions in central eastern Africa countries, and this confirms the findings of Cowan et al. (2014) and Salahuddin and Gow (2014). Our results also indicate that electricity consumption unidirectionally causes carbon emissions for the full sample and this result confirms the empirical results of Ang (2007), Apergis and Payne (2010), and Cowan et al. (2014). On the other hand, no causal relationship exists between carbon emissions and electricity consumption across the regions. The results also suggest that economic institutions unidirectionally cause carbon emissions for the full samples, southern and central eastern Africa countries, while no causality exists between economic institutions and carbon emissions in west Africa countries.

3.5 Forecast error variance and stability tests

We estimate the forecast error variance decomposition[2] using 1000 Monte Carlo simulations. We present the contribution of economic institutions to the variations that occur in carbon emissions, electricity consumption, and economic growth over 10 years. Our forecast results suggest that economic institutions explain 0.7%, 0.3%, 2%, and 2% of the variations in economic growth for the full sample, west Africa, southern Africa, and central eastern Africa countries, respectively, for the next 10-year period. Additionally, economic institutions explain about 0.09%, 0.5%, 3%, and 0.5% of the variations in electricity consumption for the full sample, west Africa, southern Africa and central eastern Africa countries, respectively, for the next 10-year period. Finally, economic institutions explain about 0.46%, 0.1%, 1%, and 1% of the variations in carbon emissions for the full sample, west Africa, southern Africa, and central eastern Africa countries, respectively, for the next 10-year period ahead. Fig. 4.5 shows that the results for the full sample and the subsamples are stable and good for policy implications.

4. Policy analysis

The results from the empirical analysis could inform economic, environmental, and energy policies, as well as institutional change for the SSA countries. The causality analysis suggests that economic growth unidirectionally causes economic institutions and the impact is positive. The implication is that increasing economic growth will enhance the efficacy of economic institutions. Therefore, sufficient and efficient investment in both human and physical capital development to boost economic growth could improve the quality of economic institutions in SSA. Additionally, to boost economic growth to improve economics institutions, policymakers should embark on structural policies that will provide the enabling environment for sub-Saharan Africa countries to make the effective transition toward industrialization rather than depending on agriculture and exportation of raw materials. It is also found that economic growth unidirectionally causes carbon emissions and the causal effect is negative. The implications are that promoting economic growth will not conflict with the environment; therefore, policymakers should embark on structural policies that could boost the economic development in SSA.

Additionally, the causality results indicate that economic institutions unidirectionally cause carbon emissions and the impact is negative. Development and transfer of technologies are needed to speed the carbon emissions mitigation. In this regards, promoting economic institutions that ensures property right protection and lower transaction costs could attract investors to invest in the development of environmentally friendly technologies that could facilitate the reduction in carbon emissions. In other words, promoting quality economic institutions would increase investment, generation, and share of renewable energy in the total energy mix, thereby reducing the overreliance on the fossil fuel and consequently reducing carbon emissions. Generally, we argue that promoting economic institutional quality will result in higher economic growth and make the region more environmentally sustainable.

The results also indicate that electricity consumption unidirectionally causes carbon emissions and the causal effect is positive. The implication is that electricity consumption in sub-Saharan Africa, which is generated using fossil fuel, contributes to the rise of carbon emissions. The policy implication is that reducing electricity consumption is necessary for carbon emissions mitigation. Reducing

---

[2]Because of space, the variance decomposition tables are not reported but available upon request.

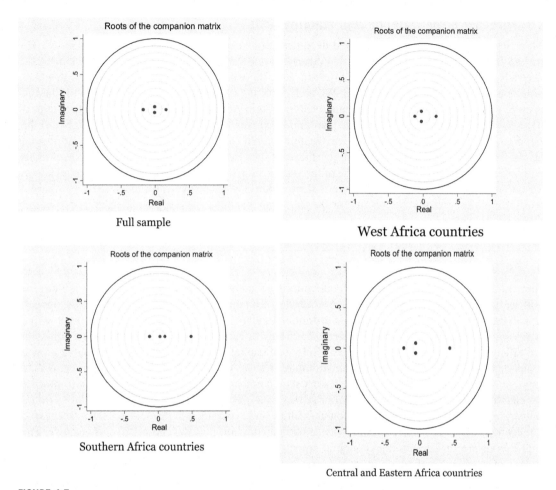

FIGURE 4.5

Stability graphs.

electricity consumption, the electricity being generated using fossil fuel, requires the government to collaborate with the private sector to increase the investment and generation of renewable energy. Increasing the generation and use of renewable energy would ease the overreliance on using fossil fuel to generate electricity. Additionally, formulating and implementing energy policies with clear goals and incentives for investment in renewable energy are also needed to promote renewable energy. The results suggest that no causal relationship exists between economic growth and electricity consumption—thus supporting the neutrality hypothesis. The implication is that electricity consumption does not contribute to economic growth while sustaining higher economic growth has no impact on electricity consumption. Thus, neither expansive nor conservative policies concerning electricity consumption make any significant contribution to SSA economic growth.

5. Concluding remarks

Achieving energy efficiency and economic growth while reducing carbon emissions has been the policy goal of most economies. In this regard, some emerging literature has recently employed multivariate models to study the causal link between energy consumption, economic growth, and carbon emissions. However, these studies failed to recognize the role of economic institutions when discussing energy consumption, economic growth, and carbon emissions nexus. Although the role of economic institutions, which involves property right protection, in economic growth has increasingly attracted scholarly attention, the extent to which economic institutions are shaping the global move toward sustainable energy consumption and carbon emissions mitigation has received less attention in the literature. This study utilizes GMM-PVAR, which is an advanced econometric technique, developed by Love and Zicchino (2006) to examine the dynamic links between economic institutions, electricity, carbon emissions, and economic growth for 45 SSA countries using annual data for the period 1960–2017.

The estimated coefficients from the GMM-PVAR system show that economic growth improves economic institutions by 0.783% while reducing carbon emissions by 0.49%. Electricity consumption increases carbon emissions by 0.141% while economic institutions reduce carbon emissions by 0.054%. Additionally, the Granger causality results indicate that economic growth causes economic institutions and carbon emissions without feedback effect while electricity consumption and economic growth are independent of each other. Electricity consumption unidirectionally causes carbon emissions while electricity consumption and economic institutions are also independent of each other. These results differ across regions within SSA. This study recommends that SSA countries should embark on expansionary structural policies that will contribute and sustain the region's economic growth, which in turn will improve domestic institutional quality while reducing carbon emissions. Additionally, policymakers should promote institutions which support property right protection and lower transaction cost; this could boost the development and transfer of environmentally friendly technologies that will contribute to carbon emissions mitigation while promoting economic growth. To lessen the environmental degrading effect of electricity consumption, policymakers should increase investment in renewable energy to ease the pressure on the overreliance on fossil fuel to generate electricity.

6. Appendix table
6.1 Countries included in the study

Angola, Benin, Botswana, Burkina Faso, Burundi, Cabo Verde, Cameroon, Central African Republic, Chad, Comoros, Congo, Dem. Rep., Congo, Rep., Cote d'Ivoire, Equatorial Guinea, Eritrea, Ethiopia, Gabon, Gambia, Ghana, Guinea, Guinea-Bissau, Kenya, Lesotho, Liberia, Madagascar, Malawi, Mali, Mauritania, Mauritius, Mozambique, Namibia, Niger, Nigeria, Rwanda, Sao Tome and Principe, Senegal, Seychelles, Sierra Leone, South Africa, Swaziland, Tanzania, Togo, Uganda, Zambia, Zimbabwe.

References

Abid, M., 2016. Impact of economic, financial, and institutional factors on CO_2 emissions: evidence from sub-Saharan Africa economies. Util. Pol. 41, 85–94. https://doi.org/10.1016/J.Jup.2016.06.009.

Abid, M., 2017. Does economic, financial and institutional developments matter for environmental quality? A comparative analysis of EU and MEA countries. J. Environ. Manag. 188, 183–194. https://doi.org/10.1016/J.Jenvman.2016.12.007.

Abrigo, M.R.M., Love, I., 2015. Estimation of panel vector autoregression in stata: a package of programs. Paper presented at the Panel data conference.

Acemoglu, D., Johnson, S., Robinson, J.A., 2001. The colonial origins of comparative development: an empirical investigation. Am. Econ. Rev. 91 (5), 1369–1401.

Acemoglu, D., Johnson, S., Robinson, J.A., 2005. Institutions as a fundamental cause of long-run growth. In: Aghion, P., Durlauf, S.N. (Eds.), Handbook of Economic Growth, 1A, pp. 385–472.

Acemoglu, D., Robinson, J., 2010. The role of institutions in growth and development. Rev. Econ. Inst. 1 (2), 1–33. https://doi.org/10.5202/Rei.V1i2.1.

Acheampong, A.O., 2018. Economic growth, CO_2 emissions and energy consumption: what causes what and where? Energy Econ. 74, 677–692. https://doi.org/10.1016/J.Eneco.2018.07.022.

Acheampong, A.O., 2019. Modelling for insight: does financial development improve environmental quality? Energy Econ. 83, 156–179. https://doi.org/10.1016/j.eneco.2019.06.025.

Adams, S., Klobodu, E.K.M., 2018. Financial development and environmental degradation: does political regime matter? J. Clean. Prod. 197, 1472–1479. https://doi.org/10.1016/j.jclepro.2018.06.252.

Ahmad, N., Du, L., Lu, J., Wang, J., Li, H.-Z., Hashmi, M.Z., 2017. Modelling the CO_2 emissions and economic growth in Croatia: is there any environmental Kuznets Curve? Energy 123, 164–172. https://doi.org/10.1016/j.energy.2016.12.106.

Akarca, A.T., Long, T.V., 1980. On the relationship between energy and GNP: a reexamination. J. Energy Dev. 5 (2), 326–331.

Akinlo, A.E., 2008. Energy consumption and economic growth: evidence from 11 sub-Sahara African countries. Energy Econ. 30 (5), 2391–2400. https://doi.org/10.1016/j.eneco.2008.01.008.

Al-Mulali, U., Lee, J.Y.M., Hakim Mohammed, A., Sheau-Ting, L., 2013. Examining the link between energy consumption, carbon dioxide emission, and economic growth in Latin America and the caribbean. Renew. Sustain. Energy Rev. 26 (Suppl. C), 42–48. https://doi.org/10.1016/j.rser.2013.05.041.

Alonso, J.A., Garcimartín, C., 2013. The determinants of institutional quality. More on the debate. J. Int. Dev. 25 (2), 206–226.

Ang, J.B., 2007. CO_2 emissions, energy consumption, and output in France. Energy Pol. 35 (10), 4772–4778. https://doi.org/10.1016/j.enpol.2007.03.032.

Ang, J.B., 2008. Economic development, pollutant emissions and energy consumption in Malaysia. J. Pol. Model. 30 (2), 271–278. https://doi.org/10.1016/j.jpolmod.2007.04.010.

Antonakakis, N., Chatziantoniou, I., Filis, G., 2017. Energy consumption, CO_2 emissions, and economic growth: an ethical dilemma. Renew. Sustain. Energy Rev. 68, 808–824. https://doi.org/10.1016/j.rser.2016.09.105.

Apergis, N., Ozturk, I., 2015. Testing environmental Kuznets Curve hypothesis in Asian countries. Ecol. Indicat. 52 (Suppl. C), 16–22. https://doi.org/10.1016/j.ecolind.2014.11.026.

Apergis, N., Payne, J.E., 2009. CO_2 emissions, energy usage, and output in Central America. Energy Pol. 37 (8), 3282–3286. https://doi.org/10.1016/j.enpol.2009.03.048.

Apergis, N., Payne, J.E., 2010. The emissions, energy consumption, and growth nexus: evidence from the commonwealth of independent States. Energy Pol. 38 (1), 650–655. https://doi.org/10.1016/j.enpol.2009.08.029.

Arellano, M., Bond, S., 1991. Some tests of specification for panel data: Monte Carlo evidence and an application to employment equations. Rev. Econ. Stud. 58 (2), 277−297.

Arouri, M.E.H., Ben Youssef, A., M'Henni, H., Rault, C., 2012. Energy consumption, economic growth and CO_2 emissions in Middle East and North African countries. Energy Pol. 45, 342−349. https://doi.org/10.1016/j.enpol.2012.02.042.

Asafu-Adjaye, J., 2000. The relationship between energy consumption, energy prices and economic growth: time series evidence from Asian developing countries. Energy Econ. 22 (6), 615−625. https://doi.org/10.1016/s0140-9883(00)00050-5.

Asafu-Adjaye, J., 2014. The economic impacts of climate change on agriculture in Africa. J. Afr. Econ. 23 (Suppl. 1_2), Ii17−Ii49. https://doi.org/10.1093/jae/eju011.

Baliamoune-Lutz, M., Boko, S.H., 2012. Trade, institutions, income and human development in African countries. J. Afr. Econ. 22 (2), 323−345. https://doi.org/10.1093/jae/ejs037.

Baltagi, B., 2008. Econometric analysis of panel data. John Wiley & Sons.

Beckerman, W., 1992. Economic growth and the environment: whose growth? whose environment? World Dev. 20 (4), 481−496. https://doi.org/10.1016/0305-750X(92)90038-W.

Bhattacharya, M., Awaworyi Churchill, S., Paramati, S.R., 2017. The dynamic impact of renewable energy and institutions on economic output and CO_2 emissions across regions. Renew. Energy 111, 157−167. https://doi.org/10.1016/j.renene.2017.03.102.

Blundell, R., Bond, S., 1998. Initial conditions and moment restrictions in dynamic panel data models. J. Econom. 87 (1), 115−143. https://doi.org/10.1016/s0304-4076(98)00009-8.

Čeh Časni, A., Dumičić, K., Tica, J., 2016. The panel VAR approach to modelling the housing wealth effect: Evidence from selected European post-transition economies. Naše gospodarstvo/Our economy 62 (4). https://doi.org/10.1515/ngoe-2016-0021.

Cong, W., Aidong, W., Chongqi, W., 2011. Analyze the relationship between energy consumption and economic growth in China. Energy Proc. 5, 974−979. https://doi.org/10.1016/j.egypro.2011.03.172.

Cowan, W.N., Chang, T., Inglesi-Lotz, R., Gupta, R., 2014. The nexus of electricity consumption, economic growth and CO_2 emissions in the BRICS countries. Energy Pol. 66, 359−368. https://doi.org/10.1016/j.enpol.2013.10.081.

Dagher, L., Yacoubian, T., 2012. The causal relationship between energy consumption and economic growth in Lebanon. Energy Pol. 50, 795−801. https://doi.org/10.1016/j.enpol.2012.08.034.

Dergiades, T., Martinopoulos, G., Tsoulfidis, L., 2013. Energy consumption and economic growth: parametric and non-parametric causality testing for the case of Greece. Energy Econ. 36, 686−697. https://doi.org/10.1016/j.eneco.2012.11.017.

Dickey, D.A., Fuller, W.A., 1979. Distribution of the estimators for autoregressive time series with a unit root. J. Am. Stat. Assoc. 74 (366), 427−431. https://doi.org/10.2307/2286348.

Eggoh, J.C., Bangake, C., Rault, C., 2011. Energy consumption and economic growth revisited in African countries. Energy Pol. 39 (11), 7408−7421. https://doi.org/10.1016/j.enpol.2011.09.007.

Esso, L.J., 2010. Threshold cointegration and causality relationship between energy use and growth in seven African countries. Energy Econ. 32 (6), 1383−1391. https://doi.org/10.1016/j.eneco.2010.08.003.

Farhani, S., Ozturk, I., 2015. Causal relationship between CO_2 emissions, real GDP, energy consumption, financial development, trade openness, and urbanization in Tunisia. Environ. Sci. Pollut. Res. 22 (20), 15663−15676. https://doi.org/10.1007/s11356-015-4767-1.

Fosu, A., Bates, R., Hoeffler, A., 2006. Institutions, governance and economic development in Africa: an overview. J. Afr. Econ. 15 (Suppl. 1_1), 1−9. https://doi.org/10.1093/jae/ejk004.

Fosu, A.K., 2012. Growth of African economies: productivity, policy syndromes and the importance of institutions. J. Afr. Econ. 22 (4), 523−551. https://doi.org/10.1093/jae/ejs034.

Góes, C., 2016. Institutions and growth: a GMM/IV panel VAR approach. Econ. Lett. 138, 85–91. https://doi.org/10.1016/j.econlet.2015.11.024.

Grossman, G.M., Krueger, A.B., 1995. Economic growth and the environment. Q. J. Econ. 110 (2), 353–377. https://doi.org/10.2307/2118443.

Halicioglu, F., 2009. An econometric study of CO_2 emissions, energy consumption, income and foreign trade in Turkey. Energy Pol. 37 (3), 1156–1164. https://doi.org/10.1016/j.enpol.2008.11.012.

Im, K.S., Pesaran, M.H., Shin, Y., 2003. Testing for unit roots in heterogeneous panels. J. Econom. 115 (1), 53–74. https://doi.org/10.1016/s0304-4076(03)00092-7.

International Energy Agency, 2014. A focus on energy prospects in Sub-Saharan Africa. International Energy Agency IEA. Retrieved from. https://www.icafrica.org/en/knowledge-hub/article/africa-energy-outlook-a-focus-on-energy-prospects-in-sub-saharan-africa-263/. on 13th August 2018.

Jahangir Alam, M., Ara Begum, I., Buysse, J., Van Huylenbroeck, G., 2012. Energy consumption, carbon emissions and economic growth nexus in Bangladesh: cointegration and dynamic causality analysis. Energy Pol. 45, 217–225. https://doi.org/10.1016/j.enpol.2012.02.022.

Karekezi, S., 2002. Poverty and energy in Africa—a brief review. Energy Pol. 30 (11), 915–919. https://doi.org/10.1016/S0301-4215(02)00047-2.

Kraft, J., Kraft, A., 1978. On the relationship between energy and GNP. J. Energy Dev. 3 (2), 401–403.

Lean, H.H., Smyth, R., 2010a. CO_2 emissions, electricity consumption and output in ASEAN. Appl. Energy 87 (6), 1858–1864. https://doi.org/10.1016/j.apenergy.2010.02.003.

Lean, H.H., Smyth, R., 2010b. Multivariate granger causality between electricity generation, exports, prices and GDP in Malaysia. Energy 35 (9), 3640–3648. https://doi.org/10.1016/j.energy.2010.05.008.

Lean, H.H., Smyth, R., 2010c. On the dynamics of aggregate output, electricity consumption and exports in Malaysia: evidence from multivariate granger causality tests. Appl. Energy 87 (6), 1963–1971. https://doi.org/10.1016/j.apenergy.2009.11.017.

Lee, C.-C., 2006. The causality relationship between energy consumption and GDP in G-11 countries revisited. Energy Pol. 34 (9), 1086–1093. https://doi.org/10.1016/j.enpol.2005.04.023.

Lee, C.-C., Chang, C.-P., 2007. Energy consumption and gdp revisited: a panel analysis of developed and developing countries. Energy Econ. 29 (6), 1206–1223. https://doi.org/10.1016/j.eneco.2007.01.001.

Levin, A., Lin, C.-F., Chu, J.C.-S., 2002. Unit root tests in panel data: asymptotic and finite-sample properties. J. Econom. 108 (1), 1–24. https://doi.org/10.1016/s0304-4076(01)00098-7.

Love, I., Zicchino, L., 2006. Financial development and dynamic investment behavior: evidence from panel VAR. Q. Rev. Econ. Finance 46 (2), 190–210. https://doi.org/10.1016/j.qref.2005.11.007.

Lutkepohl, H., 2005. New introduction to multiple time series analysis. Springer, New York.

Meadows, D., Randers, J., Meadows, D., 1992. The Limits to Growth. Universe Books, New York.

Mekonnen, A., 2014. Economic costs of climate change and climate finance with a focus on Africa. J. Afr. Econ. 23 (Suppl. 1_2), Ii50–Ii82. https://doi.org/10.1093/jae/eju012.

Mirza, F.M., Kanwal, A., 2017. Energy consumption, carbon emissions and economic growth in Pakistan: dynamic causality analysis. Renew. Sustain. Energy Rev. 72 (Suppl. C), 1233–1240. https://doi.org/10.1016/j.rser.2016.10.081.

Mishra, V., Smyth, R., Sharma, S., 2009. The energy-GDP nexus: evidence from a panel of pacific island countries. Resour. Energy Econ. 31 (3), 210–220. https://doi.org/10.1016/j.reseneeco.2009.04.002.

North, D.C., 1989. Institutions and economic growth: an historical introduction. World Dev. 17 (9), 1319–1332.

North, D.C., 1990. Institutions, Institutional Change and Economic Performance. Cambridge University Press.

North, D.C., 1992. Transaction Costs, Institutions, and Economic Performance. ICS Press, San Francisco, CA.

Omri, A., 2013. CO_2 emissions, energy consumption and economic growth nexus in MENA countries: evidence from simultaneous equations models. Energy Econ. 40, 657–664. https://doi.org/10.1016/j.eneco.2013.09.003.

Phillips, P.C., Perron, P., 1988. Testing for A unit root in time series regression. Biometrika 75 (2), 335−346.

Sachs, J.D., 2003. Institutions Don't Rule: Direct Effects of Geography on Per Capita Income. NBER. Working Paper, (w9490).

Salahuddin, M., Gow, J., 2014. Economic growth, energy consumption and CO_2 emissions in Gulf cooperation Council countries. Energy 73 (Suppl. C), 44−58. https://doi.org/10.1016/j.energy.2014.05.054.

Simbanegavi, W., Arndt, C., 2014. Climate change and economic development in Africa: an overview. J. Afr. Econ. 23 (Suppl. 1_2), Ii4−Ii16. https://doi.org/10.1093/jae/eju010.

Sims, C., 1980. Macroeconomics and reality. Econometrica 48 (1), 1−48. https://doi.org/10.2307/1912017.

Soytas, U., Sari, R., 2009. Energy consumption, economic growth, and carbon emissions: challenges faced by an EU candidate member. Ecol. Econ. 68 (6), 1667−1675. https://doi.org/10.1016/j.ecolecon.2007.06.014.

Soytas, U., Sari, R., Ewing, B.T., 2007. Energy consumption, income, and carbon emissions in the United States. Ecol. Econ. 62 (3), 482−489. https://doi.org/10.1016/j.ecolecon.2006.07.009.

Stern, D.I., 2004. The rise and fall of the environmental Kuznets Curve. World Dev. 32 (8), 1419−1439. https://doi.org/10.1016/j.worlddev.2004.03.004.

Stern, D.I., Common, M.S., 2001. Is there an environmental Kuznets Curve for sulfur? J. Environ. Econ. Manag. 41 (2), 162−178. https://doi.org/10.1006/jeem.2000.1132.

Tamazian, A., Bhaskara Rao, B., 2010. Do economic, financial and institutional developments matter for environmental degradation? Evidence from transitional economies. Energy Econ. 32 (1), 137−145. https://doi.org/10.1016/j.eneco.2009.04.004.

Wand Ji, Y.D.F., 2013. Energy consumption and economic growth: evidence from Cameroon. Energy Pol. 61 (Suppl. C), 1295−1304. https://doi.org/10.1016/j.enpol.2013.05.115.

Zhang, X.-P., Cheng, X.-M., 2009. Energy consumption, carbon emissions, and economic growth in China. Ecol. Econ. 68 (10), 2706−2712. https://doi.org/10.1016/j.ecolecon.2009.05.011.

Linkage between energy use, pollution, and economic growth—a cross-country analysis

5

Soumyendra Kishore Datta[1], Tanushree De[2]

[1]*Department of Economics, The University of Burdwan, Burdwan, West Bengal, India;* [2]*Department of Economics, Vivekananda Mahavidyalaya, Burdwan, West Bengal, India*

Chapter outline

1. Introduction

The present state of unbridled economic progress throughout the world cannot be dissociated from intensive use of energy basically linked with consumption of nonrenewable fossil fuels. Although efforts are afoot to undertake innovative drive for deriving renewable alternatives with possibility of large-scale commercial applications, the success in this direction still remains less than what is warranted. This is mainly because of two reasons (i) it involves huge expenditures toward innovation with uncertainties in success and possible time overrun (ii) problems involved in making the energy commercially viable. As a result, there has hardly been substantial replacement of nonrenewable energy sources by renewable alternatives like solar energy, wind energy, geothermal energy, etc. The concomitant aftermath of massive reliance on nonrenewable energy sources for economic growth is manifest in two adverse forms. First, its rampant use pollutes the environment with attendant

Environmental Sustainability and Economy. https://doi.org/10.1016/B978-0-12-822188-4.00001-4

ingredients for global warming, and second, unfettered use of fossil fuel gradually depletes its geological stock leaving less and less for future generations with reflections in its rising user cost. Nonrenewable energy like coal or oil is available easily in the market at comparatively lower prices than that of nonconventional energy which also involves substantial infrastructural cost component. Hence average consumer in low income condition usually prefers to use fossil fuel or articles made with conventional technology having relatively lower price than renewable alternatives or relatively costly articles made with state-of-the-art technology which saves energy and emits less pollution. As individual income rises, people can afford to access captive tapping of costly renewable energy sources or buy more and more improved articles produced by energy efficient technology. The dependence on fossil fuel is assumed to be relatively high on the part of low income countries. Increased fossil fuel energy consumption has been a major driver in raising the volume of carbon dioxide (CO_2) emission, which triggered severe ecological setbacks. However according to Environmental Kuznets curve (EKC) hypothesis, environmental degradation is assumed to decline with growth in income, resulting in a sort of inverted U-shaped relation between these two variables. With the process of growth, emissions begin to fall beyond a threshold level (the turning point) with resulting improvement in environmental quality which eventually gives rise to an inverse U-pattern. In the context of developing countries, validity of this hypothesis might have promising implications for sustainable future development (Wang et al., 2016). However in case of developed countries an N-shaped pattern is often noticed reflecting that beyond a second threshold level pollution again takes an upturn with rise in income. It is likely that economic growth increases the possibility of introduction of more modern and less pollution-intensive manmade capital and technology (Grossman and Krueger, 1995). While pollution per unit of output might go down, absolute pollution levels might very well go up as economic growth increases. So the effect of technological change on pollution is in principle often ambiguous (Lopez, 1992). Hence the problem arises as to the pattern of emission of diverse type of pollutants attendant with growth in different income group countries of the world.

Energy-environment-growth studies have assumed diverse forms reflected in global studies (Sinhababu and Datta, 2013; Datta and Sarkar, 2013), group of country studies (Arouri et al., 2012; Menegaki, 2013), individual country (Shahbaz et al., 2018; Rauf et al., 2018; Alkhathlan and Javid, 2013) or individual pollutant (Ozgur-Kayalica and Kacar, 2014; Perman and Stern, 2003; Omri, 2013) study. Some use short run while some use long-run relationship. There is of course overlapping across these studies, for example, group of country studies corresponding to an individual pollutant. There has also been diversity across the studies based on incorporation of additional explanatory variables in the assumed broader form of EKC relationship. Depending on data availability and focus of the analysis, the additional variables in different studies have assumed diverse types covering factors like financial development, industrial production, energy consumption, trade openness, urbanization, service sector development, agricultural production, etc.

As an example of global study, Sinhababu and Datta (2013) considered 22 countries spread over three continents, viz., Asia, Latin America, and sub-Saharan Africa. An attempt was made to analyze the variation of environmental degradation as associated with certain indices of economic growth. They visualized degradation as a composite of multidimensional components consisting of vitiation at all the layers including water, land, forest, and air. Fixed effect with dummy variable panel regression was used. According to them the observed N pattern relation calls for government policies toward control of emissions, devising cost-effective bioenergy sources, and increasing peoples' awareness to protect the environment. Further, an all-out effort is recommended to increase the carbon pool by undertaking extensive afforestation programs and enhancing renewable energy production.

As an instance of group of country study, Arouri et al. (2012) employed cointegration technique on data related to economic growth, energy consumption, and emissions of CO_2 for 12 countries in MENA region during the period 1981–2005. Their main objective was to test for the EKC hypothesis in MENA region for CO_2, to investigate the existence of EKC for each country and to explore the nature of the causality relationship between economic growth, energy consumption, and emissions of CO_2.

As a country-specific study, Shahbaz et al. (2018) explored the determinants of carbon emissions in France by accounting for the role played by foreign direct investment (FDI), economic growth, financial development, energy consumption, and energy research innovations. They employed unit root test on French time series data (1955–2016) to examine the order of integration and established cointegration among the time series by applying bootstrapping bounds testing approach.

In terms of individual pollutant study, Omri (2013) examined the nexus between CO_2 emissions, energy consumption, and economic growth using simultaneous equations models with panel data of 14 MENA countries over the period 1990–2011.

There is another strand of separating the studies relating to EKC into two categories. One relates to cross-section studies (Ang, 2007; Ben and Ben, 2015a; Halicioglu, 2009; Jalil et al., 2009; Jayanthakumaran et al., 2012), while the other relates to panel studies. The panel studies mostly concentrate on a number of countries in a specified continent/specified group of countries, e.g., MENA countries (Acaravci and Ozturk, 2010; Al-Mulali et al., 2015; Jaunky, 2011; Ozcan, 2013), Asian countries (Taguchi, 2012), OECD countries (Moomaw and Unruh, 1997), etc. Panel cointegration as well as short and long-run causality analyses are undertaken in most of such studies, involving mostly a single pollutant (CO_2) or at the most two pollutants (CO_2 and SO_2).

In the literature there has been little panel data studies involving countries of the world categorized into different income groups, e.g., high, higher middle, lower middle, and low income group and spread over the four continents of the world. Further the present study has focused on considering the existence of EKC corresponding to three major GHGs, viz., CO_2, N_2O, and CH_4 for all these groups of countries. The study assumes great importance in the context of relevant literature, since there remains much concern about sustainable development practices in the so-called low, lower-middle and higher-middle income countries (relative to high income group countries) which are aspiring for more development without adequate concern for pollution problem. Further, these countries are charaterized by wide variation across their level of urbanization, trade volume, industrial development which is likely to have differential impact on the relation between economic growth and pollution emission. Cost of inaction in reducing use of fossil fuels in such countries might greatly supersede the cost of action. With rising dependence on fossil fuel consumption/imports, there is dire need of a government interference (strong political system) and sound institutional mechanism (Deacon and Mueller, 2004). Introduction of incentives, enforcement of property right system, avoiding market failure and PPP mode of production might lead to efficient use of natural resources, technological innovation, and better conservation of environment.

While a considerable number of studies relating to CO_2 and SO_2 have been undertaken, studies relating to concurrent focus on the stated three pollutants in different income group countries are few and far between. In this backdrop it seems imperative to undertake a theoretical analysis with regard to the spatial concentration pattern of pollutant emission as decomposed into different components and analyze the pattern of variation of the three types of GHGs corresponding to changes in income in different income group countries of the world.

The relevant objectives include

i. Development of a theoretical approach to analyze the geographical concentration of pollutants, based on decomposition analysis coherent with application of the notion of Kaya identity.
ii. Testing the existence of EKC or otherwise relating to the stated three gases in each of the income group countries.
iii. Analysis of short- and long-run both-way causality across GDP and different pollutants.

Besides the introduction, the paper consists of the following sections. Section 2 focuses on an extended literature survey. Linked with the surveyed literature, Section 2.1 is devoted to analyzing some aspects relating to GHG emission and patterns of EKC. Section 3 covers data source and methodology. Section 3.1 deals with data source while Section 3.2 is devoted to discussion about the methodology. Again Section 3.2 is divided into two parts. Estimation method of EKC is considered in Section 3.2.1, while PVECM is alluded to in Section 3.2.2. Theoretical discussion about elasticity approach to the notion of EKC is discussed in Section 4. Results and discussion are covered in Section 5. Section 5.1 is devoted to discussing the panel regression results while Section 5.2 analyses the causality issues. Finally Section 6 provides the conclusion and policy suggestions.

2. Extended literature survey

An allusion to some of the broader dimensions of existing studies pertinent to the present analysis is provided as under.

Jalil et al. (2009) examined the long-run relationship between carbon emissions and energy consumption, income and foreign trade in case of China by employing time series data over the period 1975–2005. Particular focus was put on testing the existence of any Environmental Kuznets curve (EKC) relationship between CO_2 emissions and per capita real GDP in the long run. For empirical analysis auto regressive distributed lag (ARDL) method has been applied. The data over the sample period confirmed the existence of EKC relationship by validating a quadratic relation between income and CO_2. The results of Granger causality tests supported one way causality running from economic growth to CO_2 emissions. It was observed from the study that income and energy consumption had the major impact on CO_2 emission in the long run.

Hongbo et al. (2019) carried out a cross-country regression analysis of international trade and carbon emission based on data related to export diversification across 125 countries during the period 2000 to 2014 at the HS4 digit of disaggregation. Export diversification is broken into vertical and horizontal diversification in order to validate its correlation with pollution emission in terms of scale effect, technique effect, and composition effect. In order to stave off likely problems of heteroskedasticity and autocorrelation issues, the regression equation is estimated by incorporating Driscoll and Kraay standard errors which are robust to very general forms of cross-sectional and temporal dependence with a large time dimension. This nonparametric technique of estimating standard errors does not place any restrictions on the limiting behavior of the number of panels. Further, insertion of interaction terms between economic development and export diversification eases the comparison across different income group countries: low-income countries reveal U-shaped relationship between economic development and CO_2 emissions, while OECD countries maintain an inverted U-shaped EKC curve which is broadly common in case of the outcome across 125 countries.

Zambrano-Monserrate and Fernandez (2017) analyzed the relationship between nitrous oxide (N_2O) emissions, economic growth, agricultural land used, and exports in Germany. Using time series data over the period 1970 to 2012 and Autoregressive Distributed Lag (ARDL) model, they attempted to test for cointegration in the long run. Their results suggested the existence of a quadratic long-run relationship between N_2O emissions and economic growth, corroborating the existence of an EKC. It was also observed by them that agricultural land area influenced N_2O emissions positively, whereas exports had a negative impact on such emission.

Kaika and Zervas (2012) have in their study pointed out some controversial aspects with regard to the theoretical basis of EKC. A great debate revolves round the emergence of a pollution-free environment linked with rising income. It is often taken for a high positive income elasticity of demand. This may even turn out to be greater than one, indicating a proportionally higher concern in environmental quality than growth in income.

Ozgur-Kayalica and Kacar (2014) examine the relationship between economic growth and sulfur emission by testing the validity of the EKC hypothesis for selected countries under three different econometric models. The first model traces the impact of both log value of GDP per capita and its square on the log value of sulfur dioxide (SO_2) per capita. Model 2 is an extension of Model 1 by including the impact of trade intensity or openness variable on sulfur dioxide emission. Finally, the study also introduces population density in order to estimate whether it has statistically significant impact on sulfur dioxide emission. Estimation results based on random effect model have established the existence of an inverted U-shape pattern between economic growth and sulfur emission per capita. Further, one important result is that both openness and population density play a significant role in sulfur emission.

In a very detailed review of the empirical literature, Stern (1998) points out the relevance of the inverted-U relationship corresponding to only a subset of environmental elements, viz., air pollutants such as suspended particulates and sulfur dioxide. According to Grossman and Krueger (1993), as suspended particulates decline monotonically with income, Stern's subset is subject to challenge. In an allied work, Stern et al. (1998) used both fixed effects and random effects models with both country and time effects. Both dependent (emissions per capita) and independent (PPP GDP per capita) variables were considered in natural logarithms. The results show that estimating an EKC using data for only the OECD countries, as has often been the case, leads to estimates where the turning point is at a much lower level than when the EKC is estimated using data for the world as a whole. The paper explores possible explanations of these results using Monte Carlo analysis and other statistical tests. It is concluded that the simple EKC model is fundamentally misspecified and that there are omitted variables which are correlated with GDP.

Mandal and Madheswaran (2010) in their paper studied the existence of and direction of the causal relationship between energy consumption and output growth of Indian cement industry covering the period 1979–1980 to 2004–2005. They used recently developed panel unit root test, panel cointegration, and panel-based error correction models in a multivariate framework including capital stock, labor, and material apart from energy and output. A positive long run cointegrated relationship between output and energy consumption was confirmed from their study when heterogeneous state effect was considered in the analysis. They also found a long-run, bidirectional relationship between energy consumption and output growth in the Indian cement industry for the study period, implying that an increase in energy consumption directly affects the growth of this sector and that growth stimulates further energy consumption. These empirical findings imply that energy consumption and output are jointly determined and affect each other.

The study by Apergis and Ozturk (2015) investigated into the existence of EKC for 14 Asian countries over 1990−2011. They focused on how income, sociodemographic variables, and policies and their interaction have impact on GHG emission. For this purpose they considered variables like CO_2 emission, GDP per capita, land, population density, industry share in GDP and four indicators reflecting quality of institutions in the form of political index, government policy, policy induced regulatory activity, and corruption index. After undertaking panel unit root test, they established the long-run cointegration relation across CO_2 emission level and both economic and policy variables through use of Nyblom−Harvey, Fisher-Johansen, Pedroni and Kao test. Finally the multivariate equation was estimated by using four methods: FMOLS (Fully Modified OLS), DOLS (Dynamic OLS), PMGE (Pooled Mean Group Estimator), and MG (Mean Group) method. The results point to the validity of existence of EKC across income and CO_2 emission.

Ahmad et al. (2017) probed into the existence of EKC in connection with CO_2 emission in Croatia during the period of 1991−2011. Based on application of Autoregressive Distributed Lag (ARDL) and VECM method for this study they showed the validity of existence of a long-run EKC relation between CO_2 emission and economic growth. DOLS and FMOLS results confirm the robustness of long-run results. By applying VECM approach, they found bidirectional Granger causality between CO_2 emissions and economic growth in short run and unidirectional causality from economic growth to CO_2 emissions in the long run.

Al-Mulali et al. (2016) investigated the existence of EKC hypothesis in Vietnam by considering data over the period 1981−2011. For this purpose they employed an Autoregressive Distributed Lag (ARDL) method. According to the results of the study, pollution haven hypothesis is vindicated since capital, energy-intensive imports, as well as fossil fuel are found to significantly increase the level of pollution. However, exports have no significant effect on pollution. Again, labor force reduces pollution as most of Vietnam's labor force is mostly employed in less energy-intensive agricultural and services sectors. The derived results suggest that the EKC relation is not valid, since both in the short and long run, GDP and pollution are observed to vary positively.

Acheampong (2018) in his study applies panel vector autoregression (PVAR) technique together with a system-generalized method of moment (System-GMM) to examine the dynamic causal relationship between economic growth, carbon emissions, and energy consumption for 116 countries covering the period 1990−2014. Using multivariate model, some basic results are established which have bearing on relevant policies. At the global and regional levels, economic growth is not observed to cause significant energy consumption. Again while economic growth was observed to have negative impact on carbon emissions at the global level and the Caribbean-Latin America, in other regions no such causal impact was identified. Energy consumption positively causes economic growth in sub-Saharan Africa while it negatively affects economic growth at the global, Middle East and North Africa (MENA), Asia-Pacific, and Caribbean-Latin America. Moreover, energy consumption positively causes carbon emissions in MENA but negatively impacts carbon emissions in sub-Saharan Africa and Caribbean-Latin America. The impulse response function shows evidence of EKC at the global scale and sub-Saharan Africa. The respective policy implications have been discussed.

Adams and Acheampong (2019) in their paper, tried to assess the impact of democracy and renewable energy on carbon emissions by using unbalanced data related to 46 sub-Saharan African countries covering the period 1980−2015. They used an instrumental variable generalized method of the moment and found inverse relation between carbon emission and democracy and renewable energy. Apart from this, foreign direct investment, trade openness, population, and economic growth were identified as factors causing carbon emissions in this region. But in the presence of democracy, economic growth is observed to lessen the intensity of carbon emissions. Hence no systematic evidence of EKC could be traced. According to the study when assessing the impact of these variables on carbon emissions, the level of economic development is found to matter. Recommendation is made in the study for assigning priority to democracy and investment in renewable energy toward toning down climate change in the said African context.

Acheampong and Dzator (2020) studied the impact of institutional quality on carbon emissions covering 45 sub-Saharan African countries between the period of 2000 and 2015. Avoiding the assumption of homogeneity in the sample and categorizing the countries based on their institutional origin, they found that institutions play a significant role in mitigating carbon emissions. The result was however not significant in the absence of categorization. Hence, according to them, pursuit of better institutions of governance is basic toward achieving Sustainable Development Goal (SDG) 13.

Based on the data covering the period of 1971−2010, Shahbaz et al. (2014) explored in finding possible existence of EKC in case of Tunisia. They applied ARDL bounds testing approach to cointegration for examining long-run relationship in the presence of structural breaks. In order to detect the causality among the variables, both vector error correction method (VECM) and innovative accounting approach (IAA) were used to verify the robustness of causality. Their findings corroborated long-run relationship between economic growth, energy consumption, trade openness, and CO_2 emissions. Further the existence of EKC had been confirmed by both the VECM and IAA approaches. The authors advocated lessening of release of toxic pollutants and implementing environmentally favorable regulations for attaining sustainable development in the country.

Awad and Abugamos (2017) tried to probe into existence of EKC in MENA region by using data on CO_2 emission (CE), income (A), population (P), energy intensity (EI), urbanization (UR) spread over the time period 1980−2014.

Accordingly, within the EKC framework, the augmented model is estimated as

$$\ln CE_{it} = \alpha_i + \beta_1 \ln P_{it} + \beta_2 \ln A_{it} + \beta_3 \ln EI_{it} + \beta_4 UR_{it} + \beta_5 \ln A_{it}^2 + T_i + \varepsilon_{it}$$

Apart from *UR* and time-specific effect (*T*), other variables are taken in natural logarithm form in order to interpret the coefficients as elasticities.

They tried to explore the income-CO_2 emission connection by using panel data together with a semiparametric panel fixed effects regression. Their result suggests the existence of an inverted U-shaped relation between income and CO_2 emissions in the region. With the initiatives of renewable energy projects taken in the region, they pin hope on reversibility of environmental degradation and its improved quality together with economic growth process.

Using data spanning over 1982−2016, Sarkodie and Strezov (2019) examined the effect of foreign direct investment inflows, economic development, and energy consumption on greenhouse gas emissions for the top five GHGs in the context of developing countries like China, India, Iran, Indonesia, and South Africa. They employed a panel data regression with Driscoll-Kraay standard errors, U-test estimation approach, and panel quantile regression with nonadditive fixed-effects. The pollution haven hypothesis is validated based on strong positive impact of energy consumption on GHG emission. The EKC hypothesis is observed to be valid for only China and Indonesia. According to them FDI inflows with clean technological transfer and improved labor and environmental management practices together with enhanced energy efficiency, adoption of clean and modern energy are likely to help developing countries march toward attainment of the sustainable development goals.

2.1 Aspects of GHG emission and cross-country variations with GDP: relevance of EKC

Based on a number of literature reviews made above, it can be stated that as GDP growth in any country is conditioned by consumption of energy attendant with GHG emission, it seems obvious that pollution/degradation of the environment and global warming are caused by energy-intensive productive system. The release of GHGs into the atmosphere is also greatly linked with the efficiency of functioning of public utilities like electricity, coal, natural gas, sewage, telephone, etc., in controlling the polluting impact of their action. The release of pollutants by electricity generating power plants run on fossil fuel may have far reaching environmental and localized impact. Apart from the impacts of power generation, there are also impacts of CO_2 emission, connected with extracting, producing, and transporting certain fuels such as coal, oil, and natural gas. Developing countries should endeavor to devise innovative ways to produce power more efficiently with less environmental impact. In this context it may be pointed out that special focus should be put on lessening of flare, associated with burning of hydrocarbon gases exuded during oil extraction. This can be viewed as an exercise in promoting safety measures or means of disposal. Again emission of considerable amount of greenhouse gases (GHGs) is linked with meat production in both intensive (industrial) and nonintensive (traditional) form. Apart from this, deforestation is a major source of release of CO_2 into the atmosphere. When trees are felled to produce goods, they release the carbon that is normally stored for photosynthesis. This process releases nearly a billion tons of carbon into the atmosphere per year, according to the 2010 Global Forest Resources Assessment. This can be somewhat offset through forestry and other land-use practices, according to the EPA.

Human activity has increased the amount of methane in the atmosphere, contributing to climate change. Methane (CH_4) is particularly dangerous as it has 34 times greater impact compared to CO_2 over a 100-year period, as per the latest IPCC (Intergovernmental Panel for Climate Change) Assessment Report. A considerable source of human-made methane emissions is fossil fuel production. For instance, methane is a basic byproduct in the process of rapidly rising global extraction and processing of natural gas. Rearing of livestock also leads to release of methane resulting from enteric fermentation and generation of livestock manure. Again there is released nitrous oxide (N_2O) from excreted nitrogen, apart from the use of chemical nitrogenous (N) fertilizers employed to produce the feed meant for the animals. N_2O emission from soil also contributes in a large way to the total volume of GHG pollution. There is also a general need to modify the cropping pattern in such a way as leads to decreased amount of release of toxic greenhouse gases per unit of production matching with mitigation of climate change policies and adapted to emerging sustainability needs of the productive sphere. Thus CO_2, CH_4, and N_2O comprise some of the important components of GHGs, the emission of which when exceeds the resilience of the ecosystem can cause severe damages to human, plant, and animal life. Unbridled emission of these gases can have far-ranging environmental and health impacts. They cause climate change by trapping heat, and also lead to aggravation in respiratory disease from smog and air pollution. Unpredictable fluctuations in weather, extreme climatic conditions, intensified natural calamities, food supply disruptions, and increased wildfires are some of the other effects of climate change caused by such greenhouse gases.

In this chapter, study has been carried out to focus on the emission pattern of CO_2, CH_4, and N_2O with respect to changing GDP per capita for the below stated countries divided into four income groups. Worldwide concern has been put on controlling the emission of GHGs in the form of various consortiums, fora, protocol, etc. Depending on the length of time the molecules of these gases remain in the atmosphere, the steadiness and subsequent decline in the emission have a bearing on the stabilization of their level of concentration in the atmosphere.

Besides water vapor, which has a residence time of about 9 days, major greenhouse gases take many years to leave the atmosphere. Although it is not easy to know with precision how long it takes for greenhouse gases to leave the atmosphere, there are estimates for the principal greenhouse gases. Jacob (1999) defines the lifetime τ of an atmospheric species X in a one-box model as the average time that a molecule of X remains in the box. Mathematically, it can be defined as the ratio of the mass "m" (in kg) of X in the box to its removal rate, which is the sum of the flow of X out of the box (Fout), chemical loss of X(L) and deposition of X(D), all in kg/s). $\tau = m/(Fout + L + D)$. If output of this gas into the box died down, then after time τ, its concentration would decrease by about 63%.

Hence, commensurate with development imperatives as well as hardly any grip over executing speedy removal and monitoring of emission of these gases from the atmosphere, it seems imperative to analyze the rate of release of such gases into the atmosphere.

In this context, it seems coherent to carry out a panel data regression analysis of several countries in the world in order to understand the pattern of variation in emission of GHGs like CO_2, CH_4, and N_2O alongside the levels of development. The most pertinent method to examine the pattern of relationship that holds across such pollutants and GDP seems to fit the following regression (1) on the selected panel data corresponding to 54 countries spread over low income, lower middle income group, higher-middle income group, and high income group.

$$Z_{it} = \alpha_i + \beta_1 Y_{it} + \beta_2 Y_{it}^2 + \beta_3 Y_{it}^3 + \beta_4 P_{it} + \beta_5 T_{it} + U_{it} \tag{5.1}$$

where the subscript i stands for country index, t is time index, Z indicates pollutant type, Y indicates GDP per capita, P reflects urban population, while T indicates trade volume. U is the normally distributed error term.

Eq. (5.1) allows for testing the various forms of environmental-economic relationships.

1. $\beta_1 > 0$ and $\beta_2 = \beta_3 = 0$ reveal a monotonically increasing linear relationship, indicating that rising incomes are associated with rising levels of emissions; (Fig. 5.1A)
2. $\beta_1 < 0$ and $\beta_2 = \beta_3 = 0$ reveals a monotonically decreasing linear relationship (Fig. 5.1B);
3. $\beta_1 > 0$, $\beta_2 < 0$ and $\beta_3 = 0$ reveal a quadratic relationship, representing the EKC. The turning point of this representation of the inverted U curve is obtained by setting the derivative of (i) equal to zero, which yields: $Y_t = -\beta_1/2\beta_2$ (Fig.5.1C).
4. $\beta_1 > 0$, $\beta_2 < 0$, and $\beta_3 > 0$ reveal a cubic polynomial, representing the N-shaped figure (Fig. 5.1D).
5. $\beta_1 < 0$, $\beta_2 > 0$, and $\beta_3 < 0$ reveal a cubic polynomial, representing a reverse N-shaped or flat S-shaped figure (Fig. 5.1E).
6. $\beta_1 < 0$, $\beta_2 > 0$, and $\beta_3 = 0$ reveal a quadratic relationship, representing a U-shaped figure (Fig. 5.1F).

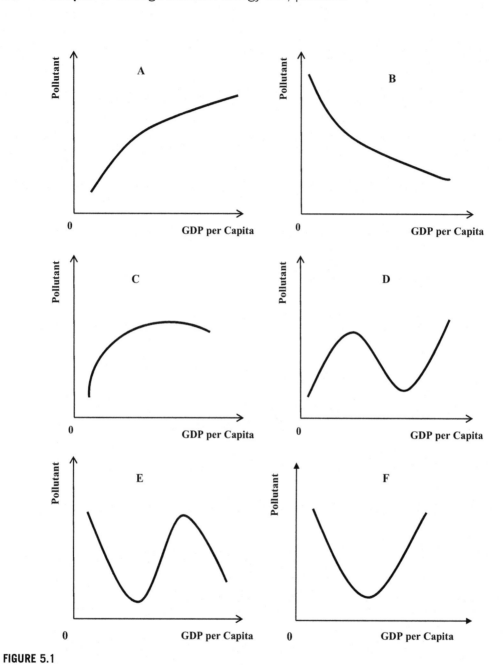

FIGURE 5.1

Relationship between pollutant and GDP per capita.

The fifth possibility (Fig. 5.1E) has been referred to by Hauff and Mistri (2015). In their study they attempted to unravel the existence of EKC in case of access to safe drinking water, ground water resource use in relation to waterborne diseases covering 32 states in India during the period 2001−12. The findings do not indicate any kind of EKC relationship in terms of the selected indicators. On the contrary, they found that with rising income at initial stages, the problem of the access to safe drinking water declines quickly, but again begins to rise at higher income levels. The reason might partially be explained by the inability to properly manage the effects of regional climatic and geomorphological variety. According to them with further rising income, there could be investment in environmental degradation abatement technologies which have a favorable effect in removing hurdle in access to safe drinking water. Again Sarkodie and Strezov (2019) found a U-shaped relationship valid for India and South Africa at a turning point of US\$ 1476 and US\$ 7573 with respect to GHG emission, which accords well with Fig. 5.1F.

The corresponding graphical exposition of the aforesaid six types is given below.

3. Data source and methodology
3.1 Data source

Capturing the aspects of environmental pollution in a theoretical frame is not an easy job. It involves a host of component factors, systematic data for which also is not always available. Further choosing important components of GHGs based on availability of consistent time series data requires careful scrutiny. In this context, the present study is wholly based on secondary data which have been collected from the World Bank database. This data source provides a motley group of data on different aspects during the 1990−2016 time zone, and it also categorizes all countries in this world into different groups on the basis of income level. In this study, we consider four income groups like "Low Level Income Group," "Lower-Middle Level Income Group" "Higher-Middle Level Income Group," and "High Level Income Group." And from each group we have considered 25% of the countries during 1960−2016 depending on data availability for the considered indicators of GHG, e.g., CO_2, CH_4, and N_2O. The details have been presented with the help of following Table 5.1.

3.2 Methodology
3.2.1 Model for fitting the EKC

Panel data regression has been run for examining the issue of existence of EKC across these countries. For estimating the panel regression, we need to first focus on both fixed and random effect modeling technique. The use of fixed effect model (FEM) and random effect model (REM) is also indicated in a study on India by Kumar and Agarwal, (2003), who separately regressed cropped area, removal of forest cover (deforestation), and pasture area on per capita income, yield, and population density.

In both the fixed and random effect panel regression model the heterogeneity among the cross-sectional units is explicitly considered. In the fixed effect model the intercept term is allowed to vary across cross-sectional category, but for each cross-sectional category the intercept is assumed to remain constant over time, i.e., it is time invariant. This is done by regarding the intercept as a variable and using dummy variable to account for differences among individual categories with regard to the intercept. The fixed effect model is estimated by applying ordinary least squares method. In the

Table 5.1 List of selected countries in different income groups.

Low level income group	Lower-middle income group	Higher-middle income group	High income group
Afghanistan	Bangladesh	Algeria	Australia
Central African Republic	Bhutan	Belize	Belgium
Ethiopia	Ghana	Brazil	Canada
Malawi	India	China	Denmark
Nepal	Indonesia	Colombia	France
Tanzania	Myanmar	Guyana	Germany
Uganda	Nigeria	Iran, Islamic Rep.	Hong Kong
	Pakistan	Iraq	Ireland
	Philippines	Malaysia	Italy
	Vietnam	Maldives	Japan
	Zambia	Mexico	Korea, Rep.
	Zimbabwe	South Africa	Kuwait
		Sri Lanka	New Zealand
		Thailand	Poland
		Turkey	Portugal
			Spain
			Sweden
			Switzerland
			United Kingdom
			United States

Author's selection based on World Bank Database.

random effect model no dummy variable is considered to capture the presence of individual effect. On the contrary the individual effect is considered as a random variable with a specific mean value. The individual differences in the intercept values across cross-sectional units are recognized by random error term ε. Thus in terms of Eq. (5.1) $\alpha_i = \alpha_1 + \varepsilon_i$. So the combined error term becomes $w_{it} = \varepsilon_i + U_{it}$. The model derives its name by considering individual effect α_i as a random variable. The following assumptions are made with regard to U_{it} and ε_i

$$U_{it} \sim N\left(0, \sigma_{U^2}\right)$$
$$\varepsilon_i \sim N\left(0, \sigma_{\varepsilon^2}\right)$$
$$E(U_{it}, \varepsilon_i) = 0$$
$$E\left(\varepsilon_i, \varepsilon_j\right) = 0$$

$E(U_{it}, U_{is}) = E\left(U_{ij}, U_{jt}\right) = E\left(U_{ij}, U_{js}\right) = 0$ for $i \neq j$ and $t \neq s$. Here $\sigma_{\varepsilon^2} = \text{var}(\varepsilon)$ and $\sigma_{U^2} = \text{var}(U_{it})$.

It can be observed that composite error term w_{it} has zero mean and constant variance. The random effects model is estimated by using the Generalised Least Square technique (GLS).

Generally, Hausman test play a very important for selecting the appropriate model between FEM and REM. Sometimes, it is observed that if there be any correlation between explanatory variables and composite error term, then the REM model cannot be recommended. In this situation Hausman test is most appropriate to resolve this type of realistic problems. Hausman based the test on the idea that if there be no correlation between w_{it} and explanatory variable(s), both OLS and GLS are consistent but OLS is inefficient. However, if such correlation exists, OLS is consistent while GLS is not. To be more specific Hausman assumed that there exist two estimators $\widehat{\beta}^{FE}$ and $\widehat{\beta}^{RE}$ of the parameter vector β and added two hypothesis testing procedures. They are

H_N : both $\widehat{\beta}^{FE}$ and $\widehat{\beta}^{RE}$ are consistent, but $\widehat{\beta}^{FE}$ is inefficient

H_A : both $\widehat{\beta}^{FE}$ is consistent and efficient, but $\widehat{\beta}^{RE}$ is inconsistent.

We test here H_N (random effect are consistent and efficient) against H_A (random effects are inconsistent, as the fixed effects will always be consistent). Hausman takes $\widehat{q} = \left(\widehat{\beta}^{FE} - \widehat{\beta}^{RE} \right)$ as the basis for the relevant test statistic. The Hausman test statistic is given by

$$H = \widehat{q}' \left[Var\left(\widehat{\beta}^{FE} \right) - Var\left(\widehat{\beta}^{RE} \right) \right]^{-1} \widehat{q} \sim \chi^2(k)$$ where "k" stands for degree of freedom. Hausman test result recommends the application of random effect model instead of fixed effect model when the computed value of chi-sq appears to be statistically insignificant (Bhaumik, 2015).

Based on Hausman test, the random effect model appears to be suitable and hence its results have been presented in the analysis. Culas (2012) also used random effects model to find the inverted U-shaped EKC for nine Latin American countries.

The significance of the coefficients of estimated equations help identify the existence of inverted U-shaped or N-shaped relation or otherwise between per capita GDP and emission of respective indicators for GHGs. Further, panel vector error correction model has been fit to analyze the possibility of causality across GDP and emission of different GHG pollutants.

This basic model involves

$$Z_{it} = \alpha_i + \beta_1 Y_{it} + \beta_2 Y_{it}^2 + \beta_3 Y_{it}^3 + \beta_4 P_{it} + \beta_5 T_{it} + U_{it}$$

$Z_{it} =>$ CO_2 emission or CH_4 emission or N_2O emission depending on the specific GHG the regression seeks to address.

$Y_{it} =>$ GDP per capita.

$Y_{it}^2 =>$ Square of GDP per Capita.

$Y_{it}^3 =>$ Cubic of GDP per capita.

$P_{it} =>$ Urban population.

$T_{it} =>$ Volume of trade.

$U_{it} =>$ Error term

$t =>$ 1990, 1991, 1992, … … … …. 2016.

Based on application of Hausman test, the superiority of random effect model is established in the present context.

3.2.2 Panel vector error correction model (PVECM) for testing causality

In order to check short- and long-run (both way) causality between pollutant and GDP, PVECM is utilized for the analysis. Before running the VEC model, it is necessary to check the existence of unit root in the variables reflecting different pollutant and GDP per capita. In this test there are three methods (Levin, Lin and Chu method, ADF, and PP-Fisher technique), which help in accepting or rejecting unit root at different levels. If both variables are integrated of same order, then we can process for Fisher-Johansen test of cointegration. Once cointegration is established, VEC model is applied to test for existence of short as well as long run causality across considered variables.

4. Theoretical development of elasticity approach to the notion of EKC

Decomposition of pollution intensity of GDP and attendant analysis and exposition in terms of elasticity coefficient have been considered in some studies (Wang et al., 2015; Datta and Sarkar, 2013). In order to understand the geographical concentration of pollutants (or pollution intensity of specific area), it is deemed pertinent to undertake a decomposition analysis of the same based on somewhat modified application of the notion of Kaya identity. Further, an analysis of variation of elasticity of area-specific pollution concentration with respect to GDP is also carried out corresponding to decomposition method. Thus we may consider

Pollution/Area = Pollution/Energy $*$ Energy/GDP $*$ GDP/Population $*$ Population/Area

i.e, $E_P/A = E_P/E * E/Y_D * Y_D/P * P/A$

i.e, $\log(E_P/A) = \log(E_P/E) + \log(E/Y_D) + \log(Y_D/P) + \log(P/A)$

Taking first difference

$$\text{i.e, } \Delta\log(E_P/A) = \Delta\log(E_P/E) + \Delta\log(E/Y_D) + \Delta\log(Y_D/P) + \Delta\log(P/A)$$

$$\text{or, } \Delta\log(E_P/A)/\Delta\log Y_D = \Delta\log(E_P/E)/\Delta\log Y_D + \Delta\log(E/Y_D)/\Delta\log Y_D$$

$$+ \Delta\log(Y_D/P)/\Delta\log Y_D + \Delta\log(P/A)/\Delta\log Y_D$$

$$\text{or, } \epsilon_{ASPC} = \epsilon_{PIE} + \epsilon_{EIGDP} + \epsilon_{PCGDP} + \epsilon_{PD}$$

where E_P/A = Area Specific Pollution Concentration (ASPC).

E_P/E = Pollution Intensity of Total Energy Use (PIE).

E/Y_D = Energy Intensity of Gross Domestic Product (EIGDP).

Y_D/P = Per Capita GDP (PCGDP).

P/A = Population Density (PD).

Since pollution is influenced by energy consumption relative to output (GDP) and drive to such output growth emerges from improving peoples' wellbeing through its enhanced availability, it seems worthwhile to account for such factors contributing to pollution concentration in a defined geographical region and related concept of elasticity of spatial concentration of pollution with respect to output. In this context we consider relevant concepts of spatial concentration of pollution, GDP, energy, population etc. In a bid to square up these aspects with the notion of EKC, we consider some decomposition analysis in terms of explaining the variation in elasticity of spatial concentration of pollution with respect to changes in output.

In this context it seems important to explain the relevant ratios and their specific elasticities. The ratio E_P/A refers to the intensity of environmental pollution in a specified geographical area or in other words it indicates the spatial concentration of pollutants arising out of domestic economic activities. As energy is considered as an important input in resource-intensive production and its use involves release of pollutants in diverse forms, the pollution intensity of energy use (PIE) seems to be an important key contributing element in the process of environmental pollution. Again as production in all three spheres like agriculture, industry, and service activity (comprising the GDP) involves energy use in various forms, considerable volume of pollution is likely to emerge out of the energy use in the production process. Substantial amount of energy use in the form of fossil fuel burning and use of electricity are likely to be concomitant in the production process of GDP. The less the energy use per unit of GDP, the less is the environmental pollution. and implicit betterment of the environment. Hence energy intensity of GDP (EIGDP) indicated by E/Y_D is very likely to indicate the contributory aspect toward pollution. Again since with the growth process of an economy, per capita GDP (PCGDP) captured by Y_D/P is likely to rise, reflecting flourishing of activities in all the three sectors of an economy with attendant rise in energy component of input use, it is likely to signal toward the generation of increasing volume of waste and pollutants in an economy. Finally population density (PD) captured by the term P/A indirectly reflects the contribution as well as susceptibility of population in a defined geographical space toward pollution concentration. It also indicates the direct and indirect pressure on available resource use in a given area and hence captures the concern for associated release of pollutants and waste elements. ϵ_{PIE} captures the aspect of elasticity of pollution intensity of energy use with respect to GDP. This value is expected to be negative in case of developed countries where with rise in output, proportion of pollution in relation to energy use tends to decline unlike in developing countries. The underlying cause of such expected pattern is that developed countries can nurture the tempo of sustained adoption of green and efficient technology that results in release of less pollution and even less energy in specific cases. The speed of diffusion of improved green technology is likely to be rather slow in case of developing economies compared to that in developed countries. Again increased awareness among people about environmental norms and regulations, implementation of stern measures in the form of emission tax, tradable pollution permits, carbon banks, etc., have been instrumental in developed economies in putting a brake on release of pollutants with rise in production together with increasing concern among people for protection of environment.

Again more resources are at present being spent by developed countries in the pursuit of innovative technologies capable of undertaking production requiring less fossil fuel−based energy in domestic production sphere with gradual shift to use of renewable type energy. This process is yet to make deep inroads in case of production in developing countries. With huge funds devoted to the sphere of devising alternative nonconventional sources of energy and land specifically assigned for farming of jatropha/veranda or other biofuels, advanced countries are likely to experience a negative value of elasticity of EIGDP (ϵ_{EIGDP}) with respect to GDP contrary to that in developing economies. But in cases where biofuel hardly acts as a good substitute for fossil fuel−related energy, the stated elasticity may take a positive value with low magnitude resulting from innovative efforts. Again with proportional change in GDP, there is likely to be unidirectional proportional change in per capita GDP. This is likely to have impact on pollution resulting from possible technological changes or scale effect of rising input use both of which are conditioned by enhanced use of fossil fuel and electricity. Even if it be assumed that in the case of developed countries alternative energy sources are more and more used to churn out increased GDP, the production of alternative energy or its apparatus also often requires electricity which is based on fossil fuel. Hence the elasticity of per capita GDP (ϵ_{PCGDP}) is likely to be positive for all types of economies. With increase in GDP there is likely to be decline in infant mortality and better health facility resulting in enhanced population density with concomitant pressure on pollution intensification. Hence the elasticity of population density with respect to GDP growth (ϵ_{PD}) is likely to be positive leading to enhanced area-specific pollution concentration.

5. Results and discussion

Large-scale economic activities coherent with satisfying the development and consumption needs is conditioned by using diverse type of inputs with attendant production of GHGs. The degradation of the environment jeopardizes the economic activities. In order to derive a steady-state economy and to ensure high environmental quality, economic development or growth of the economy should match with the issues of the environment. However, due to variation in different macroeconomic variables and level of growth attained in a country, the dimensions of environmental quality and the EKC hypothesis vary from place to place and country to country.

Two macroeconomic variables have been considered in the study, viz., trade and urban population. It is often held that trade promotes economic growth but degrades the environment. Usually import reduces carbon emissions, while export increases it. However, open trade has either a positive or negative impact, as higher trade does not always necessarily mean higher emissions and vice versa. It depends on the composition and volume of output traded. Hence trade may have a mixed impact on the type of emission, and it varies with each country's environmental situation.

Several studies have observed a positive relationship between urbanization and CO_2 emissions (Wu et al., 2016; Katircioglu and Katircioglu, 2017). Aggregate energy consumption and greenhouse emissions are positively correlated with the urbanization process resulting from higher urban living standards. Urban citizens are usually prone to consume high energy-intensive goods. Hence spread of urbanization is likely to lead to increased direct and indirect energy consumption with consequential global warming impact.

There are four groups of countries as already indicated.

(1) High income group countries, (2) Upper-middle income group countries, (3) Lower-middle income group countries, (4) Low income group countries.

In order to fit the Eq. (5.1), Hausman test has first been carried out in order to choose between random effect and fixed effect model.

5.1 Discussion of panel regression results

The use of Hausman test invalidated the use of fixed effect model for all the GHGs across the four groups of countries. Accordingly the random effect model results are displayed in the following Table 5.2.

The figures reveal that for both high-middle income and high income group countries, the sign of the estimated coefficient for Y^2 is negative and significant while that for Y^3 is positive and significant corresponding to all the three considered GHGs like CO_2, CH_4, and N_2O. The sign of Y is positive and significant for all the cases excepting that for methane in high income countries, where it is positive but insignificant. Further, as expected, there is observed to be significant positive impact of urban

Table 5.2 Results of the emission pattern of different GHGs across different income group countries.

Variable	CO$_2$				N$_2$O				CH$_4$			
	Low income group	Lower-middle income group	High-middle income group	High income group	Low income group	Lower-middle income group	High-middle income group	High income group	Low income group	Lower-middle income group	High-middle income group	High income group
GDP	−5.46E-07	0.000123	4.20E-07	5.69E-07	−108.2568	−7.065014	32.42223	1.455566	−66.98747	−16.28538	101.5831	0.535294
	(0.0000)	(0.7393)	(0.0000)	(0.0000)	(0.0542)	(0.5544)	(0.0000)	(0.0006)	(0.1935)	(0.4089)	(0.0000)	(0.4622)
GDP^2	4.44E-10	4.62E-09	−2.27E-11	−1.11E-11	0.064174	0.000970	−0.002355	−5.64E-05	0.043876	0.008970	−0.007240	−3.81E-05
	(0.0000)	(0.9523)	(0.0004)	(0.0000)	(0.0930)	(0.6946)	(0.0000)	(0.0000)	(0.2083)	(0.0283)	(0.0000)	(0.0141)
GDP^3	−8.99E-14	−1.44E-12	5.32E-16	6.84E-17	−1.39E-05	−1.04E-07	4.76E-08	3.81E-10	−9.75E-06	−7.28E-07	1.46E-07	2.75E-10
	(0.0000)	(0.7641)	(0.0009)	(0.0000)	(0.0864)	(0.4958)	(0.0000)	(0.0000)	(0.1849)	(0.0040)	(0.0000)	(0.0040)
Urban population	−5.71E-08	0.000440	0.000103	7.50E-05	2,033.835	1,307.490	3,053.363	−38.38819	1,465.646	−1,121.488	9,702.261	63.12474
	(0.9339)	(0.9705)	(0.0000)	(0.0000)	(0.0000)	(0.0621)	(0.0000)	(0.4819)	(0.0000)	(0.3469)	(0.0000)	(0.5031)
Trade	5.77E-08	−0.004103	−1.71E-06	−3.92E-05	−200.7811	−81.34545	−33.23273	138.1255	−153.4770	−120.8504	−88.12016	85.06564
	(0.6708)	(0.1027)	(0.0680)	(0.0000)	(0.0003)	(0.5365)	(0.2309)	(0.0000)	(0.0057)	(0.5864)	(0.3465)	(0.0879)
C	0.000247	−0.076633	−0.004458	−0.001657	49,194.46	20,812.94	−228,221.5	34,831.23	39,217.88	146,552.6	−761,729.7	74,699.47
	(0.0002)	(0.8901)	(0.0000)	(0.3614)	(0.0770)	(0.4551)	(0.0000)	(0.0437)	(0.1340)	(0.0229)	(0.0000)	(0.0075)
R^2	0.635278	0.013261	0.738072	0.143552	0.389447	0.027174	0.612518	0.241804	0.187045	0.081093	0.599622	0.106993
F	63.75027	0.854708	224.8637	17.90109	23.34566	1.776567	126.1452	34.06059	8.420916	5.612692	119.5114	12.79593
	(0.000000)	(0.511920)	(0.000000)	(0.000000)	(0.0000)	(0.117219)	(0.0000)	(0.0000)	(0.0000)	(0.00005)	(0.0000)	(0.0000)

**Figure within first parentheses indicates the P-value of "t-statistic."*

Author's calculation based on World Bank database.

population on the emission status of the three gases pervading in almost both these groups of countries excepting in one case. The sign of the coefficient of the said variable corresponding to N_2O in case of high income group country appears to be negative which is however insignificant. So the intensity of urbanization is found to be directly related to vitiation of the environment through greater impact on ecological footprint. Accordingly the significant positive coefficients vindicate this hypothesis.

The more export-oriented a country is it is likely to have a greater trade openness and volume. Depending on the nature of goods traded, the impact on pollution is likely to be different. The impact of trade on emission of N_2O and CH_4 is found to be positive and significant while that on CO_2 is observed to be negative as well as significant. For CO_2, the result may be due to the fact that in many cases the major polluters are not the major exporters of industrial goods and vice versa. The enhanced volume of trade resulting from increased economic activity usually gets associated with three effects, viz., scale effect, composition effect (through structural changes), and technique effect. When the combined intensity of composition and technique effects dominates the scale effect, there may emerge an inverse relation between pollution emission and trade volume as is evinced in case of CO_2 for these two groups of countries.

Thus, on the whole the entire results for these two groups of countries seem to be fitting and consistent with category four of the aforesaid type of Kuznets curve. This shows that for all these considered pollutant gases, the curve takes the shape of cubic polynomial representing N-shaped figure corresponding to high and high-middle income nations. The aforesaid findings seem perfectly consistent with the increasing pressure on energy consumption needs arising out of never ceasing developmental cravings of these countries, which experience saturation in technological development after a threshold income level. Observed decreases in emission with a rise in GDP per capita could only be a temporary phenomenon. In the absence of measures to improve the technology and maintain technical advances with increasing returns, there appears a point at which decreasing technical returns force such economies back to a state of increasing environmental destruction (Lorente and Alvarez, 2016). These findings also find some support in a study by Sengupta (1997), who observed an N-shaped curve for aggregate CO_2 emissions in case of a sample 16 OECD countries covering data over the period 1971–1988. Similarly Sinha and Sengupta (2018) tested the EKC hypothesis for N_2O emissions in APEC (Asia Pacific Economic Cooperation) countries (most of which fall in the above two group of countries) over the period of 1990–2015. Their results also confirmed the existence of N-shaped curve. In such conditions energy innovation might play a significant positive effect to improve their environmental quality. Under such circumstances increased budget allocations to R&D related to energy might help avoid a scale effect that would otherwise trigger a return to rising GHG emissions.

In case of low income countries the sign of Y^2 is found to be positive and significant while that of Y^3 is negative and significant corresponding to CO_2 and N_2O. However, in case of CH_4 both these coefficients have aforesaid type of signs but are found to be insignificant. The possible reasons for this reverse type of signs may be explained as follows. In case of low income countries, people at the lower rung of income ladder tend to use twigs, small branches of trees, cowdung cakes, etc. for fuel purposes, which produce rather moderately low volume of pollution. Hand carts, bullock carts, or cycles are often used across countryside for peoples' movement. Further, they also do not have easy access to domestic gadgets or cars. Most of their income is spent on subsistence consumption with little saving. As income starts slightly rising, some income tends to be saved and dependence on collected fuel gets reduced with gradual shift to cleaner fuel/electricity, etc., for domestic uses. Hence pollution initially tends to fall with slightly increasing income. But with further rising income, beyond a threshold level, people can better afford different types of appliances that create pollution, can buy fossil-fuel driven vehicles, and to cater to their demand industrial production starts taking an upswing. So the relation between GDP per capita and stated pollutants first assumes a U-shape pattern. This result is akin to that of Hongbo et al. (2019) who found a U-pattern relation between economic development and CO_2 emission for low income countries. However, again after a second threshold level of income, people become more and more aware of environmental problems and try to replace fossil fuel with renewable substitutes as far as practicable. Thus CFL lamps, solar light/pumps, CNG-driven cars, etc., partly lessen the intensity of pollution. Improved literacy and better awareness among people together with adoption of superior method of pollution control devices tend to reverse the growth in pollutant level. As a result, beyond a certain level of income, pollution begins to decline and takes a downturn; thus, finally the relation between GDP and pollutants looks like a reverse N pattern.

In case of lower-middle income countries similar significant results are obtained only in case of methane.

5.2 Analysis of causality across GDP and pollutants

One of the interesting debatable issues that have surrounded development and environment over the past few decades is the linkage between output growth and environmental pollution. Environmentalists argue that expanding output in the domestic sphere leads to emission of pollutants vitiating the ambient atmosphere. More output requires more use of fossil fuel and adverse intervention in environmental resources thus generating pollutants. They claim that the real costs of enhanced output are depleting natural resources and deteriorating environmental quality, while growth economists argue that pollutant emissions act as input in the sphere of domestic production. According to them, pollution inherent in rising industrial and improved agricultural activities is an important instrumental adjunct in triggering growth. In this section an investigation is made about the existence of a long-run/short-run causal relationship between environmental pollutants and output growth using cointegration technique, as well as Panel Vector Error Correction Model (PVECM), which is applied to examine the dynamics of the system. PVECM helps examine whether a variable X that evolves over time Granger causes another evolving variable Y in case predictions of the value of Y based on the past values of X and its own past values are better than predictions of Y based only on its own past values.

Based on the literature, three types of operations have been performed for examining the existence of unit root, cointegration, and causality analysis. First, a panel unit root test has been carried out to examine whether or not the variables in our model are stationary. Second, we test for cointegration among the variables employing the trace test and max-Eigen test which are inherent in the Johansen-Fisher cointegration test. Third, once cointegration relationship is established, we investigate the existence of both way/one way/no causality across GDP and different types of pollutants.

Using Eviews software, panel unit root test is performed by Levin, Lin and Chu method, ADF, and PP-Fisher Chi-sq technique. It may be so that at the level of the variable there may not exist any unit root indicating stationarity of the data. In case the existence of unit root and hence nonstationarity of the data is ascertained, first difference/second difference of the respective variable is performed and again unit root test is carried out based on the aforesaid three methods. Once the existence of unit root is not observed with desired degree of significance, the data is considered as stationary. Next cointegration test is carried out by following Johansen-Fisher cointegration test. It is important to note the

Johansen-Fisher test for cointegration is applicable only when the concerned variables are integrated of the same order. In case of the test technique, both trace test and max Eigen test are performed based on the Fisher statistic. When at least one cointegated equation encompassing the two variables is established, based on the stated tests, we confirm the existence of cointegration between the respective variables. When this is established, it is possible to run the panel vector error correction model to examine causality across the two variables. In this case first variable put in the software is called as the dependent variable and second variable is the explanatory one. The desirable lag is also chosen for both the variables. We assume lag two as the optimum for running the PVECM depending on the state of uniform integration of both the variables. The model is run by taking each of the pollutants in turn as the first variable and then GDP as the first variable.

The results based on system model for reflecting short- and long-run causality are reflected in the following Table 5.3. The system models are estimated by OLS technique. There are two system models, one relevant to Causality from Pollutant to GDP and the other relevant to Causality from GDP to Pollutant. The respective system models are indicated as follows:

The first modelis

$$D(Pollu \tan t_i) = C(1) * \{Pollu \tan t_i(-1) + \alpha_i * GDP(-1) + \beta_i\} + C(2) * D(Pollu \tan t_i(-1))$$
$$+C(3) * D(Pollu \tan t_i(-2)) + C(4) * D(GDP(-1)) + C(5) * D(GDP(-2)) + C(6)$$

where, $i = 1,2,3$ ($1 = CO_2$, $2 = N_2O$, $3 = CH_4$).

Similarly the second model stands as

$$D(GDP) = C(1) * \{GDP(-1) + \alpha_j * Pollu \tan t_j(-1) + \beta_j\} + C(2) * D(GDP(-1))$$
$$+C(3) * D(GDP(-2)) + C(4) * D(Pollu \tan t_j(-1)) + C(5) * D(Pollu \tan t_j(-2)) + C(6)$$

where, $j = 1,2,3$ ($1 = CO_2$, $2 = N_2O$, $3 = CH_4$).

α, β stand for coefficients in respective models that can be estimated from data.

"D" stands for difference and bracketed negative terms indicate order of lag in both the cases.

It may be noted that in case of Lower-Middle Income Group countries and High Income Group countries, the respective variables stated in the left hand side (LHS) are not integrated of the same order as that of GDP. Hence existence of cointegration of the respective pollutants with that of GDP is not validated by the Johansen test. So, for these countries PVECM cannot be run for the respective pollutants.

C(1) as reflected in the above equations actually refers to the error correction term. When C(1) is negative in coefficient and significant, then we say that there is a long run causality running from the GDP to the respective pollutant or vice versa as the case maybe. This also implies that there exists a long run equilibrium with corresponding speed of adjustment as reflected by the respective coefficient. For checking short-run causality we need to set the coefficient C(4) and C(5) equal to zero (by null hypothesis) and test it by using Wald test. When the corresponding chi-square statistic value turns out to be significant, it is settled that there does exist short-run causality, and if it appears insignificant there is no such causality. The following table helps in identifying the valid existence of long-run or short-run causality (either one-way or two-way type) across the respective pollutants and GDP for the indicated countries on the LHS.

Table 5.3 Coefficient and significant level of the respective "C values."

Country type	Pollutant type	Causality from pollutant to GDP		Causality from GDP to pollutant	
		Long run	Short run	Long run	Short run
Low income group	CO_2	C(1) = −0.040077 (0.0775)	C(4) = C(5) = 0 Chi Square = 13.00827 (0.0015)	C(1) = −0.000761 (0.0092)	C(4) = C(5) = 0 Chi Square = 9.803551 (0.0074)
	N_2O	C(1) = 0.007246 (0.6651)	C(4) = C(5) = 0 Chi Square = 0.173315 (0.9170)	C(1) = 0.017689 (0.0169)	C(4) = C(5) = 0 Chi Square = 0.216193 (0.8975)
	CH_4	C(1) = 0.001338 (0.4116)	C(4) = C(5) = 0 Chi Square = 0.301503 (0.8601)	C(1) = 0.018179 (0.0589)	C(4) = C(5) = 0 Chi Square = 0.121434 (0.9411)
Lower-middle income group	CO_2 N_2O CH_4	No cointegrating relationship established			
High income group	CO_2	C(1) = 0.000250 (0.4265)	C(4) = C(5) = 0 Chi Square = 0.046078 (0.0015)	C(1) = −0.012814 (0.0024)	C(4) = C(5) = 0 Chi Square = 1.251845 (0.5348)
	N_2O CH_4	No cointegrating relationship established			
High-middle income group	CO_2	C(1) = 0.033537 (0.0000)	C(4) = C(5) = 0 Chi Square = 0.156027 (0.9250)	C(1) = −0.000801 (0.3645)	C(4) = C(5) = 0 Chi Square = 0.367848 (0.8320)
	N_2O	C(1) = 0.027826 (0.0000)	C(4) = C(5) = 0 Chi Square = 0.207163 (0.9016)	C(1) = −0.000928 (0.2175)	C(4) = C(5) = 0 Chi Square = 0.709768 (0.7013)
	CH_4	C(1) = 0.017721 (0.0000)	C(4) = C(5) = 0 Chi Square = 0.431980 (0.8057)	C(1) = −0.000143 (0.2041)	C(4) = C(5) = 0 Chi Square = 0.867778 (0.6480)

[**]*Figure within first parentheses indicates the P-value.*
Author's calculation based on World Bank Database.

The cointegration test results suggest that the considered variables tend to move together in the long run in low income, higher-middle income, and partly for high income countries. No such movement is detected in case of lower-middle income group nations. It is observed from Table. 5.4 that for low income countries there exists both long- and short-run two-way causality across GDP and CO_2, while no such causality exists across N_2O and GDP or CH_4 and GDP. For lower-middle income

Table 5.4 Long- and short-run causality status across pollutants to GDP and GDP to pollutants across country type.

Country type	CO$_2$ to GDP		GDP to CO$_2$		N$_2$O to GDP		GDP to N$_2$O		CH$_4$ to GDP		GDP to CH$_4$	
	Long run	Short run	Long run	Short run	Long run	Short run	Long run	Short run	Long run	Short run	Long run	Short run
Low income	Causality	No causality	Causality	Causality	No causality	No causality	No causality	No causality	No causality	No causality	No causality	No causality
Lower-middle income	CO$_2$, N$_2$O, and CH$_4$ are stationary at 1st difference but GDP is stationary at 2nd difference. So, these variables are not integrated of same order invalidating cointegration test and PVECM.											
High income	No causality	No causality	Causality	No causality	N$_2$O, CH$_4$ are stationary at level but GDP is stationary at 1st difference. So, N$_2$O and GDP and CH$_4$ and GDP are not integrated of same order. Thus, cointegration test is not valid.							
Higher-middle income	No causality	No causality	No causality	No causality	No causality	No causality	No causality	No causality	No causality	No causality	No causality	No causality

Author's calculation based on World Bank Database.

countries no such model could be fit because of stated reasons disallowing checking of causality. In case of high income countries, only one-way long-run causality is found to exist from GDP to CO_2. The results suggest that CO_2 has a predictive power for economic growth or feedback impact with GDP growth in low income countries. On the basis of speed of adjustment, it is observed that there exists bidirectional causality between CO_2 emission and economic growth in such low income countries. Further based on long-run unidirectional causality running from GDP to CO_2, it may be stated that GDP has a predictive power for release of CO_2 in case of high income countries.

6. Concluding remarks and policy prescription

This study aims at analyzing the dynamic relationship between emission of GHGs like CO_2, CH_4, and N_2O, and real GDP covering panel data for individually seven low income countries, 12 lower-middle income countries, 15 higher-middle income countries, and 20 high income countries across the period 1990−2016. Revised form of Kaya identity has been applied to explain area-specific Pollution Concentration through an interactive relation across Pollution Intensity of Total Energy Use, Energy Intensity of Gross Domestic Product, Per Capita GDP, and Population Density. Existence of EKC or otherwise has been probed for each individual category of countries. It is observed that the relevant curve takes shape of cubic polynomial representing N-shaped figure corresponding to upper and upper-middle income nations. As all possible cost and energy saving technological innovations reach the brink of saturation, technical coefficients of carbon emissions approach some lower bounds, with possible positive relinking between GDP and emission. Again pervading huge demand for tourism and travel in such countries involves pressure on direct and indirect use of electricity and transport with attendant emission of GHGs.

In case of low income economies the sign of Y^2 is found to be negative and significant while that of Y^3 is positive and significant corresponding to CO_2 and N_2O, thus confirming the existence of a reverse N-shaped pattern. Similar pattern has been detected corresponding to only methane in case of lower-middle income countries. Low income countries and to some extent lower-middle income nations are gradually becoming a victim to the assault by the waves of globalization with emulation of the lifestyle of advanced economies. With gradual progress of time, expansion of industrial production, mechanized agriculture, and transport and communication are likely to threaten the ambient environment through release of CO_2 and N_2O arising from consumption of fossil fuel.

Both long-run and short-run causality across CO_2 to GDP and GDP to CO_2 have been found to exist only in case of low income economies. And for high income nations only long run causality has been recorded from GDP to CO_2 for N_2O and CH_4, no such causality has been recorded for any type of countries. Thus, the absence of causality from emissions of N_2O or CH_4 to growth or in reverse direction suggests that the respective type countries can control their N_2O or CH_4 emissions without troubling their economic growth. Since CO_2 is the most important pervading gas in atmosphere, the finding of two-way causality for low income countries suggests that unless immediate protective measures are taken in the form of reduced power consumption by switching off electronic devices instead of unnecessarily keeping them on, by walking or cycling short distances instead of using motor-bikes or bus, switching to renewable sources of energy, adopting recycling waste materials, etc., the environment might be vitiated early allowing point of no immediate return. High income countries also should put a brake on unbridled output growth; otherwise, the N pattern of relation between output and CO_2 would reflect a steep rise in its emission after certain point of time. Quick switch to biodiesel producible from renewable energy sources such as sugar beet, rape seed, palm oil, and sunflowers cannot wait any more. Again ethanol also serves as a biological fuel substitute for petrol and is made from renewable energy sources. Thus proper care should be taken to protect the ecosystems, air quality, and sustainability of our resources and focusing on the elements that lighten the pressure on the environment. Concern should be placed on maintaining a greener future with a proper innovative blend of engineering technology and biotechnology.

References

Acaravci, I., Ozturk, I., 2010. On the relationship between energy consumption, CO_2 emissions and economic growth in europe. Energy 35, 5412–5420.

Acheampong, A.O., 2018. Economic growth, CO_2 emissions and energy consumption: what causes what and where? Energy Econ. 74, 677–692. https://doi.org/10.1016/j.eneco.2018.07.022.

Acheampong, A.O., Dzator, J., 2020. Managing environmental quality: does institutional quality matter? In: Hussain, C.M. (Ed.), Handbook of Environmental Materials Management. Springer, Cham, pp. 1–24. https://doi.org/10.1007/978-3-319-58538-3_215-1.

Adams, S., Acheampong, A.O., 2019. Reducing carbon emissions: the role of renewable energy and democracy. J. Clean. Prod. 240 https://doi.org/10.1016/j.jclepro.2019.118245.

Ahmad, N., Du, L., Lu, J., Wang, J., Li, H.Z., Hashmi, M.Z., 2017. Modelling the CO_2 emissions and economic growth in Croatia: is there any environmental Kuznets curve? Energy 123, 164–172.

Alkhathlan, K., Javid, M., 2013. Energy consumption, carbon emissions and economic growth in Saudi Arabia: an aggregate and disaggregate analysis. Energy Pol. 62, 1525–1532.

Al-Mulali, U., Ozturk, I., Solarin, S.A., 2016. Investigating the environmental Kuznets curve hypothesis in seven regions: the role of renewable energy. Ecol. Indicat. 67, 267–282. https://doi.org/10.1016/j.ecolind.2016.02.059.

Al-Mulali, U., Weng-Wai, C., Sheau-Ting, L., Mohammed, A.H., 2015. Investigating the environmental Kuznets curve (EKC) hypothesis by utilizing the ecological footprint as an indicator of environmental degradation. Ecol. Indicat. 48, 315–323.

Ang, J.B., 2007. CO_2 emissions, energy consumption and output in France. Energy Pol. 35, 4772–4778.

Apergis, N., Ozturk, I., 2015. Testing environmental Kuznets curve hypothesis in Asian countries. Ecol. Indicat. 52, 16–22. https://doi.org/10.1016/j.ecolind.2014.11.026.

Arouri, M.E., Youssef, A., Mhenni, H., Rault, C., 2012. Energy consumption, economic growth and CO_2 emissions in Middle East and North African countries. Energy Pol. 45, 342–349.

Awad, A., Abugamos, H., 2017. Income-carbon emissions nexus for Middle East and North Africa countries: a semi-parametric approach. Int. J. Energy Econ. Pol. 7, 152–159.

Ben, J.M., Ben, Y.S., et al., 2015a. The environmental Kuznets curve, economic growth, renewable and non-renewable energy and trade in Tunisia. Renew. Sustain. Energy Rev. 47, 173–185. https://doi.org/10.1016/j.rser.2015.02.049.

Bhaumik, S.K., 2015. Principles of Econometrics, A Modern Approach Using EVIEWS. Oxford University Press.

Culas, R.J., 2012. REDD and forest transition: tunneling through the envrionmental Kuznets curve. Ecol. Econ. 79, 44–51. https://doi.org/10.1016/j.ecolecon.2012.04.015.

Datta, S.K., Sarkar, K., 2013. An enquiry into the existence of environmental Kuznets curve and issue of convergence related to CO_2 emission—a panel data analysis. Int. J. Glob. Environ. Issues 13 (1).

Deacon, R., Mueller, B., 2004. Political Economy and Natural Resource Use. Department of Economics Working Paper, UC Santa Barbara. Series No. 5-IV.

Grossman, G.M., Krueger, A.B., 1993. Pollution and growth: what do we know? In: Goldin, I., Winters, L. (Eds.), The Economics of Sustainable Development. MIT Press, Cambridge, MA.

Grossman, G.M., Krueger, A.B., 1995. Economic growth and the environment. Q. J. Econ. 110 (2), 353−377.

Halicioglu, F., 2009. An econometric study of CO_2 emissions, energy consumption, income and foreign trade in Turkey. Energy Pol. 37, 1156−1164.

Hauff, M.V., Mistri, A., 2015. Economic Growth, Safe Drinking Water and Ground Water Storage: Examining Environmental Kuznets Curve (EKC) in Indian Context. University Library of Munich, Germany. MPRA Paper 61656.

Hongbo, L., Hanho, K., Justin, C., 2019. Export diversification, CO_2 emissions and EKC: panel data analysis of 125 countries. Asia-Pac. J. Reg. Sci. 3 (2), 361−393.

Jacob, D., 1999. Introduction to Atmospheric Chemistry. Princeton University Press, ISBN 978-0691001852, pp. 25−26.

Jalil, A.A., Syed, F., Mahmud, S.F., 2009. Environment Kuznets curve for CO_2 emissions: a co-integration analysis for China. Energy Pol. 37, 5167−5172.

Jaunky, V.C., 2011. The CO_2 emissions-income nexus: evidence from rich countries. Energy Pol. 39, 1228−1240.

Jayanthakumaran, K., Verma, R., Liu, Y., 2012. CO_2 emissions, energy consumption, trade and income: a comparative analysis of China and India. Energy Pol. 42, 450−460.

Kaika, D., Zervas, E., 2012. Theoretical examination of the environmental Kuznets curve (EKC) concept: possible driving forces and controversial issues. In: Proceedings of 3rd International Conference on Development, Energy, Environment, Economics., Paris, France.

Katircioglu, S., Katircioglu, S., 2017. Testing the role of urban development in the conventional environmental Kuznets curve: evidence from Turkey. Appl. Econ. Lett. 25, 741−746.

Kumar, P, Agarwal C, S, et al., 2003. Does an environmental Kuznets curve exist for changing land use? Empirical evidence from major states of India. Int. J. Sustain. Dev. 6 (2), 231−245.

Lopez, R. (Ed.), 1992. The Environment as a Factor of Production: The Economic Growth and Trade Policy Linkages. World Bank, Washington D.C.

Lorente B, D, Alvarez, A, et al., 2016. Approach to the effect of energy innovation on environmental Kuznets curve: an introduction to inflection point. Bull. Energy Econ. 4 (3), 224−233.

Mandal, S.K., Madheswaran, S., 2010. In: Causality between Energy Consumption and Output Growth in Indian Cement Industry: An Application of Panel Vector Error Correction Model, vol. 238. The Institute for Social and Economic Change, Bangalore, ISBN 81-7791-194-5. Working paper.

Menegaki, A.N., 2013. Growth and renewable energy in europe: benchmarking with data envelopment analysis. Renew. Energy 60, 363−369.

Moomaw, W.R., Unruh, G.C., 1997. Are environmental Kuznets curves misleading us? The case of CO2 emissions. Environ. & Dev. Econ. 2 (4), 451−463.

Omri, A., 2013. CO_2 emissions, energy consumption and economic growth nexus in MENA countries: evidence from simultaneous equations models. Energy Econ. 40, 657−664.

Ozcan, B., 2013. The nexus between carbon emissions, energy consumption and economic growth in Middle East countries: a panel data analysis. Energy Pol. 62, 1138−1147.

Ozgur-Kayalica, M., Kacar, S.B., 2014. Environmental Kuznets curve and sulfur emissions: a comparative econometric analysis. Environ. Econ. 5 (1), 8−20.

Perman, R., Stern, D., 2003. Evidence from panel unit root and cointegration tests that the environmental Kuznets curve does not exist. Aust. J. Agric. Resour. Econ. 47 (3), 325−347.

Rauf, A., Zhang, J., Li, J., Amin, W., 2018. Structural Changes, Energy Consumption and Carbon Emissions in China: Empirical Evidence from ARDL Bound Testing Model Structural Change and Economic Dynamics. https://doi.org/10.1016/j.strueco.2018.08.010.

Sarkodie, S.A., Strezov, V., 2019. Effect of foreign direct investments, economic development and energy consumption on greenhouse gas emissions in developing countries. Sci. Total Environ. 646, 862–871. https://doi.org/10.1016/j.scitotenv.2018.07.365.

Sengupta, R., 1997. CO_2 emission-income relationship: policy approach for climate control. Pac. Asian J. Energy 7 (2), 207–229.

Shahbaz, M., Khraief, N., Uddin, G.S., Ozturk, I., 2014. Environmental Kuznets curve in an open economy: a bounds testing and causality analysis for Tunisia renewable and sustainable. Energy Rev. 34, 325–336. https://doi.org/10.1016/j.rser.2014.03.022.

Shahbaz, M., Nasir, M.A., Roubaud, D., 2018. Environmental degradation in France: the effects of FDI, financial development, and energy innovations. Energy Econ. 74, 843–857. https://doi.org/10.1016/j.eneco.2018.07.020.

Sinha, A., Sengupta, T., 2018. Impact of energy mix on nitrous oxide emissions: an environmental Kuznets curve approach for APEC countries. Environ. Sci. Pollut. Control Ser. 26 (8).

Sinhababu, S., Datta, S.K., 2013. The relevance of environmental Kuznets curve (EKC) in a framework of broad-based environmental degradation and modified measure of growth - a pooled data analysis. Int. J. Sustain. Dev. World Ecol. 20 (4), 309–316.

Stern, D.I., 1998. Progress on the environmental Kuznets curve? Environ. Dev. Econ. 3, 173–196.

Stern, D.I., Auld, A., Common, M.S., Sanyal, K.K., 1998. Is There an Environmental Kuznets Curve for Sulfur? Working Papers in Ecological Economics, 9804 Center for Resource and Environmental Studies, Australian National University, Canberra.

Taguchi, H., 2012. The environmental Kuznets curve in Asia: the case of Sulphur and carbon emissions. Asia Pac. Dev. J. 19 (2).

Wang, Y., Han, R., Kubota, J., 2016. Is there an environmental Kuznets curve for SO_2 emissions? A semi-parametric panel data analysis for China. Renew. Sustain. Energy Rev. 54, 1182–1188.

Wang, Z., Zhao, L., Mao, G., Wu, B., 2015. Factor decomposition analysis of energy-related CO_2 emissions in tianjin, China. Sustainability 7 (8), 9973–9988.

Wu, J., Wu, Y., Guo, X., Cheong, T.S., 2016. Convergence of carbon dioxide emissions in Chinese cities: a continuous dynamic distribution approach. Energy Pol. 91, 207–219.

Zambrano-Monserrate, M.A., Fernandez, M.A., 2017. An environmental Kuznets curve for N_2O emissions in Germany: an ARDL approach. Nat. Resour. Forum 41 (2), 119–127.

Environmental pollution and sustainability

6

Priyanka Yadav[1], Jyoti Singh[1], Dhirendra Kumar Srivastava[2], Vishal Mishra[1]
[1]*School of Biochemical Engineering, IIT (BHU), Varanasi, Uttar Pradesh, India;* [2]*Council of Science and Technology, Lucknow, Uttar Pradesh, India*

Chapter outline

1. Introduction

The key problems of the environmental policy pursued by any government include the rectification of tradeoffs between local freedom to adapt to environmental standards and the implementation of local values, revenues, and physical environments (Howe, 1996; Hills and Roberts, 2001). Environmental sustainability problems are emerging rapidly as one of the most valuable topics for strategic business, manufacturing, management, and decision of product development. This increases the appreciation of the natural environment which was reflected in environmental-cum-innovative conscious products recommended to consumers in recent years. Green manufacturing may lead to lower raw material costs, improved corporate image, and diminished environmental and occupational safety expenses coupled with production efficiency gains. The relation between green practices and performance results has been subjected to several studies, but results are not convincing (Sezen and Cankaya, 2013). Demands on global energies have been increasing rapidly due to the technology and economic developments (Sorgulu and Dincer, 2018). Sustainable development means addressing existing needs without undermining future generations to meet their demands (Goodland and Daly, 1996).

Sustainability science requires connecting scientific research to political agenda. From the very beginning, this connection had been a core aspiration of sustainability science (Rockstrom et al., 2018). Environmental pollutants have dominant toxic effects leading to respiratory disorders, perinatal dysfunctions, cardiovascular dysfunction, endothelial dysfunction, malignancies, increase in oxidative stress, and mental issues (Kelishadi, 2012). Science may contribute to the sustainability transition by

111

giving guidance and knowledge for navigating the journey from contemporary patterns to a sustainable future (Miller et al., 2014). Sustainable development may not only carry an intergenerational dimension but may also include features like spatial allocations, and these spatial dimensions may range from global to local scale. Sea level rise is an example of global sustainability challenge mainly caused by climatic changes, ozone depletion, global warming, desertification, deforestation, as well as species extinction. On the other side, the example of local sustainability challenge is pollution in areas which face exposure to different toxic chemicals, healthy urban development, construction of roads, radioactive industries, and local catastrophes (Nijkamp and Opschoor, 1997).

This chapter focuses on how the environment can be managed to achieve sustainability, followed by a brief discussion on sustainability. It also addresses various environmental sustainability challenges and the sustainability myth, accompanied by various sustainability-related activities.

2. Environmental management

The green supply chain includes analysis of total environmental effects of products from its whole life cycle of products and services. The management philosophy may influence business to search for improvements in our environment that may yield parallel economic benefits. Product design may integrate the environmental considerations with aspects of product designs. Waste minimization is also considered to be a cleaner production. There are two main waste strategies; the first deals with the waste after the mitigative and generative effect on the environment and the second one is to reduce the amount of waste generated in the first step. Under defensive compliance (Fig. 6.1), firms were reactive in environmental management and simply comply with existing regulations (AlKhidir and Zailani, 2009). Essential practices are required to attain objectives regarding the increase of ecoefficiency in different economy sectors (Juknys, 2003). Sustainability is termed as a key to recent economic, ecological, as well as developmental issues (Midilli and Dincer, 2010). For a long time, it has been recognized that resource depletion and environmental pollution are the byproducts of industrial productions (Levy, 1997).

FIGURE 6.1

Shift of environmental management.

Table 6.1 Benefits linked with final ecosystem services with its functions.

Ecosystem services	Benefits	Intermediate ecosystem functions
Providing natural areas for human use like wildlife viewing	*Information functions* Outdoor recreations	Biodiversity, primary productivity
Provision of aesthetic views	*Information functions* Residential amenities	Primary productivity
Habitat as well as refugia provision for wildlife and humans	*Information functions* Property value premiums	Primary productivity, biodiversity
Production of fruits, nuts, grains, as well as seeds	*Production functions* Food harvests	Primary productivity, pollution, soil productivity, nutrient cycling, and disease regulation
Woody biomass production	*Production functions* Fuel production	Primary productivity
Woody biomass production	*Production functions* Wood products	Primary productivity
Aquifer as well as surface water availability, as well as decrement in storm water runoff	*Regulation functions relevant to pollution mitigation* Drinking water provision—lessens water transport costs as well as pumping	Soil quality, hydrologic cycle
Aquifer as well as surface water quality (sediment as well as nutrient removal)	*Regulation functions applicable to pollution mitigation* Drinking water provision—avoids treatment costs	Nutrient cycling, soil quality as well as hydrologic cycle
Tree shades	*Regulation functions relevant to pollution mitigation* Cost reduction in heating or cooling	1° productivity

Table 6.1 gives a set of ecosystem functions that are reported in urban forests studies to improve urban life quality (Escobedo et al., 2011).

In Britain, there was an environmental justice conflict which showed a gap between the perceptions of people and current activities. In the United Kingdom, it was reported that for a lot of people the issue of environmental justice is not very much clear. The words 'environmental' and 'justice' do not agree with them. At best it stimulates a recollection of distant news reports and a documentary on how poor communities in America face disproportionately harmful problems. In comparison, when applied to white middle-class populations, environmental justice failed to record a signal (Agyeman and Evans, 2004).

There are several environmental issues globally, for instance, Malaysia. Fig. 6.2 illustrates the graph of an environmental survey conducted in Malaysia in 2019.

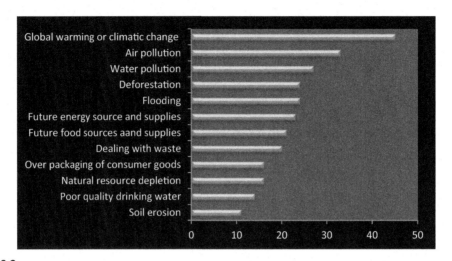

FIGURE 6.2

Most concerning environmental problems according to citizens of Malaysia as of March 2019.

According to a survey, 45% of global warming and climate change was a significant concern for the environment, followed by 33% for air pollution. Other destructive effects due to several environmental issues have also been shown in Fig. 6.2, as per the citizens of Malaysia.

3. Sustainability: a practical approach

Fig. 6.3 shows a practical approach to the validation of issues and perspectives on the sustainability of social, human, and global systems (Komiyama and Takeuchi, 2006). Energy demand may depend on demographic profile, age distribution, size of population, expectations, and habits, and the economic activities as well (Howard et al., 2009). There are many pieces of evidence which indicate that humanity is at a threshold, or crossed it already, where some of the hard-wired earth systems have changed with global and unavoidable consequences.

This may harm the upcoming potential for social benefits that sequentially need careful rethinking and a change in our research agenda regarding plausible desired future in Anthropocene (the time in which human activity has been the main or at least a significant reason for the change in the human environment). Sustainability science is interdisciplinary, which integrates the natural and social sciences in anthropogenic inquiries on pathways to sustainable development and sounds straightforward but is nothing short of a scientific revolution (Rockstrom et al., 2018).

In 1981, Brown (1981) applied two main guidelines, per capita gross world product and per capita consumption of fossil fuel, to determine economic growth trends. It suggested that when the costs of production and consumption increase and climate changes, forest destruction and soil erosion exceed the benefits provided, society would become unsustainable (Brown, 1981). Clark and Munn (1986) raised the issue of 'sustainable development of the biosphere'" together with a series of density measures (Clark and Munn, 1986). Brown (1981) applied guidelines on agricultural production density, population density, and energy consumption density to compare the sustainability of regions

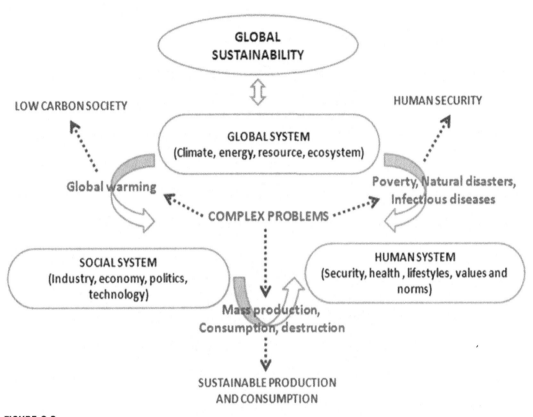

FIGURE 6.3

Addressing the sustainability science through the lens of three systems as well as the links among them.

and nations (Brown, 1981). Heilbroner (1991) suggested a precedence guideline for evaluating up-coming trends such as oil prices, economic growth rates, as well as environmental issues (Heilbroner, 1991). Brubaker (1972) gave a classification system to measure environmental issues, which included climate change, pollution, and overpopulation (Brubaker, 1972). World Conservation Strategy (IUCN, 1980) reported destruction of the topsoil along with lack of farmland, water supply disruption, deforestation, as well as fisheries contamination and reduction as guidelines of nonsustainable resource utilization (Brubaker, 1972). Environmental pollution leads to several health effects like infant mortality, respiratory disorders, cardiovascular disorders, perinatal disorders, allergies, endo-thelial dysfunctions, malignancies, and some other harmful effects from early life (Kelishadi, 2012). Sustainability issues are often characterized by complexity, irreducible uncertainty, as well as con-tested values (Miller et al., 2014).

Fig. 6.4 demonstrates an equation of life that indicates the key role of soil in catalyzing carbon cycle, nutrients, as well as energy in the ecosphere and environmental interface to maximize plant and human conditions (Doran, 2002).

Fig. 6.5 illustrates the three-legged stool that balances agricultural sustainability to achieve long-term sustainability (Doran, 2002).

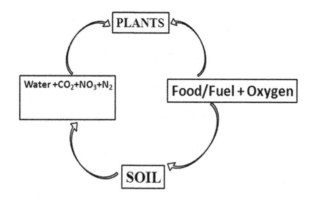

FIGURE 6.4

Simplified form of "equation of life."

FIGURE 6.5

Agricultural sustainability is analogous to a three-legged stool, main elements of which must be in stabilized to get long-term sustainability.

4. Challenges to environmental sustainability

Fig. 6.6 illustrates the four main challenges to environmental sustainability, namely human action, first order environmental changes, second order environmental consequences, and human development impacts. As shown in Fig. 6.6, these challenges are interlinked, which means that human actions, such as food production, will bring first-class environmental change that helps in the distribution of species.

Many human actions have an impact on what people value. One way in which actions can bring about change is by recognizing the consequences from a decadal scale to centuries. This aspect makes people more likely to be involved in taking measures on the present time to protect the core values of those who would be affected by environmental changes in future. Nonetheless, because of the confusion of how worldwide environmental processes operate, and because the people affected live in very different situations and may have different beliefs, it is difficult to know how current policies would impact them.

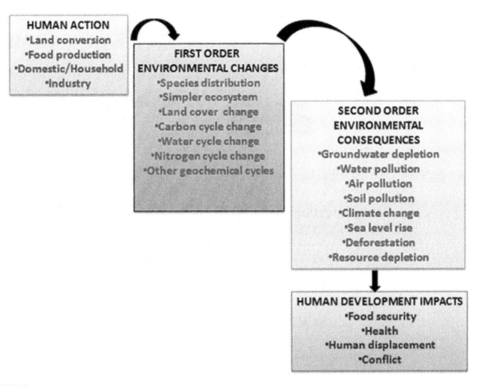

FIGURE 6.6

Challenges to environmental sustainability.

If environmental changes are not positive with nature, it will result in several environmental consequences, as stated by the environmental consequences of the second order. This will also involve the development of multiple human impacts, such as conflict, health issues, food safety, and human displacement.

The term "environmental sustainability" means the creation and maintenance of conditions under which both nature and humans can exist in productive harmony to meet current and future generations' natural, social, and economic needs (Fig. 6.7). In 2000 a report was published by the Royal Commission addressing problems in climatic change as well as its relationship to the supply of energy. The crucial principle recommendation of the study was that the CO_2 emission from anthropogenic activities in the United Kingdom should be decreased by about 60% from 1998 levels by 2050 (Clift, 2007). Throughout the world exposure to environmental pollution remains the main cause for the rise of health issues, and risks have been reported to be commonly high in developing countries. Deprivation, the deficit of advancement, or replacement of the old technology, as well as weaker environmental legislation combine to result in high pollution levels (Briggs, 2003).

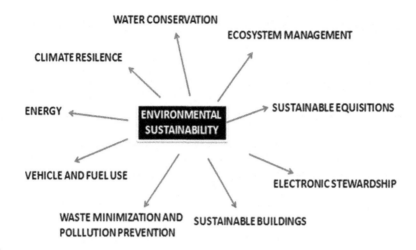

FIGURE 6.7

Environmental sustainability.

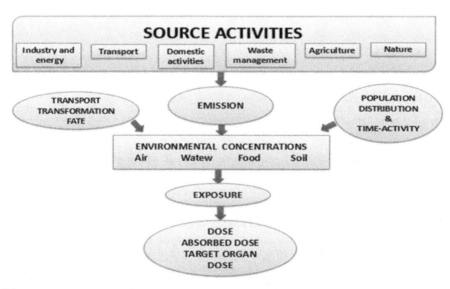

FIGURE 6.8

Source-effect chain.

5. Activities for attaining sustainability

Fig. 6.8 illustrates the source-effect chain. Emission to atmosphere tends to be more closely modeled as well as analyzed and generally listed, than any other media, partly due to its higher significance for environmental pollution as well as human health and also due to the presence of better-established policy as well as regulation. The amount of biological pollutants is often not the focus of such reports. Biological pollutants comprise the viable and living organisms like bacteria but may also include a vast array of the endotoxins, which arise from the protoplasm of the organism after death (Briggs, 2003).

Stockholm Declaration in 1972 aimed to develop a reference to sustainability in higher educations that would recognize an interdependency between humanity and the environment and also suggests various ways to achieve the environmental sustainability (Alshuwaikhat and Abubakar, 2008). United Nations Development Program's (UNDP) Human Development Index (HDI) focused toward higher comprehensive measures of human development and not on economic development alone. The UNDP does not negate that per capita income is a significant determinant of the country's level of social development but it also includes other socioeconomic indicators, like access to clean energy, food, and basic human needs like sanitation, health, and many others (Neumayer, 2001). It is easier to implement and design effective research and development based on this index, and some of the successful international programs exhibit several characteristics that have already been developed to address issues ranging from the increasing agricultural productivity to human health and environmental impacts of pollution (Strigl, 2003).

Based on World Bank data, CO_2 emissions have increased in the past 50 years from 3.09 to 4.99 in the world. In the past few years, sustainable hydrogen-based power generation system has become a major component of clean energy paradigm and sustainability. The most important reason is its low environmental impact. Varieties of hydrogen energy resources give a folding array of possibilities for its application (Midilli and Dincer, 2010).

6. Conclusion

For sustainable development and the current economic as well as political issues, it is mandatory to use a sustainable supply of fuels like hydrogen; also there is a need for practical efforts like ecofriendly policies. For global networking, research and training are considered to be best practice examples used to encourage the multidisciplinary research on environmental and human interactions that are affecting and are being affected by global changes. A sustainable society uses its human, economic, and financial capital to meet their current needs while ensuring that future generations have sufficient resources. Sustainability is essential for quite a simple and obvious reason: If we accept it, we could sustain our quality of life as human beings, the nature of life on earth. There are signs from all fields and from the smallest to the largest level that we have to address sustainability.

References

Agyeman, J., Evans, B., 2004. 'Just sustainability': the emerging discourse of environmental justice in Britain? Geogr. J. 170 (2), 155–164.

AlKhidir, T., Zailani, S., 2009. Going green in supply chain towards environmental sustainability. Global J. Environ. Res. 3 (3), 246–251.

Alshuwaikhat, H.M., Abubakar, I., 2008. An integrated approach to achieving campus sustainability: assessment of the current campus environmental management practices. J. Clean. Prod. 16 (16), 1777–1785.

Briggs, D., 2003. Environmental pollution and the global burden of disease. Br. Med. Bull. 68 (1), 1–24.

Brown, L.R., 1981. Building a Sustainable Society. WW Norton & Company, Inc., New York, NY, 500 Fifth Avenue, 10110.

Brubaker, S., 1972. To Live on Earth. Man and His Environment in Perspective.

Clark, W.C., Munn, R.E., 1986. Sustainable Development of the Biosphere. Cambridge University Press.

Clift, R., 2007. Climate change and energy policy: the importance of sustainability arguments. Energy 32 (4), 262–268.

Doran, J.W., 2002. Soil health and global sustainability: translating science into practice. Agric. Ecosyst. Environ. 88 (2), 119–127.

Escobedo, F.J., Kroeger, T., Wagner, J.E., 2011. Urban forests and pollution mitigation: analyzing ecosystem services and disservices. Environ. Pollut. 159 (8–9), 2078–2087.

Goodland, R., Daly, H., 1996. Environmental sustainability: universal and non-negotiable. Ecol. Appl. 6 (4), 1002–1017.

Heilbroner, R.L., 1991. An Inquiry into the Human Prospect: Looked at Again for the 1990s. WW Norton & Company.

Hills, P., Roberts, P., 2001. Political integration, transboundary pollution and sustainability: challenges for environmental policy in the Pearl River Delta Region. J. Environ. Plann. Manag. 44 (4), 455–473.

Howard, D.C., Wadsworth, R.A., Whitaker, J.W., Hughes, N., Bunce, R.G., 2009. The impact of sustainable energy production on land use in Britain through to 2050. Land Use Pol. 26, S284–S292.

Howe, C.W., 1996. Making environmental policy in a federation of states. In: Braden, J.B., Folmer, H., Ulen, T.S. (Eds.), Environmental Policy with Political and Economic Integration. The European Union and the United States, Cheltenham, pp. 21–34.

International Union for Conservation of Nature and World Wildlife Fund (IUCN), 1980. World Conservation Strategy: Living Resource Conservation for Sustainable Development. Gland, Switzerland.

Juknys, R., 2003. Transition period in Lithuania—do we move to sustainability? Energy 4 (26), 4–9.

Kelishadi, R., 2012. Environmental pollution: health effects and operational implications for pollutants removal. J. Environ. Public Health 2012.

Komiyama, H., Takeuchi, K., 2006. Sustainability Science: Building a New Discipline.

Levy, D.L., 1997. Environmental management as political sustainability. Organ. Environ. 10 (2), 126–147.

Midilli, A., Dincer, I., 2010. Effects of some micro-level exergetic parameters of a PEMFC on the environment and sustainability. Int. J. Glob. Warming 2 (1), 65–80.

Miller, T.R., Wiek, A., Sarewitz, D., Robinson, J., Olsson, L., Kriebel, D., Loorbach, D., 2014. The future of sustainability science: a solutions-oriented research agenda. Sustain. Sci. 9 (2), 239–246.

Neumayer, E., 2001. The human development index and sustainability—a constructive proposal. Ecol. Econ. 39 (1), 101–114.

Nijkamp, P., Opschoor, H., 1997. Urban environmental sustainability: critical issues and policy measures in a third-world context. In: Regional Science in Developing Countries. Palgrave Macmillan, London, pp. 52–73.

Rockstrom, J., Bai, X., DeVries, B., 2018. Global sustainability: the challenge ahead. Global Sustain. 1.

Sezen, B., Cankaya, S.Y., 2013. Effects of green manufacturing and eco-innovation on sustainability performance. Proc. Soc. Behav. Sci. 99, 154–163.

Sorgulu, F., Dincer, I., 2018. A renewable source based hydrogen energy system for residential applications. Int. J. Hydrogen Energy 43 (11), 5842–5851.

Strigl, A.W., 2003. Science, research, knowledge and capacity building. Environ. Dev. Sustain. 5 (1–2), 255–273.

The 2030 agenda and the push for electrification in Africa: a tale of two countries

7

Dan Banik
Centre for Development and the Environment, University of Oslo, Oslo, Norway

Chapter outline

1. Introduction

The idea of "sustainable development" has witnessed a resurgence in recent years following the adoption of the 2030 Agenda and its accompanying 17 Sustainable Development Goals (SDGs) by 193 heads of state in 2015 (United Nations, 2015). By closely linking "sustainability" with "development" through the principles of "universality," "integration," and "leave no one behind," the 2030 Agenda has been much celebrated in academic, activist, business, and policy circles as a means to stimulate a radical shift in world affairs (Banik and Miklian, 2017; Banik, 2018). But it has also been roundly criticized for its unrealistic ambitions and lack of focus (The Economist, 2015; Easterly, 2015). Despite the initial euphoria, progress on achieving the SDGs has thus far been mixed. Indeed, although there is growing awareness on sustainable development, the emerging consensus in most current academic and policy discussions is that the pace of action in achieving the SDGs is slow and that far too many people of the world continue to be excluded from the development process (United Nations, 2019b).

Environmental Sustainability and Economy. https://doi.org/10.1016/B978-0-12-822188-4.00011-7

There is now widespread agreement among scholars and policymakers that ensuring access to affordable, reliable, sustainable, and modern energy for all is a prerequisite for progress on all SDGs. And the world has witnessed some success in this regard, with the proportion of people in least developed countries with access to electricity more than doubling between 2000 and 2016 (United Nations, 2019a). Despite such progress, however, recent evidence clearly shows that the world is falling short of sustainable energy goals (ESMAP, 2019). The problem is particularly acute in sub-Saharan Africa (SSA), where sustainable and reliable power remains in short supply and over 570 million people continue to lack access to electricity (Chingwete et al., 2019).

This chapter examines the push for sustainable energy transitions in SSA as part of the growing interest on the implementation of the 2030 Agenda in general and SDG 7 in particular, which promotes sustainable energy.[1] The empirical focus is on the recent attempts at boosting electricity generation and access in Kenya and Malawi through coal-fired power plants. I examine how domestic energy policies balance social, economic, and environmental objectives. The key question is as follows: How and to what extent can governments promote policy coherence with the aim of integrating energy concerns with other sustainable development objectives?

The study builds on a range of sources and includes a review of national development plans, energy policy documents, media reports, project evaluations, and feasibility studies. These secondary sources were supplemented by over 40 interviews with key informants conducted in the period 2019–20. The majority of these interviews were undertaken in Malawi. One set of key informants included current and former officials involved in formulating and implementing energy policies. Another broad set of informants included representatives from the National Planning Commission, the judiciary, United Nations agencies, major foreign aid donors, the media, private sector actors, and the academic community. In March 2020, I visited two major sources of hydroelectric power in Malawi—Nkula and Tedzani power plants—that are run by the Electricity Generation Company of Malawi (Egenco). This facilitated in-depth discussions with engineers and technicians on water levels in the Shire river, the costs of dredging, and the viability of water transportation. These semi-structured interviews and informal discussions with stakeholders were conducted in Lilongwe, Blantyre, Zomba, Nkula, and Tedzani.

In Kenya, I interacted with journalists, civil society organizations, policymakers, and academicians working on issues of sustainable development in September 2018 and May 2019. Follow-up fieldwork in Kenya was not possible due to the ongoing coronavirus crisis. However, with the help of a research assistant based in Nairobi, I was able to review the major news reports, policy documents, and evaluations related to a major coal-fired power plant project that has been the subject of considerable controversy. Due to the politically sensitive nature of this research, all interviews have been anonymized.

1.1 Background

Despite the growing rhetoric in national and international circles on the urgency of solving Africa's energy woes, all available evidence points to slow and uneven progress. The abysmal record of energy poverty in SSA has major implications not just for economic growth but also the ability of the state to

---

[1]Cf. SDG 7: "Ensure access to affordable, reliable, sustainable and modern energy for all", see United Nations. Transforming our World. The 2030 Agenda for Sustainable Development. A/Res/70/1.

provide quality social services. SSA not only lags behind other world regions in terms of access rates but "the total number of people without electricity has increased in recent decades as population growth has outpaced growth in electrification" (World Bank, 2019).

While most SSA countries are struggling to boost production and access to electricity, a few have enjoyed notable success. An illustrative example is Rwanda, which despite its small and landlocked status, is often highlighted in the international discourse as a success story that others could emulate. It is one of the fastest electrified countries in the world despite having low access rates, and has recently been ranked fifth in making the transition to clean power (Bloomberg NEF, 2018). For the past 5 years, Rwanda has also been actively promoting renewable energy projects, including solar power plants and off-grid solutions, often through successful private-public partnerships. At the other end of the SSA electrification spectrum is Malawi, which is heavily dependent on foreign aid and continues to have one of the lowest electrification rates in the world (10% connected to the grid). While electricity only accounts for 3% of energy use in the country, 89% of Malawi's total energy supply consists of unsustainably sourced biomass, which has caused widespread deforestation (Government of Malawi, 2017). Thus, while both Rwanda and Malawi are small, poor, and landlocked, public policies aimed at promoting energy security in these two contexts greatly vary. And both appear to find it challenging to provide electricity access in rural areas.

A particularly vexing problem for many SSA countries relates to the environmental impacts of their energy policy decisions. In recent years, such challenges have received media attention in regard to societal protests over coal-powered power plants. Although supporters claim that coal is the "cleanest least costly option," critics wonder why Africa should construct such projects when coal is being phased out in large parts of the world.

1.1.1 Kenya

Kenya, like many others on the African continent, suffers from energy poverty. The government's Kenya Vision 2030 strategy (Government of Kenya, 2019) is aimed at transforming the country "into a newly industrializing, middle-income country providing a high quality of life to all its citizens by 2030 in a clean and secure environment" (Government of Kenya, 2007). In recent years, the government has made substantial investments in electricity generation. There is also considerable political interest in pursuing green energy initiatives such as the Lake Turkana Wind Power Project, the largest wind power farm in Africa. Despite this growing interest in green energy, the government has nonetheless pursued more traditional fossil fuel—powered projects. The goal of building east Africa's first coal-fired plant adjacent to the Lamu Archipelago on the northern coast of Kenya (a group of six inter-connected islands) with a $2 billion Chinese loan is a good example in this context (Source Watch, 2019). The 1050 MW Lamu Coal Power Station was proposed in the country's Least Cost Power Development Plan (LCPDP) in 2005 (Government of Kenya, 2010). After a public tender process, the contract was awarded to Amu Power Company in 2014 with plans for the construction process (estimated to last 21 months) to begin in 2015. The initial reaction in the country was generally positive and the Kenyan government claimed that the project would strengthen national energy security in addition to creating jobs and reducing poverty in the region. The coal plant, together with the development of Lamu port, was also sold to the public as an important part of the US$29.2 billion Lamu Port South Sudan Ethiopia Transport Corridor (Lapsset) program[2] which would promote

---

[2]http://www.lapsset.go.ke/.

economic development in Lamu and the entire coastal region of Kenya. The coal plant project, however, soon began facing resistance from environmental activists and members of the local community. It is currently suspended following an order by Kenya's National Environmental Tribunal which decided to stop the project in 2019 citing environmental concerns and a failure to adequately consult the public (Human Rights Watch, 2019).

1.1.2 Malawi

With an estimated 10.8% of the population connected to the grid, Malawi has one of the lowest electrification rates in the world. It mainly relies on hydroelectric power generated along the Shire river, which provides 384 MW of a total national power generation capacity of 438 MW (Power Africa, 2018). Over the years, Malawi has struggled to attract further investments that are required to upgrade existing hydropower plants as well as to build newer plants. The lack of power generation capacity in the country is a frequent source of frustration among the population. Even when one is connected to the grid in mainly urban areas, it does not mean one has continued access to electricity. In rural areas, access to electricity is even more limited. There is also considerable national debate on the adverse impacts the lack of electricity has had on the manufacturing sector and on economic growth and social development. Since over 96% of current electricity capacity originates from hydropower plants on just one river source—the Shire—and with the increased advent of droughts and lack of adequate rainfall, the country's ability to produce adequate power for its needs is severely restricted (Government of Malawi, 2017). Power cuts and loadsheddings are common for large parts of the day. The lack of adequate electricity generated from traditional sources such as hydropower has resulted in policymakers prioritizing short-term and ad-hoc solutions such as the introduction of diesel generators in recent years which provide around 55 MW. While several coal plants have been planned in Malawi over the years, most have either been canceled or shelved (Bloomberg NEF, 2018). Coal power plants have been promoted by the government as an alternative to expensive and unavailable investments in hydroelectricity and solar power. The Kam'mwamba Power Station in Neno district of southern Malawi, with a planned installed first phase capacity of 300 MW and with assistance from the China Energy Engineering Group Ltd. (CEEC), has been under planning since 2013. However, the project has faced numerous obstacles, most importantly related to lack of finances, and construction is yet to begin.

1.2 Achieving integration and coherence in policy-making on sustainable development

The 2030 Agenda and the 17 SDGs have revived interest in the idea of sustainable development understood as meeting the needs of the current generation without sacrificing the needs of future generations. The SDGs apply to all countries and are considered by many to offer a much-needed framework for promoting economic development while addressing environmental concerns. Although the SDGs are now in their sixth year of implementation, and overall global progress has been mixed, a particularly challenging feature facing most countries relates to achieving policy integration aimed at promoting one goal (e.g., reduced hunger, economic growth, poverty reduction, energy sufficiency, consumption or infrastructure) without compromising another (e.g., biodiversity, environment and climate-related goals). And while there is a growing international rhetoric which claims that achieving the SDGs will be impossible unless countries adopt an integrated approach toward policy formulation and implementation, the 2030 Agenda is less clear on how exactly governments can achieve such integration.

The term "policy integration" was popularized by the World Bank, organisation for economic cooperation and development (OECD), and UN agencies in the 1990s and has been variously understood as the creation of interdependence, cooperation, and coordination between actors representing various sectors and domains that may, on occasion, even result in shared decision-making (Metcalfe, 1994; Braun, 2008). There are numerous integration-related concepts in the government-centered and governance-centered literature. Such concepts in the policy studies literature include "joined-up government" with a focus on administrative cooperation (Christensen et al., 2014), and "horizontal governance" and "holistic governance" that highlight the role of networks in government units and application of technology to improve coordination (Bolleyer, 2011). Still others distinguish between "spatial" (policymakers connecting in a nonhierarchical manner across jurisdictions) and "temporal" features of integration (which entail making policymakers aware of the long-term impacts of their decisions) (Stevens, 2018).

A much talked-about concept in recent years has been that of "policy coherence," popularized by the OECD in the early 1990s. The concept has achieved considerable impact in advocating closer synergies between foreign aid and other national policies of affluent nations, in addition to shaping the discourse around the predecessor to the SDGs—the Millennium Development Goals (MDGs) (Tosun and Lang, 2017). Ensuring policy coherence, particularly in low income country contexts, entails a commitment to creating enabling environments, minimizing overlaps between administrative agencies, ensuring coherence of actions at all levels of government, and assessing the impact of policies (OECD, 2016). Thus, one must prioritize the creation of "institutional mechanisms and processes to harmonize and manage often competing policy objectives and interests" (OECD, 2016: 15). An emphasis on policy coordination is particularly important in order to "resolve conflicts or inconsistencies between policies" in addition to creating and strengthening "systems for monitoring, analysis and reporting on the impacts of policies to provide evidence to inform decision-making" (Tosun and Lang, 2017). Thus, policy coordination contributes to resolving energy insecurity concerns from governance and institutional perspectives.

The 2030 Agenda highlights the integrated nature of the 17 SDGs, which puts even greater responsibility on governments "to be able to work across policy domains, actors and governance levels" and requires "breaking out of sectoral silos and adopting integrated approaches to consider more systematically complex inter-linkages (such as the water-energy-food nexus), trans-boundary and intergenerational impacts, and trade-offs" (OECD, 2016: 20—21). Thus, governments must be capable of understanding how their policies and sectoral policies influence and impact on the various dimensions of energy security. These include access to electricity and other forms of energy (related to poverty reduction measures and focus on vulnerable groups and rural/urban residents), availability and stability of energy supply (policies that promote production and generation of electricity), and affordability issues (ensuring that household incomes are resilient to price shocks).

An interest in policy integration or policy coherence is not deeply engrained in the political leadership of most countries. Why is this the case? Le Blanc (2015: 186) argues that thinking about integration "across sectors and policy advice represents a challenge to the way development work is usually conducted." Indeed, most political leaders and the institutions they head, are accustomed to working in silos that further one's own interests even though it may hinder the work of a colleague or another institution working for the same government. This, in turn, requires, a reorganization of national administrations and government systems (Bierman et al., 2017). Some suggest that integration on sustainable development is best achieved if focused at the local community level (e.g., integrating agriculture, nutrition, and health of smallholder farmers) (Canavan et al., 2016).

Related to lack of political interest in furthering policy integration is the administrative architecture and competition among government units, and rivalry and distrust between officials on the one hand and local nongovernmental organizations and international agencies on the other, which create obstacles for shared vision and decision-making. Many practitioners as well as local officials I have interacted with in recent years in countries as diverse as India, China, Rwanda, Bangladesh, Kenya, and Malawi cite lack of capacity and expertise in addition to the lack of political commitment. They cite the problem of weak or nonexistent institutional setups that leave the planning, monitoring, and reporting to individual ministries rather than strengthening policy coordination (and thereby integration) at the highest political level (e.g., at the office of the prime minister or president). There are, of course, some exceptions. For example, Bangladesh has appointed a senior official as chief coordinator for SDG affairs in the prime minister's office, tasked with coordinating policies across government ministries. And there is strong political support for the SDGs in China, India, and Rwanda—where nodal agencies have been established at national, and sometimes at provincial, levels, although the policy impact such institutions will exert in such national contexts is still unclear. There are also concerns that not all societal sectors (e.g., including civil society) are included in policy dialogues. Similarly, it is often unclear how and to what extent new institutional arrangements aimed at promoting integrated policies on sustainable development (e.g., Niti Aayog in India) have broad political support and whether they will be continued with when a new political regime assumes power.

1.3 Formulating and implementing energy policy

There is growing talk in international forums on the need to intensify efforts to achieve sustainable development and the 17 SDGs, particularly those related to sustainable energy. But most countries in SSA are yet to begin mapping existing policies and programs against the 169 cross-cutting SDG targets. When engaging in energy policy-making and implementation, political and administrative leaders in such contexts often do not publicly mention "trade-offs." But in private many insist that they prefer prioritizing a few key goals among the 17 SDGs that they can fund rather than try to integrate all goals related to sustainable development. Indeed, while the global discourse on sustainable forms of energy increasingly highlights the importance of pursuing an integrated approach to economic, social, and environmental aspects of development, it often overlooks issues of local justice and messy local politics including competition between groups for control over scarce resources.

Most low-income countries continue to plan activities over a 5−10 year period, often guided by documents variously promoted as visionary, ambitious, and forward-looking. The day-to-day policy environment for ministers and civil servants, however, requires quick decisions and routine adjustments to such long-term plans. In addition to domestic pressures for income generation, efficient revenue collection, and the provision of quality social services, policymakers must be well-prepared to respond to international events and trends, including fluctuating prices of major imports and exports, and changing trade arrangements that can have a major impact on GDP growth. Added to these concerns is the growing threat of economic progress being adversely affected by climate change.

In recent years, the Malawian government has passed new legislation and undertaken a set of reforms in the energy sector. For example, the Malawi Energy Policy (2018) aims to "increase access to affordable, reliable, sustainable, efficient and modern energy for every person in the country." And the main goal of the Sustainable Energy for All (SEforAll) Action Agenda (2017) is to "provide access to modern energy services for all by 2030, through on- and off-grid electrification and improved cookstoves." And finally, the provision of "sufficient sustainable energy for industrial and socio-economic development" is a crucial goal of the revised Malawi Growth and Development Strategy III (2017), which includes a combination of hydroelectricity and solar power. The government has undertaken a set of reforms in the energy sector, which has included the restructuring of the national utility company—Electric Supply Company of Malawi (ESCOM) which was split into two companies in 2017. The newly established Electricity Generation Company of Malawi (EGENCO) is tasked with the operation, maintenance, and improvement of state-owned power plants. ESCOM is now responsible for procurement, transmission, and distribution of electricity to the consumer.

Similarly, focus on energy has been a key feature of Kenya's Vision 2030 plan. And in recent years, public policy on energy generation and distribution have highlighted the growing need not just for any type of energy, but for clean and affordable sources aimed at mitigating climate change. The government has involved several agencies including the Ministry of Energy, Kenya Electricity Generating Company (KenGen), the Electricity Regulatory Commission (ERC), Kenya Electricity Transmission Company Limited (KETRACO), and Kenya Power & Lighting Company (KPLC). It has also formulated a set of overarching plans and documents to guide the process forward. In addition to Vision 2030, this includes the Least Cost Power Development Plan 2010–2030 (LCPDP), which provides overarching guidelines for the strategic capacity development for the energy sector. Kenya has also expressed considerable interest in facilitating a transition to green energy, and renewable energy constitutes almost 70% of the country's current installed electric power capacity. At the 2018 Paris Peace Forum, President Uhuru Kenyatta expressed his government's commitment to transition to 100% green energy sufficiency within a short period. The completion of the Lake Turkana Wind Power Project—the largest wind power farm in Africa with the capacity to generate 310 MW—is a major step in realizing the goal of addressing Kenya's energy needs through renewable sources.

Thus, on paper, at least, both countries have displayed considerable interest in formulating energy policies aimed at addressing national energy deficits. Overall national plans and visions, together with specific policies related to energy, have been formulated and organizational reforms have been undertaken to address inefficiencies. Despite these efforts, however, electricity generation remains extremely limited in Kenya and Malawi. There are several explanations to this, which include the lack of policy space, inadequate funding mechanisms, inadequate consideration of environmental concerns, and a lack of adequate consultation with citizens and relevant stakeholders. I discuss each of these broad sets of issues below.

1.3.1 Policy space

The Global North has, in recent decades, been lukewarm to funding major infrastructure projects in low-income countries with foreign aid. As Milanovic (2017) has argued, affluent nations of the world have until recently basically withdrawn from the "hard" aspects of development and focused their development aid in recent decades on so-called "soft" areas such as direct budget support to governments and funding initiatives that focus on governance, transparency, and empowerment. A persistent complaint among many world leaders is the poor state of their roads and ports and the lack of interest in the international community to fund infrastructure projects. China has filled this gap through its ambitious Belt and Road Initiative (BRI), launched in 2013 and estimated to cost over \$5 trillion.[3] And Beijing has actively financed development projects in both Malawi and Kenya. But even before the introduction of the BRI, Beijing wielded great influence in many African countries through its investments and aid, which largely consists of grants and concessional loans. It touts its aid and investment policies as representative of a set of interrelated general principles, such as win-win outcomes, noninterference in the domestic affairs of other countries, mutual respect, and friendship.

However, Chinese aid differs significantly from Western aid. China does not abide by the Paris principles, which aim to achieve better impacts by formulating aid policies around five pillars.[4] Unlike aid and development assistance from most Western countries, Chinese aid does not come with stringent conditions for improving local governance, strengthening women's rights, or combating corruption. Indeed, China's overall strategy is to present itself as a country driven by the conviction that state-to-state relations ought to benefit both parties (Banik, 2019). When Beijing makes decisions about aid and investment projects around the world, it strives to present itself as not questioning the legitimacy of the governments in recipient countries. However, despite the growing prominence and novel approach of Chinese foreign aid, it suffers from challenges and limitations that constrain its effectiveness and can distort its economic impact in some respects. One major concern that the citizens of some recipient countries have expressed is that certain Chinese-funded projects may not be suitable for or in accordance with local needs. Some also criticize the one-size-fits-all model that Beijing employs in large parts of the African continent despite the continent's wide range of different sociopolitical conditions and cultural traditions. Another issue is the lack of transparency surrounding the terms of Chinese investment decisions.

China has also been stepping-up its activities in the energy sector and has provided loans and helped build several energy projects including coal-fired power plants and hydroelectric projects. It has been involved in the energy sector in both Malawi and Kenya, and the promotion of coal-fired power plants has been at the forefront of this push by Beijing to achieving energy security in these countries. This preference for coal, rather than renewal energy, has been criticized in some circles in both Kenya and Malawi as in other SSA countries. For example, in my interactions with several civil society activists and environmental journalists in Malawi and Kenya, Beijing was accused of transferring its coal technology and equipment to Africa while it phases out the use of coal within its own territory. They also pointed to the fact that Kenya and Malawi have been specifically selected by Beijing for coal-related investment projects. Others, especially journalists and scholars, blamed their own national government for failing to stand up to Chinese interests. They claimed that in their desperation to improve the production of and access to electricity, politicians in Kenya and Malawi were unwilling to press for clean power solutions.

---

[3] The State Council of the People's Republic of China (undated), The Belt and Road Initiative, http://english.www.gov.cn/beltAndRoad/.

[4] These pillars include: greater national ownership over aid recipients' self-identified development priorities, increased alignment between donor goals and recipients' national policy priorities, harmonization of aid from different donors to improve coordination, better management of aid to achieve measurable results, and greater accountability on the parts of both donors and recipients for failed interventions.

Political leaders in Malawi have, over the years, complained to me of the lack of policy space for them to make independent decisions when they are reliant on the generosity of external actors such as bilateral foreign aid donors and international development agencies. Ensuring national and local "ownership" to policies and projects, they argued, is difficult because they do not control the purse strings. And without the ability to attract investments from alternative sources, for considerably more expensive projects related to hydroelectricity, these politicians believed that their options were limited. Moreover, several Malawian officials and engineers told me that coal plants were not only cheaper, but could also be built far quicker than hydroelectric power stations. In addition, they highlighted the difficulty of attracting investments from Western capitals for environment-friendly power projects, which in turn makes Malawi increasingly dependent on the Chinese in relation to policies aimed at strengthening energy security.

1.3.2 Funding

Although some estimates find that achieving the SDGs globally will require at least USD 45 trillion (OECD 2017), a key question that has dogged the 2030 Agenda since its inception relates to funding ambitious programs in low-income countries. Three main sources of funding for the SDGs are typically identified: own funding by poor countries through taxation, foreign aid from developed countries, and private sector finance (Rudolph, 2017; Banik and Lin, 2019). But there are growing concerns that many of the poorest countries will not be able to self-finance programs and that recent threats to aid from the Global North will stall efforts to advance the SDGs—thus making a stronger case for involvement of the private sector and the trillions of dollars for business opportunities that the SDGs open up (Business Commission, 2017). Thus, the availability of funds and investments for power generation and distribution is crucial. Securing domestic financing for major energy infrastructure projects is, of course, important. However, the private sector in many low-income countries is often small and without technical expertise or the financial muscle to undertake major energy projects. This is certainly the case for Malawi. And although Kenya has a much larger economy and a thriving private sector, very few businesses have the capital or the technical knowhow to undertake major infrastructure projects. Attracting investments from external actors with proven expertise is thus crucial.

Neither of these tasks—securing domestic and foreign investments—are however easy. They are not easy because they may come with conditionalities, some of which may be politically unpalatable or when conditionalities are detrimental to the livelihoods of local populations. An illustrative example is a loan, which although concessional in nature, would nonetheless have to be repaid at some point of time. Whether an investment worth several hundred million dollars will pay off in the long-run is often difficult to ascertain with precision given the unpredictability associated with future demand, potential spillover effects in the economy, and costs related to maintenance and expansion. Consequently, thorough homework in the form of impact assessments and feasibility studies must be undertaken by impartial actors and institutions.

Let us consider the case of the proposed coal-fired Kam'mwamba Power Station in Neno district of southern Malawi with a planned installed first phase capacity of 300 MW. The Government of Malawi and the China Energy Engineering Group Ltd. (CEEC) began negotiations in 2013 for the power plant that would be built by a subsidiary of CEEC—the China Gezhouba Group International Engineering Company. The plan included securing partial funding for the $667 million project through domestic revenue generation in addition to a loan from the Export-Import Bank of China. The Malawian government's ownership of the project was set at 85%.[5] The initial plan was to use local sourced coal. However, news reports subsequently claimed that coal would be sourced from the Moatize coalfields in Mozambique. The project failed to get started. According to several news reports, a major hurdle was that Malawi has not been able to generate the resources (an estimated $104 million) required under the collaboration agreement. Additionally, the project was delayed as a proper feasibility study was not undertaken and a Power Purchase Agreement was not in place.

In April 2019, the Chinese ambassador to Malawi told journalists that the project was "still alive" and that "we are doing some feasibility studies on some infrastructure like roads, railways systems and other things to enable easy transportation of coal as well as distribution."[6] Thus, it came as a major surprise when the Malawian government announced in October 2019 that it had withdrawn from its agreement with the China Gezhouba Group International Engineering Company and had "relocated the project to Energy Generation Company (Egenco) to implement it with government as the sole financier."[7] The official reason for this decision, according to the former CEO of the power station, was "lack of financial closure" as Malawi did not receive a firm commitment on how, and the extent to which, the project would be financed. No mention was made by Malawian officials, however, of any problems or challenges in relation to their Chinese counterparts. The construction was yet to begin as of May 2020 and it is unclear to what extent the Malawian government has generated financial resources to fund this expensive project on its own. Indeed, despite repeated queries to Malawian journalists, scholars, civil society activists, and politicians, it was not possible to elicit further information regarding the status of this project.

The Kenyan government's proposed construction of the Lamu coal power station was anchored in the 20-year Least Cost Power Development Plan (LCPDP) that it had formulated in 2005 (Government of Kenya, 2010). After tenders were invited for this public-private partnership project in 2014 and in 2015, the government selected Amu Power Company—a Kenyan, Omani, American, and Chinese consortium—to construct the power plant estimated to cost at least $2 billion. Financing for the project was secured after the consortium signed a credit agreement for $900 million with the Industrial and Commercial Bank of China (ICBC) and with local lenders such as the Standard Bank of South Africa. The project was deemed by the Chinese authorities to be an important part of its BRI with the goal of solving "the power shortage for millions in the region."[8] Construction was planned to begin in September 2015 and expected to be complete in less than 3 years.[9] The actual task of construction was assigned to the Power Construction Corporation of China. The consortium intended to recover its investment by selling electricity until the year 2050 based on the build-own-operate model (Wang, 2019).

Despite a few hiccups along the way (such as the withdrawal of the Standard Bank of South Africa from the project in 2017), a Power Purchase Agreement was eventually put in place. The Kenyan president met his Chinese counterpart in Beijing in May 2017 and an agreement to begin construction

[5]"Kam'mwamba project to start Nov", The Nation, 29 September 2017.

[6]Malawi: Chinese Ambassador says Kam'mwamba 200 MW Coal-Fired Power Plant Still On", 26 April 2019, https://www.iea-coal.org/malawi-chinese-ambassador-says-kammwamba-300mw-coal-fired-power-plant-project-still-on/.

[7]"Govt to go solo on Kam'mwamba", The Nation, 26 October 2019.

[8]"Kenyan court blocks China-backed power plant on environment grounds", Financial Times, 17 June 2019, https://www.ft.com/content/9313068e-98dc-11e9-8cfb-30c211dcd229.

[9]"Construction of coal-powered plant to begin in September", The East African, 28 March 2015, https://www.theeastafrican.co.ke/business/Construction-of-coal-powered-plant-to-begin-in-September/-/2560/2668674/-/bd73kmz/-/index.html.

in September 2017 was signed with the Power Construction Corporation of China.[10] However, a protest movement by local members of the community and environmental activists soon began to gather momentum. The Kenyan government also faced international criticism for causing damage to the UNESCO world heritage site in Lamu and for going ahead with a highly polluting project.

Both Malawi and Kenya continue to be increasingly dependent on Chinese finance for such projects. Unlike in Malawi, where there was no "financial closure" for the coal plant, a complex financing system together with a Power Purchase Agreement was in place in the Kenyan case. Although it is not entirely clear what exactly resulted in Malawi withdrawing from its planned collaboration with Chinese partners in the Kam'mwamba Power Station project, the Malawian government's inability to provide its share of financial resources appears to be a major cause for the disruption of the original plan. It is unclear to what extent the Kenyan government had other options available, particularly since two of the world's most influential actors in the region—China and the United States—believe that coal is the best solution for addressing Kenya's energy needs.[11] Moreover, the African Development Bank has recently decided not to provide loans to Kenya and other member countries for any future coal-fired plants.[12]

In recent months, there have been several new funding pledges by agencies such as the African Development Bank and the Export-Import Bank of the United States aimed at promoting electricity access on the continent. While some of this funding is earmarked for small-scale solar systems and minigrids, others are aimed at boosting access to electricity in rural areas. But more is needed. Not just money, but also technology and sound advice that makes a persuasive case for clean energy.

1.3.3 Environmental impacts and civil society activism

While civil society organizations in Malawi remain largely free to operate and are influential actors in the country's political discourse, the voices of those championing environmental issues, especially related to energy policy, remain muted. The general societal agreement is that the government ought to do all it can to promote electricity access and coverage in addition to increasing production to ensure uninterrupted power supply. As such, environmental concerns in relation to power generation do not necessarily receive the kind of social and political attention they deserve. However, this does not mean that there is no interest in promoting the environment. Malawi has enacted legislation that bans single-use plastics, and there is growing attention in the media on the perils of climate change. However, since the country mainly relies on clean hydropower projects for electricity, environmental activism in the energy sector has been limited. Most activists tend to focus on issues such as food security and economic growth, in addition to the strengthening of democratic freedoms. The proposed Kam'mwamba coal power station project did receive some criticism for sourcing coal from neighboring Mozambique. However, even in this case, the criticism was not related to the pollution dimension but pointed to increased costs.[13]

---

[10]"Row over Chinese coal plant near Kenya World Heritage site of Lamu", BBC News, 05 June 2019, https://www.bbc.com/news/uk-48503020.

[11]"The Trump Administration Protested when Kenya Halted a Coal-Fired Power Plant", https://bit.ly/2PbKD8z.

[12]"African Development Bank decides not to fund Kenya coal project", Reuters, 13 November 2019, https://reut.rs/36qVAsM.

[13]Malawi: Chinese Ambassador says Kam'mwamba 200 MW Coal-Fired Power Plant Still On", 26 April 2019, https://www.iea-coal.org/malawi-chinese-ambassador-says-kammwamba-300mw-coal-fired-power-plant-project-still-on/.

In contrast, the Lamu coal power station project in Kenya has elicited substantial pushback from civil society organizations and environmental activists. Ever since the initial reports of the project emerged, there has been a wave of protests citing concerns over revenue-sharing arrangements and the real threat of increased pollution, endangered marine species, destruction of mangroves, air pollution, destruction of a UNESCO World Heritage Site, and a 700% increase of Kenya's annual emissions of greenhouse gases. Sustained protests by a diverse group of actors and organizations have marred the construction of the power plant. In 2009, various individuals and groups joined forces to create a coalition under the *Save Lamu* banner. As this movement grew in momentum, the coalition grew from 12 to over 40 local organizations, including development organizations, environmental groups, and organizations working for the welfare of women and the youth. Over the years, *Save Lamu* has raised awareness on the environmental and social challenges facing the region and the importance of policy-making that pays adequate attention to community knowledge, practices, and rights over natural resources. Initiatives such as *Save Lamu* and other environmental and social justice organizations became a part of the deCOALonize campaign—a movement that—with the help of community engagement, public activism, and legal advocacy—is opposed to coal-related industrialization in Kenya and the region. The overall focus of the efforts of such coalition groups was to promote sustainable development. The role of international organizations in partnership with national and local organizations was also crucial to the work of the protest movement.

The concerns expressed by civil society organizations and activists related to numerous aspects of the project. One set of issues related to the lack of community involvement in the project, the disruption of local livelihoods, and destruction of the Lamu heritage. Thus, a community representative from *Save Lamu* listed, in October 2018, the numerous grievances of the local community: "a lack of public participation, a lack of environmental assessment and management plans, and failure to take into account traditional fishing rights and the right to the protection of our cultural heritage" (Ahmed, 2018). Another set of issues concerned the exact terms of reference, the type of negotiations that took place between Kenyan authorities, private investors, and Chinese actors. There were concerns regarding the costs incurred and the huge financial burden on the government and the tax payer every year. Thus, *Save Lamu* and the deCOALonize Lamu coalition have argued that despite government promises of a major increase in jobs, most of this (around 1400) would take place during the construction process and only 400 jobs would be needed once the plant began its operations. Omar Mohamed Elmawi from *Save Lamu* has claimed that the loss of livelihoods of fishermen and others (around 6000) would be far higher than the number of new jobs created by this project.[14]

But the most persuasive arguments concerned the adverse environmental and health impacts of continued usage of coal. Supporters of the project, including the Kenyan and Chinese governments, have long promoted the idea of "clean coal." Organizations such as Haki Africa have, however, argued that Kenya's attempts to address climate change through mitigation should not rely on coal, as irrespective of claims to being clean, coal has devastating effects on the environment all over the world. *Save Lamu* and their allies organized protests with slogans proclaiming that "Clean coal is a lie," and "Don't be fooled by fossil fool. Coal is not cool," urging the government to put the lives of "Kenyans

---

[14]https://www.mixcloud.com/AEPEP/episode-1-omar-mohamed-elmawi-from-save-lamu-on-energy-anti-coal-resistance-alternatives/.

before coal" (Dahir, 2018). Some scholars have also questioned the impartiality and credibility of the Environment and Social Impact Assessment (ESIA) study that was conducted.[15]

Legal experts also played a key role in the Kenyan case. For example, the Katiba Institute—based in Nairobi, and which undertakes Public Interest Litigation (PIL) and promotes knowledge on constitutionalism in order to better facilitate the implementation of the country's constitution and thereby achieve social transformation—played a crucial role. With legal representation from the Katiba Institute, *Save Lamu* appealed against the National Environment Management Authority's (NEMA) approval of the proposed coal-fired power plant. The appeal contended that NEMA had violated the Constitution, the Environmental Management and Coordination Act, as well as EIA Regulations in issuing the license to Amu Power Company. Kenya's National Environmental Tribunal decided in June 2019 to stop the project citing environmental concerns and a failure to adequately consult the public.[16] The Tribunal concluded that Kenya's NEMA had failed to comply with the law in issuing the Environmental Impact Assessment license to Amu Power Company. In revoking the license issued to Amu Power Company, the Tribunal concluded that NEMA had approved the ESIA study without public consultations.[17] It also pointed to several flaws and missing elements in the ESIA study, including the extent of pollution that would result from coal, dust, and ash produced by the plant and the extent to which these would affect humans, plants, animals, and marine life (UNEP 2019). Thus, for the project to proceed, the company would have to successfully appeal the ruling and/or carry out a fresh ESIA before reapplying. While celebrating this victory, Kenyan activists do not believe they have achieved total closure, as there is a risk that construction activities will once again resume in the near future. However, the legal fight has showcased the desire of locals that their leaders should prioritize renewable sources of energy rather than coal-powered solutions.

2. Conclusion

Large parts of the African continent continue to face an acute shortage of energy and hundreds of millions of people do not have any access to electricity. As governments struggle to formulate and implement energy policies aimed at boosting power generation and making electricity more accessible, they are increasingly being forced to make hard decisions and address numerous tradeoffs. While the global discourse on sustainable development increasingly highlights the importance of pursuing an integrated approach to economic, social, and environmental aspects of development, it often overlooks issues of local justice and messy local politics including competition between groups for control over scarce resources.

The cases from Malawi and Kenya highlight the numerous challenges related to the integration of the various SDGs and achieving coherence in formulating and implementing public policy on sustainable development. While the SDGs are not legally binding, governments are expected to take ownership and establish national frameworks for the achievement of the 17 Goals. Recent evidence, however, indicates the need for greater efforts across the goals without losing focus on poverty

---

[15]"At what cost will Lamu coal power plant be to Man and sea?", Daily Nation, 09 April 2018, https://mobile.nation.co.ke/news/In-Lamu–a-coal-power-plant-faces-opposition/1950946-4378174-item-1-n9sjs0/index.html.

[16]Human Rights Watch (2019) "Tribunal Stops Kenya's Coal Plan Plans", https://bit.ly/2sge5RJ.

[17]Republic of Kenya in the National Environmental Tribunal at Nairobi Tribunal Appeal No. Net 196 of 2016, http://www.katibainstitute.org/wp-content/uploads/2019/06/Save-Lamu-Judgement.pdf.

reduction. Operationalizing the SDGs requires a clearer understanding of the interconnected yet distinct role of national governments, international agencies, and businesses. This is particularly urgent in low income countries and conflict-prone fragile states, which are confronted with the dilemmas and potential pitfalls associated with coordinating the activities of numerous competing actors while at the same time securing financing for ambitious projects. Consequently, thorough research in the form of reliable and credible participatory impact assessments and feasibility studies must be undertaken by impartial actors and institutions. But even when such studies are available, the decision to choose among a list of alternatives may not be easy, particularly when such alternatives do not differ much from one another. Achieving policy coherence in energy policy-making and implementation appears to be the key.

Another key issue relates to state capacity and ability of local public administrations to identify, articulate, coordinate, and implement development programs aligned with the national interest, while also making it sufficiently attractive for both domestic and international actors to become involved in SDG-related activities. Governments must therefore develop the capacity to identify mere profit-making initiatives that can thwart sustainable development objectives, especially as it is questionable whether low-income country governments will be able to say no to powerful business interests when they come calling with enticing offers of help. Much of the current discourse on the 2030 Agenda and the SDGs does not adequately address the role of politics that underlies the sustainable development framework. In addition to interdisciplinary research and exploration of new and effective mechanisms for policy planning and implementation, intersectoral collaboration and knowledge sharing appear to be crucial. There must also be a greater focus on better understanding the incentives for politicians to use available scientific evidence to promote an integrated approach to sustainable development.

Acknowledgment

I wish to thank Victor Onyango, Hannah Swila and Kaja Elise Gresko for research assistance and two anonymous referees for helpful comments.

References

Ahmed, R.F., 2018. Keynote Address by Raya Famau Ahmed, Community Representative from Save Lamu Organisation, Delivered during the 2018 ACCA General Assembly, Held from 10—11 October 2018 in Nairobi, Kenya. http://www.accahumanrights.org/images/ACCA_GA_2018_/Keynote_address_Ms_Raya_Ahmed.pdf.

Banik, D., 2018. Taking stock of the SDGs. In: Deepak, D., Pooran, P. (Eds.), Leaving No One behind: SDGs and South-South Cooperation. Crossbill, New Delhi.

Banik, D., 2019. Coordinating Chinese Aid in a Globalized World. Research Article. Carnegie-Tsinghua Center for Global Policy, Beijing. https://carnegietsinghua.org/2019/01/06/coordinating-chinese-aid-in-globalized-world-pub-78058.

Banik, D., Lin, K., 2019. Business and morals: corporate strategies for sustainable development in China» (with Ka Lin). Bus. Polit. 21 (4), 514—539.

Banik, D., Miklian, J., 2017. New Business: The Private Sector as a New Global Development Player. Global Policy. https://bit.ly/2rrHVmq.

Biermann, F., et al., 2017. Global governance by goal-setting: the novel approach of the UN sustainable development goals. Curr. Opin. Environ. Sustain. 26–27, 26–31.

Bloomberg NEF, 2018. Emerging Markets Outlook 2018: Energy Transition in the World's Fastest Growing Economies.

Bolleyer, N., 2011. The influence of political parties on policy coordination. Governance 24 (3), 469–494.

Braun, D., 2008. Organising the political coordination of knowledge and innovative policies. Sci. Publ. Pol. 3584, 227–239.

Business Commission, 2017. Better Business Better World. https://goo.gl/2bXZ9q.

Canavan, C.R., et al., 2016. The SDGs will require integrated agriculture, nutrition, and health at the community level. Food Nutr. Bull. 37 (1), 112–115.

Chingwete, A., Felton, J., Logan, C., 2019. Prerequisite for Progress: accessible, reliable power still in short supply across Africa. Afrobarometer Dispatch 1–23. No. 334. https://bit.ly/2PDfaLD.

Christensen, T., et al., 2014. Joined-up government for welfare administration reform in Norway. Publ. Organ. Rev. 14 (4), 439–456.

Dahir, A.L., 2018. A Chinese Coal Plant on a UNESCO-Protected Island in Kenya Is Facing Major Protests. Quartz Africa, 9 June 2018. https://qz.com/africa/1301469/photos-kenya-environmentalists-protest-china-coal-plant-in-lamu/.

Easterly, W., 2015. The SDGs Should Stand for Senseless, Dreamy, Garbled. Foreign Policy. https://bit.ly/2rzSqE8.

ESMAP, 2019. The Energy Progress Report 2019. https://esmap.org/2019_sdg7_report.

Government of Kenya, 2019. Kenya Vision 2030, Updated Least Cost Power Development Plan. Study Period 2011–2031. https://bit.ly/3b3KSKI.

Government of Kenya, 2007. https://vision2030.go.ke/.

Government of Kenya, 2010. Kenya Vision 2030, Updated Least Cost Power Development Plan, Study Period 2011–2031. https://bit.ly/3b3KSKI.

Government of Malawi, 2017. Malawi Renewal Energy Strategy. https://bit.ly/2YF29Fu.

Human Rights Watch, 2019. Tribunal Stops Kenya's Coal Plan Plans. https://bit.ly/2sge5RJ.

Le Blanc, D., 2015. Towards integration bat last? The sustainable development goals as a network of targets. Sustain. Dev. 23, 176–187.

Metcalfe, L., 1994. International policy coordination and public management reform. Int. Rev. Adm. Sci. 60 (2), 271–290.

Milanovic, B., 2017. The West Is Mired in 'soft' Development. China Is Trying the 'hard' Stuff. The Guardian, 17 May 2017. https://www.theguardian.com/global-development-professionals-network/2017/may/17/belt-road-project-the-west-is-mired-in-soft-development-china-is-trying-the-hard-stuff.

OECD, 2016. Better Policies for Sustainable Development: A New Framework for Policy Coherence. OECD, Paris.

OECD, 2017. Where to Start with the SDGs. https://oecd-development-matters.org/2017/07/20/where-to-start-with-the-sdgs/.

Rudolph, A., 2017. The Concept of SDG-Sensitive Development Cooperation: Implications for OECD-DAC Members, Discussion Paper 1/2017, German Development Institute; SDG Africa (2017) "SDG Financing for Africa: Key Propositions and Areas of Engagement. https://goo.gl/dBjFex.

Source Watch, 2019. Lamu Power Project. https://bit.ly/38s91u1.

Stevens, C., 2018. Scales of integration for sustainable development governance. Int. J. Sustain. Dev. World Ecol. 25 (1), 1–8.

The Economist, 2015. The 169 Commandments. https://goo.gl/Ma73i6.

Tosun, J., Lang, A., 2017. Policy integration: mapping the different concepts. Pol. Stud. 38 (6), 553–570.

UNEP, 2019. Lamu Coal Plant Case Reveals Tips for Other Community-Led Campaigns. https://www.unenvironment.org/news-and-stories/story/lamu-coal-plant-case-reveals-tips-other-community-led-campaigns.

United Nations, 2015. Transforming Our World: The 2030 Agenda for Sustainable Development. https://bit.ly/2PBqg3w.

United Nations, 2019a. Sustainable Development Goals Report 2019. https://bit.ly/2LJ5qho.

United Nations, 2019b. Global Sustainable Development Report 2019. https://bit.ly/35bOQP1.

Wang, C.N., 2019. Kenya's Lamu Coal Fired Power Plant: Lessons Learnt for Green Development and Investments in the BRI. Green Belt and Road Initiative Center. https://green-bri.org/kenyas-lamu-coal-fired-power-plant-lessons-learnt-for-green-development-and-investments-in-the-bri.

World Bank, 2019. Electricity Access in Sub-saharan Africa: Uptake, Reliability, and Complementary Factors for Economic Impact. https://bit.ly/2RJaNkG.

Indicators and sustainability

Economic indicators for material recovery estimation

Matan Mayer

School of Architecture and Design, IE University, Segovia, Spain

Chapter outline

1. Introduction

It is widely agreed that recycling and other material recovery techniques are essential for a global transition from linear to circular consumption patterns (MacArthur, 2013). In order to shift consumption patterns to materials that have higher recovery yields, manufacturers and policymakers need to be able to universally estimate and compare recyclability rates. Even though material recovery for recycling and energy production purposes has been extensively practiced in many industries for a number of decades now, there still seems to be little agreement about a single effective method to rank and compare materials based on their recovery potential (Hiroshige et al., 2001; Maris and Froelich, 2013; Li et al., 2018). While this situation might be driven by major manufacturers looking to maintain ambiguity regarding the actual recycling potential of their products, it might also be a result of real challenges in quantifying the efficiency of complex recycling processes in an industry-wide standardized manner (Sarja, 2002). Additionally, end-of-life solutions for materials and products may assume many forms of recovery—from direct reuse, through remanufacturing, upcycling, recycling, and downcycling, to energy recovery. The distinction between one solution and the other in terms of recovery quality is in many cases unclear and the terminology that is used in popular press to depict recovery solutions typically labels all operations as "recycling." In effect, end-of-life solutions exist on a continuous spectrum between direct reuse on one end and landfilling on the other with no established and agreed upon divisions.

Possible implementation of recyclability assessment in the construction industry adds even further complexities to these issues as many building materials, such as concrete, for example, are applied in-situ, far from the controlled environment of a production facility, and thus diminish or add to the recovery potential of the component in the future. Developing a universal method for measuring recovery potential and distinguishing between different levels of recovery are essential in order to inform material usage in circular economy scenarios. Within this context, this chapter presents the development of a recovery potential assessment method that is simple to implement in a range of geographical contexts and sectors. It begins with a conceptual background with regards to existing recovery technologies, followed by a description of the method, its implementation in practice, and lastly, a discussion regarding proposed extensions that make the method more applicable for a multidimensional assessment of building materials.

2. Background

Prior to expanding on the methodological foundations of recovery potential evaluation, it would be beneficial to provide a conceptual background about material recovery techniques and technologies. Those vary widely from one industry to another. As a general observation, industries that are tied to products with relatively short life spans (for example, beverage containers, print press, or cell phones) tend to have more efficient recovery supply chains and therefore more advanced recovery technologies. On this spectrum, the construction industry positions closer to the slower end, where long service life spans translate to inefficient recovery processes. Apart from nonferrous metals (e.g., aluminum) and ferrous alloys (e.g., steel), most of the common construction material groups in the world are highly problematic from a recycling standpoint. The molecular structure of anisotropic materials such as timber and composites such as concrete prevents them from being recycled into materials of equivalent properties to their virgin form. Such materials can either be downcycled into substances of inferior properties or incinerated for energy recovery. That said, recycled building materials play an increasingly important role in new construction. The following is a survey of available technologies and uses for six of the most prevalent material groups in the industry (Table 8.1).

Concrete—Concrete is mostly crushed into rubble to be used in two common forms: Bound, for use as virgin aggregate replacement in new concrete, and unbound, for use in road construction and landscaping. About 90% of the global concrete-recycling industry is oriented to unbound applications (Tam et al., 2006).

Masonry—One of the major challenges associated with brick recycling is contamination. Mortar tends to attach to the coarse surface of the brick in a manner that renders smooth serration almost impossible. Lime-based mortar, which was widely used until the 1930s, is a much easier binder to remove than its contemporary replacement, cement-based mortar (Nordby et al., 2009). Weathered clay facing bricks are often salvaged for their appearance and high market value. Other brick types may be crushed into aggregates to be used in new clay bricks as well as in sodium silicate brick production.

Ferrous metals—Steel lends itself well to preconsumer or postconsumer recycling. If processed in a correct manner, recycled steel can most often come close to matching the properties of virgin steel.

Table 8.1 Summary of typical end-of-life treatments and applications of building materials.

Material group	Treatment	Reuse	Recycle	Downcycle	Incineration	Applications
Concrete	Crushing			●		Bound: Concrete aggregate
				●		Unbound: Road construction
Masonry	Restoration	●				Facing brick
	Crushing			●		Sodium silicate aggregate
Ferrous metals	Melting		●			Same as virgin applications
Glass	Grinding			●		Thermal/acoustic insulation
				●		Filler
				●		Tiles and panels
				●		Paving blocks
	Crushing			●		Road and concrete aggregates
Nonferrous metals	Melting		●			Same as virgin applications
Timber	Burning				●	Energy recovery
	Hydrolysis			●		Cellulostic ethanol production
	Grinding			●		OSB and MDF production
				●		WPC production
				●		Mulch production
				●		Binderless fiberboard production

The significance of the symbol (●) is an indication of the type of material recovery for each treatment.

As is the case with the production of virgin steel, steel recycling is an energy-intensive process which releases large amounts of pollutants into the atmosphere (Ashby, 2012).

Glass—Unlike most other common construction materials, glass is widely used in consumer products as well. As a result, there are relatively well-developed recycling technologies for glass, although much like other ceramics, glass loses many of its virgin properties during the recovery process. End-of-life glass is typically crushed into either coarse or fine powder, suitable for a range of applications in the construction industry: (a) Thermal and acoustic insulation: recycled glass can be processed into fibers which are formed into insulation wool or embedded into resin. (b) Filler: glass powder can be used to strengthen concrete mixes. One example is ConGlassCrete, a product developed at the University of Sheffield (Zhu, 2004). (c) Tiles and translucent panels: glass typically loses much of its transparency during the recycling process. As a result, a common application of recycled glass is the production of wall tiles or translucent façade components. (d) Paving blocks: rough glass powder can be consolidated to create blocks which have been proven to reduce water absorption and provide enhanced compressive strength. A major challenge in the production of recycled glass blocks is the alkali-silica chemical reaction which occurs if a cement binder is used. (e) Road and concrete aggregates: recycled glass aggregates can be used with asphalt for road construction or with concrete for building construction. Cross-contamination resulting from impurities in the glass is a common problem in this application field. Finally, it should be noted that while recycled glass finds many applications in construction, its recycling is expected to continue to be increasingly challenging, primarily due to the common use of embedded films and composites in high-performance glazing products.

Nonferrous metals—Nonferrous metals are perhaps the most recyclable group of construction materials. The main nonferrous metals used in construction are aluminum, copper, and zinc. All three have been recovered in high percentages in recent decades. Much like ferrous metals, nonferrous metals can be restored their virgin properties through correct recycling.

Timber—Similarly to many other building materials, end-of-life options for timber waste are limited to downcycling and energy recovery. The one uncommon exception is refurbishment and reuse of salvaged timber, which is highly dependent on the type of wood used and the condition of the salvaged component. End-of-life timber is typically processed into pellets and used for a number of applications: (a) Energy recovery through incineration. (b) Cellulostic ethanol: molecular bonds in wood can be broken through hydrolysis to produce biofuel. (c) Ground timber and wood flour can be used in the production of oriented strand board (OSB) and medium-density fiberboard (MDF), respectively. (d) Mixed with plastics, wood flour can be used to produce wood plastic composite (WPC) for decking purposes or cladding materials. (e) Mulched timber can be used for soil enrichment and for recycled paper production. (f) Scientists at the US Department of Agriculture Forest Product Laboratory have developed a binderless recycled timber fiberboard. Wood particles are joined by steam in a high pressure oven. The major advantage of this production technology over comparable existing technologies such as MDF production, for example, is that the product in this case is fully recyclable (Hunt et al., 2006).

3. Methodology

Material recovery potential is dependent upon multiple regional and global variables. The quality of collection infrastructure defines which products or subassemblies have a better chance of being recycled (Pokharel and Mutha, 2009); availability of developed recycling technologies can determine the existence of recovery streams for given products (Nowosielski et al., 2008); the chemical makeup of a material determines how much energy needs to be invested in its recovery (Asokan et al., 2009); market demand for recycled content motivates recovery efforts around certain materials (Geyer and Blass, 2010); and so on. Accounting for these variables quantitatively in a universal manner is a challenging task. However, there might be a simple way to reflect their collective impact. A closer examination of the variables suggests that all are tied in one way or another to economic resources and could therefore be expected to be reflected in the market value of the product or the material.

Fig. 8.1 tracks market value fluctuations of a generic product as it progresses through its service life. As materials are sourced, processed, and assembled, the manufactured product gains market value—which peaks at the point of sale. From that point on, the product loses market value—first

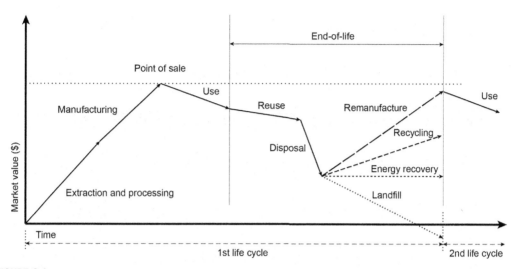

FIGURE 8.1

Value fluctuations throughout a generic product life cycle.

Adapted from Joshi, K., Venkatachalam, A., Jawahir, I.S., 2006. A new methodology for transforming 3R concept into 6R concept for improved product sustainability. In: IV Global Conference on Sustainable Product Development and Life Cycle Engineering, pp. 3—6.

through use, then reuse, and finally disposal. At that point, the product faces multiple possible end-of-life solutions that present value gains or loses. Landfilling would not only diminish the remaining market value of the disposed product but might lead to negative value due to tipping fees (Katers et al., 2009). Energy recovery (incineration) would generate some returns and therefore might maintain the product's market value or even increase it slightly. Recycling in its various forms (downcycling, upcycling, etc.) is almost certain to increase the market value of the product somewhat, especially if there is substantial demand for recycled content in that specific product category. Lastly, remanufacturing, where a reused product is restored to near-new condition, is certain to increase the market value of the disposed product (Vogtlander et al., 2017).

In addition to providing greater understanding of the value evolution of products from cradle to cradle, this sequential graph can also offer insights regarding specific performance drivers. By looking at individual segments of the graph and calculating the ratio between the beginning and the end value points, one can make determinations regarding extraction and production efficiency (processing and manufacturing phases), perceived durability and reliability of the product (use phase), preowned market (reuse phase), maturity of recycling technology (disposal phase), and more.

To illustrate this point, Fig. 8.2 shows a simplified application of the market value graph on 37 common construction materials. Even at a quick glance, certain materials stand out due to a particularly strong or a particularly weak performance. Copper, for example, loses very little of its peak value when it is sold as scrap. In contrast, glass-fiber insulation batt loses almost all of its value at the end-of-life phase. This difference implies that copper recycling is much more viable technologically and economically than glass-fiber insulation batt recycling. In broader terms, while materials may

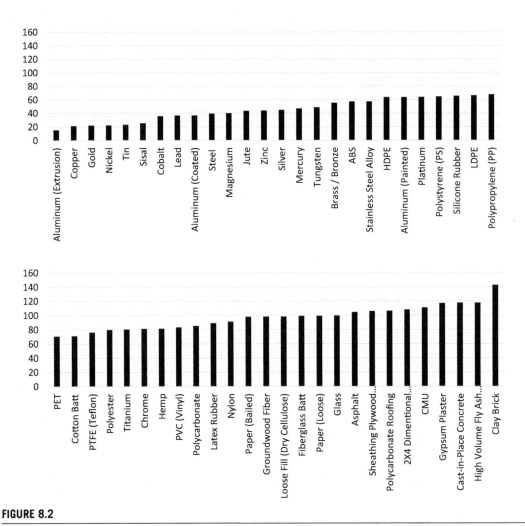

FIGURE 8.2

Market value drop (%) between point-of-sale and post-use scrap for common building materials.

differ in their properties, the market value of recycled content compared to the market value of virgin content reflects the efficiency and availability of appropriate recycling processes. The closer the market value of the recycled version to the virgin version, the more efficient the readily available recycling technologies are. In order to put this insight into practical use, the ratio between the virgin (point of sale) market value and the scrap market value of a given material or product could be defined as an index for estimating recovery potential.

To put this notion into mathematical terms, if V_V represents the market value of a virgin material in its raw manufactured form (i.e., metals in ingots, polymers in granules) and V_P represents the market value of a material after it has been recycled and it is ready for its second use, then R the recovery potential index (RPI) is defined as:

$$R = \frac{V_P}{V_V} \qquad (8.1)$$

R is unitless and as its value approaches 1, it indicates that the ability of the evaluated material to regain most of its original functionality is higher. Aluminum, which recycles well, for instance, is found to have an R value of 0.84 (the distance from 1 represents the fact that even in aluminum, the quality and application of the recycled product cannot always fully match the original properties—in the case of aluminum mainly due to the presence of tramp elements). Timber studs, which cannot be recycled to match the properties of its virgin version, are found to have an R value of -0.05. In addition to aluminum and timber, Fig. 8.3 compares recovery potential index findings for steel, high-density polyethylene (HDPE), and cement and clay bricks. Aluminum and steel rank highest with results of 0.84 and 0.58, respectively, and timber and clay rank lowest with results of -0.05 and close to 0.00, respectively. This analysis demonstrates that although all six materials are fundamentally different in chemical composition and application range, they can be compared in order to identify the best choice for recovery purposes based solely on economic indicators.

The implications of material recovery potential results are twofold. First, they can indicate desirable environmental trajectories in relation to consumption patterns. Consuming materials with low recovery potential could likely result in landfilling, while consumption of materials with high recovery potential would likely result in recycling or reuse. Second, material recovery potential findings—particularly those at the lower end—indicate necessity for action, either by governments or by the private sector. Low recovery potential results from either a lack of demand for recycled content or a lack of available technology to recycle a given material. Both conditions can be addressed through regulation (by governments) or through increased demand (by the

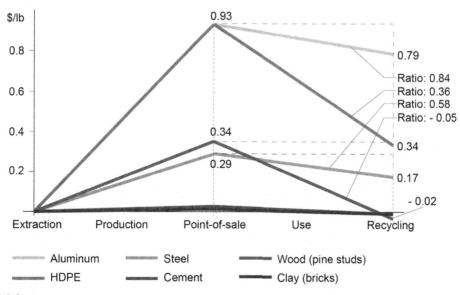

FIGURE 8.3

Recovery potential index findings for six common construction materials.

private and public sectors). An example for regulatory action could be a ban on materials with low recovery potential (e.g., plastic bags); an example for private sector action could be a corporate decision to switch from virgin to recycled stock (e.g., a transition to paper manufacturing based on paper waste rather than trees). Accordingly, maximizing the recovery potential of materials will depend on their specific market context as well as the local regulatory landscape.

4. Applications in practice

Since the recovery potential index is based on market value, it can be easily utilized to calculate disassembly depth in products as well as in building assemblies. Disassembly depth is defined as the balance between financial investment into material recovery operations and the market value of the materials recovered (Giudice, 2010). The standardized 0.00–1.00 scale allows the user to define a cutoff limit under which disassembly and retrieval efforts for components are no longer economically viable and should therefore cease. Additionally, it is possible to determine what percentage of a given assembly has a recovery potential index above a predetermined threshold. This rate can be calculated by simply summing up all material weights with the predetermined (x) or higher recovery potential index values ($R \geq x$), and then dividing this sum by the total weight of the assembly:

$$\%_{R \geq x} = \frac{\sum\limits_{i}^{n} W_{R \geq x}}{W_{\text{total}}} \cdot 100 \tag{8.2}$$

In order to expand this calculation to comprehensively evaluate the economic viability of disassembly operations for a given product, one would need to add the cost of labor and determine a profit to loss margin (PLM) value in the following manner:

$$\text{PLM}_{\text{recycle}} = \sum_{n} V_r - C_d + C_l - \sum_{n} C_r \tag{8.3}$$

where V_r represents the value of used materials before recycling; C_d represents the cost of disassembly labor; C_l represents savings of disposal cost; C_r represents the cost of recycling for each material; and n represents the number of materials in the evaluated product (Johnson and Wang 1998).

5. Discussion

Concerning the limitations of the presented approach, a narrow focus on cost as a single determining factor in the evaluation of material recovery potential could be problematic, as it may overlook other factors that may affect end-of-life strategies. For this reason, it might be beneficial to speculate on adding a number of extensions in order to reach a more multidimensional assessment of recovery potential.

5.1 Recovery energy index

Adopting the calculation method proposed in the RPI, a recovery energy index would reflect the efficiency of available recycling technologies from an embodied energy standpoint. Measuring recovery potential through embodied energy would also enable assessment of environmental impacts such as greenhouse gas emissions and pollution potential. Unlike the RPI, where the higher the value of the recovered material cost the better, embodied energy values of recovery processes should ideally be lower than the embodied energy values of the original production process: the greater the distance between the recovery energy and the production energy, the better. For this reason, in the recovery energy index, the embodied energy generated during the initial production of the product (EE_p) is divided by the embodied energy accumulated during its recovery process (EE_r). Here, as in the RPI, the higher the value the better; however in the recovery energy index values are often greater than 1.00:

$$R_{\text{energy}} = \frac{EE_p}{EE_r} \tag{8.4}$$

Most products result in a value that is higher than 1.00 due to the fact that usually primary (virgin) production requires extraction and processing operations prior to manufacturing. Primary production of a steel beam, for example, would involve mining and processing of multiple minerals prior to the extrusion of the product. In comparison, production of a recycled steel beam requires only the very last part of the process—melting and extrusion. Cases where the value of the Recovery Energy Index is smaller than 1.00 indicate an undesirable situation from an environmental impact point of view—a recovery process that is more energy-intensive than the primary production process. This kind of situation can happen in natural products that experience little processing between the extraction point and the consumer. One example in the construction industry would be natural stone tiles for flooring (granite), cladding (limestone), or roofing (slate). In this case, products that involve stone recovery and further processing might be more environmentally harmful that the original product.

5.2 Recovery carbon dioxide index

Although embodied energy levels reflect the efficiency of existing recycling and recovery technologies, they do not describe the sources used for the energy measured. Take, for example, the energy required to recycle a given amount of aluminum. It can be generated by burning fossil fuels such as coal, or it can be generated through renewable sources such as wind or solar power. These various sources have radically different environmental implications. For this reason, the environmental efficiency of recycling technologies should be measured by the amount of environmentally harmful agents released during the recovery process. The use of carbon dioxide as an atmospheric environmental impact indicator is well established and used by prominent policymakers such as the USEPA. A recovery CO_2 index would be structured similarly to the embodied energy index, with the exception of replacing embodied energy levels with carbon dioxide amounts released to the atmosphere:

$$R_{CO_2} = \frac{CO_{2p}}{CO_{2r}} \tag{8.5}$$

As can be observed in Fig. 8.4, a comparison between market value recovery index rankings and carbon dioxide recovery index rankings for common industrial materials shows that although some recycling processes might be relatively inexpensive, they produce large amounts of greenhouse gases. One example is recycling of aluminum extrusion products: a 0.84 score on the market value index and only 0.18 on the CO_2 index. Similarly, while some processes may be relatively costly, they are more environmentally benign. An example is platinum recycling: A 0.94 score on the CO_2 index and only

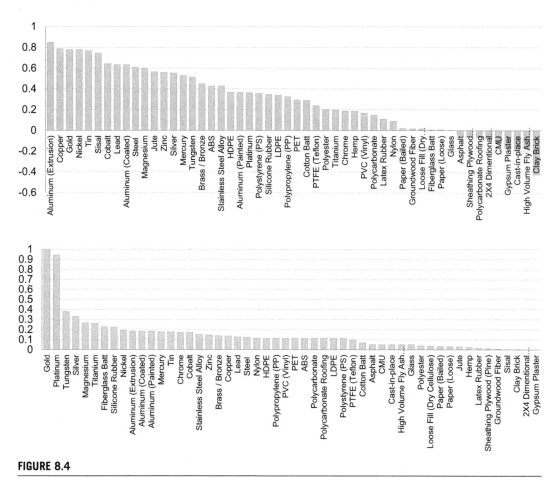

FIGURE 8.4

A comparison between a market value recovery index (top) and a CO_2 recovery index (values are normalized to 0.00–1.00) ranking commonly used materials.

0.36 on the market value index. Both indices represent a diverse set of important considerations and interests that should ideally be taken into account in concert.

5.3 Recovery resource depletion index

In addition to economic considerations and greenhouse gas emissions, the decision on whether or not to recycle some substances could be informed by their abundance in natural reserves. Lee (1998) proposes a global resource depletion index that is designed to calculate the amount of maximum extractable years left for a given material. The index is calculated in the following way:

$$\text{Maximum extractable years} = \frac{\text{Global reserves(by weight)}}{\text{Annual production rate(by weight)}} \quad (8.6)$$

5.4 Recovery yield strength index

Most materials lose some of their structural properties during the course of recovery operations. This degradation reoccurs in varying degrees of severity in every recovery cycle (Sarja, 2002). A recovery index that could describe the relative structural degradation in common materials would assist designers in avoiding the use of problematic materials in structural applications. A number of terms exist to identify the stress level in which materials begin to exhibit plastic deformation. Yield strength is perhaps the most commonly used value to describe this condition. A recovery yield strength index would compare the yield strength of a given recycled material to a virgin version of the same material in the following manner:

$$R_{fx} = \frac{f_{rx}}{f_{vx}}$$

(8.7)

where f_{rx} is the yield strength of a recycled material x, f_{vx} is the yield strength of the virgin version of material x, and R_{fx} is the recovery yield strength index of the material. Comparing the evaluated material to the original virgin material is significant, as it could provide an indication of the total structural degradation of the material rather than an indication that is relative to the last recovery cycle.

As a final point, it should be noted that the indices presented in this chapter are founded on a base assumption according to which the recycled materials are used for the same set of original applications as the virgin materials they replace. If the target application of the recovery differs from the original application, much of the comparison loses its relevance. In a similar manner, the quality of the recycling process influences the validity of this comparison as well.

6. Conclusion

This chapter presented an assessment method for material and product end-of-life recovery potential based on economic indicators. The method is based on the notion that market value fluctuations reflect recovery-related technological availability and consumer demand for recovered materials. The method is demonstrated through a comparative analysis of recovery potential of common construction materials. The chapter concludes with a discussion that proposes four extensions to the developed recovery potential index based on environmental impact and structural degradation. Future research in this context will focus on validating the developed index against collected data regarding recycling fractions in current supply for a wide range of materials and products.

References

Ashby, M.F., 2012. Materials and the Environment: Eco-Informed Material Choice. Elsevier.

Asokan, P., Osmani, M., Price, A.D., 2009. Assessing the recycling potential of glass fibre reinforced plastic waste in concrete and cement composites. J. Clean. Prod. 17 (9), 821–829.

Geyer, R., Blass, V.D., 2010. The economics of cell phone reuse and recycling. Int. J. Adv. Manuf. Technol. 47 (5–8), 515–525.

Giudice, F., 2010. Disassembly depth distribution for ease of service: a rule-based approach. J. Eng. Des. 21 (4), 375–411.

Hiroshige, Y., Nishi, T., Ohashi, T., 2001. Recyclability evaluation method (REM) and its applications. In: Proceedings Second International Symposium on Environmentally Conscious Design and Inverse Manufacturing. IEEE, pp. 315–320.

Hunt, J.F., Supan, K., 2006. Binderless fiberboard: comparison of fiber from recycled corrugated containers and refined small-diameter whole treetops. For. Prod. J. 56 (7/8), 69–74.

Johnson, M.R., Wang, M.H., 1998. Economical evaluation of disassembly operations for recycling, remanufacturing and reuse. Int. J. Prod. Res. 36 (12), 3227–3252.

Joshi, K., Venkatachalam, A., Jawahir, I.S., 2006. A new methodology for transforming 3R concept into 6R concept for improved product sustainability. In: IV Global Conference on Sustainable Product Development and Life Cycle Engineering, pp. 3–6.

Katers, J.F., Walczak, D., Burch, T., 2009. The impacts of a recycling surcharge to increase landfill tipping fees in Wisconsin. Waste Manag. Res. 27 (5), 501–511.

Lee, C.H., 1998. Formulation of resource depletion index. Resour. Conserv. Recycl. 24 (3–4), 285–298.

Li, Z., He, J., Lai, X., Huang, Y., Zhou, T., Vatankhah Barenji, A., Wang, W.M., 2018. Evaluation of product recyclability at the product design phase: a time-series forecasting methodology. Int. J. Comput. Integr. Manuf. 31 (4–5), 457–468.

MacArthur, E., 2013. Towards the circular economy. J. Ind. Ecol. 2, 23–44.

Maris, E., Froelich, D., 2013. Critical analysis of existing recyclability assessment methods for new products in order to define a reference method. In: REWAS 2013. Springer, Cham, pp. 202–216.

Nordby, A.S., Berge, B., Hakonsen, F., Hestnes, A.G., 2009. Criteria for salvageability: the reuse of bricks. Build. Res. Inf. 37 (1), 55–67.

Nowosielski, R., Kania, A., Spilka, M., 2008. Integrated recycling technology as a candidate for best available techniques. Arch. Mater. Sci. Eng. 32 (1), 49–52.

Pokharel, S., Mutha, A., 2009. Perspectives in reverse logistics: a review. Resour. Conserv. Recycl. 53 (4), 175–182.

Sarja, A., 2002. Life cycle design methods. In: Integrated Life Cycle Design of Structures. CRC Press, pp. 36–85.

Tam, V.W., Tam, C.M., 2006. A review on the viable technology for construction waste recycling. Resour. Conserv. Recycl. 47 (3), 209–221.

Vogtlander, J.G., Scheepens, A.E., Bocken, N.M., Peck, D., 2017. Combined analyses of costs, market value and eco-costs in circular business models: eco-efficient value creation in remanufacturing. J. Remanufact. 7 (1), 1–17.

Zhu, H., Byars, E., 2004. Post-consumer glass in concrete: alkali-silica reaction and case studies. In: Sustainable Waste Management and Recycling: Glass Waste. Thomas Telford Publishing, pp. 99–108.

Economic approaches to sustainable development: exploring the conceptual perspective and the indicator initiatives

André C.S. Batalhão[1,2,3], João H.P.P. Eustachio[1,3], Adriana C.F. Caldana[1,3], Alexandre R. Choupina[4]

[1]*School of Economics, Business Administration and Accounting at Ribeirão Preto, University of São Paulo, Ribeirão Preto, Brazil;* [2]*Center for Environmental and Sustainability Research (CENSE), Nova Lisbon University, Caparica, Portugal;* [3]*Global Organization Learning and Developing Network (GOLDEN) — Brazilian Chapter, University of São Paulo, São Paulo, Brazil;* [4]*School of Agronomy, Federal University of Jataí, Jataí, Brazil*

Chapter outline

Environmental Sustainability and Economy. https://doi.org/10.1016/B978-0-12-822188-4.00007-5

1. Introduction and objectives

In the early days of the English Revolution, at the end of the 18th century, classical economists made economics a discipline, worrying about the incipient conditions of industrial capitalism of the time. Adam Smith tried to explain the economic system inserted in the environment, although he considered it passive and benevolent. The classics recognized free gifts; however, they had no concern about the possible impacts of the waste and waste disposal by the economic system.

According to Adam Smith, economic growth would be the result of a process of capital accumulation resulting from increasing levels of use of a specialized workforce. The accumulation of capital would allow the expansion of the number of jobs and the the markets (adding directly higher productivity), and with that the profit would expand progressively, generating more capital accumulation (cumulative process of economic expansion). Thus, per capita production would grow more than per capita consumption, generating a surplus for investments.

Adam Smith pointed out population growth as a limit in this process, citing the steady state. In line with the scientific thinking of that historic moment, in the analysis of the relationship between economics and the environment, the classical school adopted mechanistic epistemology. They viewed the environment as neutral and passive, and the constraints it would impose on growth would stem only from the limited availability of resources—notably agricultural land.

The neoclassical school adopted a posture similar to that of the classical school, with the difference of totally ignoring natural resources. The hypothesis of free gifts and free disposal of waste from the economic system has become implicit, treating the economy as an isolated system (Friedman, 2005). The ecological economy, in turn, rejects the view of the neoclassical environmental economy, arguing that the disregard of the biophysical-ecological aspects of the economic system leads to a partial and necessarily reductionist analysis of the interfaces between economics and the environment (Ropke, 2004).

In this chapter, the research method was applied to provide an overview of how economic approaches have been working in sustainability context based on indicators initiative. We have focused on main pillars of theory, discussing the main perspectives of the indicators. Finally, we provide the contextual meaning and understanding of the economic issue toward sustainability progress, suggesting directional recommendations and further reflections.

2. Economic approaches to sustainable development
2.1 Contextualization of environmental economics

The main schools of the environment differ in relation to the effective importance of natural capital for SD. One takes the concept of weak sustainability, and the other of strong sustainability (Turner et al., 1993). The first concept defends the economic growth of capital and product in an unlimited way, justifying that natural capital can be replaced by other categories of capital. The second concept is based on not replacing natural capital with other categories of capital. These approaches are based on the premise that natural capital must be an integral part of the economic process, including allocation, distribution, and scale (Hardi and Barg, 1997). Within different sectors such as industry, governments, and the domestic market, there must be integration between the economy and the environment through decision-making at all levels (MacNeill et al., 1991).

At the end of the 1960s, with the advent of the environmental economy, the approaches on its main fronts generated two main environmental hypotheses. One is the tenuous environmental hypothesis, which admits that the environment is benign, passive, which can be annoying if attacked, but which is basically stable. In the deepened environmental hypothesis, the environment is endowed with a certain fragility, liable to undergo potentially destabilizing changes as a result of cumulative anthropic pressures, focusing on the relationship between the economic and environmental systems.

2.2 Neoclassical environmental economics

The neoclassical environmental economy develops on a mechanistic epistemological basis, assuming a neutral and passive environment, subject to the impacts of the economic system that can be reversed (Meckling and Allan, 2020). The tenuous environmental hypothesis and the concept of weak sustainability make up the burden of optimism in the thinking of neoclassical environmental economics. Neoclassical thinking considers the possibility that the depletion of nonrenewable natural resources will prevent the expansion of the economy to be minimal. Also, according to this thinking, there may be inefficient exploitation of resources (based on the Paretian criterion); however, the appropriate policies, inspired by market mechanisms, can easily solve these problems. This current of thought uses sophisticated and complex mathematical methods and models as the main analytical tool, but such models have been demanding increasingly simplified basic hypotheses.

2.3 Ecological economics

The current of thought in ecological economics, especially in its aspect of survival economics, does not accept the hypotheses of neutral environment and reversibility (Pols and Romijn, 2017). The analyses carried out, especially for the survival economy, are based on aspects of maintaining the opportunities of future generations, based on the deepened environmental hypothesis. This current of thought emphasizes the threats that the expansion of the scale of the contemporary world economy is imposing on the stability of the global ecosystem, directly implying the well-being of future generations. Its assessments are pessimistic about the future of humanity, emphasizing that the current world economy would not be sustainable.

2.4 Institutionalists

Institutionalists argue that their economic approach is more suited to addressing environmental problems than the neoclassical approach (Jespersen and Gallemore, 2018). They argue that environmental problems need a holistic approach, which has the following characteristics: (a) multidimensional and multidisciplinary nature; (b) complexity and uncertainty; (c) irreversibility; (d) conflicts of interest. These characteristics are, for institutionalists, incompatible with the instruments of unidimensional, limited, and disciplinary analysis of the neoclassicals. Another criticism by institutionalists of the neoclassical view is that the measurement of environmental impacts in monetary terms is based on static and dynamic efficiency, that is, treated only as an optimization problem. Another important divergence between currents of thought is the origin of the formulation of environmental policy. Neoclassicals are concerned with policy instruments, while institutionalists are concerned with the behavior of the individual and his socio-cultural context.

2.5 Evolutionists

In view of the economic perspective, evolutionary theories (Freeman, 1982; Lundvall; Borras, 2009) contemplate five main elements, which highlight their main concerns: (1) technological innovations: product innovations; (2) nonlinear dynamics: multiple trajectories predetermined by long-term equilibrium forces; (3) selection process: "principle of rationality"; (4) agents: several groups of agents influence the choice of technology; (5) lock-in: adjusts technology to the social environment (adaptations and behavior change). These five elements are sufficient to understand the theoretical framework of the evolutionary school in dealing with environmental problems.

The concept of externality, from the neoclassical school, is the starting point of the theoretical approach of environmental issues by evolutionists. For evolutionists, positive and negative externalities, related to production processes and patterns, must have a dynamic and long-term focus. In the evolutionary perspective, before proposing environmental policy instruments, we seek to understand the development of ecologically sustainable technology. Here, the diffusion of environmental technology must involve economic restructuring in order to limit damage to the environment.

Evolutionists treat environmental issues as an important element of dichotomy in the development path of technology, even recognizing the political relevance and its power to directly influence environmentally correct actions. However, the environmental discussions held by evolutionists are still restricted, mainly with regard to the profile of environmental policies and employment and choice of instruments. Related to environmental policies, the evolutionists suggest the emergency adoption of environmentally sound technological development trajectories. As for instruments, the choices are divided between a flexible approach and a more incisive or hard approach. Among the evolutionists, there is a consensus that the guiding presence of the government is fundamental to indicate the trajectory said to be environmentally correct.

3. Natural resources: how to classify?

There are several classifications of natural resources, but taking into account the analysis of the economy of the environment, the one that best fits distinguishes between nonrenewable resources and conditionally renewable resources. The first is characterized by the finite endowment of natural resources, encompassing two broad categories: (1) exhaustible but recyclable resources (fixed stock, with partial possibility of recycling) and (2) depletable resources (fixed amount of inventory and/or little increase in inventory worldwide). Conditionally renewable resources are resources that have at least partial replacement of what is extracted. There are three categories in this group: (1) renewable resources, but dispersed and difficult to capture (e.g., capture of solar energy); (2) renewable resources, but subject to extinction (of common property); and (3) renewable resources, but subject to degradation due to inadequate management (subject to management practices).

When dealing with nonrenewable resources, it is necessary to speak of a fundamental element in the availability of natural resources: reserves. Three concepts are highlighted, understood as inseparable from the problem of depletion, and management of natural resource reserves:

1. Concept of current reserves: the amount of current reserves of a mineral is determined by geological and economic factors;
2. Concept of potential reserves: includes the subeconomic reserves and the undiscovered portion of the reserves (but there are signs of existence);

3. Role of technology: related to environmental impacts on extraction (enabling mineral extraction), exploration (facilitating exploration), and recycling (expanding the supply without increasing extraction).

4. Models of economic regulation and greening of international trade

Environmental policy in the national and international contexts goes through determining factors that influenced its environmental policy options: the style of regulation and international environmental trends. The regulation style is characterized by a specific political choice, with its own characteristics. In international environmental trends, we can highlight trade barriers based on environmental regulations, called greening of international trade.

Vogel et al. (1986) carried out a comparative study between American and British environmental policies. The author found a strong contrast between the orientation of environmental policies between the two countries. In the United States, the regulation process is legalistic and contentious, with extremely ambitious goals (legal requirements). British regulators, on the other hand, prefer a close and cooperative relationship with industrialists, with a gradual regulation process (voluntary agreements). Vogel also revealed that the practical results of these environmental policies were very similar. That is, the improvement of environmental quality in the two countries was not considerably different. This result is evidenced with great care by the author, who points out some methodological problems in assessing the effectiveness of policies in relation to their real environmental impact, making it difficult, in his view, to make a direct comparison between the two countries.

Dealing with a national political system, the characteristics of this system overlap with the nature of a specific political area, which refers to the proposal for national styles of regulation. In this case, Vogel makes an important reservation, saying that this conclusion is valid only for two countries and for the limited number of issues covered by his study.

International trends in environmental policy can influence national environmental policies, under the practice of trade restrictions. A country's trade policy can be used as a single or complementary instrument for internal environmental control. The problem occurs when a country, through trade restrictions, forces another country to adhere to an environmental policy formulated and applicable only in one of these countries (imposition of restrictions). The main argument of countries that adopt trade restrictions for environmental purposes is that a country with a milder environmental policy practices an unfair commercial practice, keeping its prices lower, compromising competitiveness (ecological dumping). The results presented by Stevens (1993), analyzing empirical studies in developed countries, refute this argument concluding that the impacts of environmental policies on competitiveness and the trade balance are low or null (between 1% and 2% of the total costs).

A very controversial and debated topic is the use of trade restrictions as an environmental measure, linking the effectiveness of environmental policy with ecological and economic efficiency terms. The costs of environmental control are different in each country, which makes the homogeneity of environmental policies unfeasible. In the legal aspect, the use of commercial policy as an instrument of environmental control and conflicting scenarios has also been debated lately. Basically, the influence of environmental policies on international trade is exercised by two forms of restriction: product barriers (restrictions on access to the domestic market) and process barriers (imposition of commercial restrictions on the product).

Many international agreements related to environmental problems are in force and are being debated, however, none of them seem to meet conditions that ensure environmental effectiveness and economic efficiency. In many of these agreements, the use of trade sanctions has been used to enforce them.

5. Environmental economics and ecological economics: a vision of sustainability resource

The vision of resource sustainability and the way to develop indicators and metrics capable of measuring and managing processes to mitigate environmental damage affect several areas of knowledge, including the economic sciences. Within economic sciences are included Ecological Economics and Environmental Economics. Efforts to attribute monetary values to services and environmental losses, thus correcting macroeconomic accounting, are part of the ecological economy, but their main orientation is centered on the introduction of physical indicators and indices to contribute to sustainability. In turn, environmental economics, based on the models of Hotelling (1931) and Hartwick (1977), demonstrated how models based on "individual rationality" can be used to understand the long-term relationship between the resource and the behavior of agents guided by market forces and their impact on the sustainability of resource use. All of this suggests that the existing neoclassical system, in one way or another, could adequately respond to emerging concerns about environmental issues along with neoclassical tools and principles (Solow, 1974).

Environmental Economics understands that the use of natural resources can be analyzed within the total economic value and can be maintained in real terms. For this they assume that: (a) renewable resources have the capacity to regenerate in a certain period of time and (b) if a renewable resource becomes "non-renewable" then the economic structure can be handled properly by neoclassical tools and principles.

In turn, the Ecological Economy amplifies the absolute scarcity to show how the irreversibility and other biophysical limits of the "closed" ecosystem restrict the growth of the economic subsystem and cause the degradation of natural resources (Costanza et al., 1992). According to this current, the issue is not the replacement of natural capital by human capital and the theoretical frameworks of neo-classical tools to treat natural resources, but synergy between forms of capital due to environmental assumptions such as irreversibility, uncertainty, and existence services that are essential to the well-being of the biosphere (Pearce et al., 1990).

Environmental economics and ecological economics aim to understand the issues involved between natural resources and economics. While environmental economics is centered on relevant issues within the neoclassical approach in a systematic way, ecological economics evolves through a more holistic approach that transcends classical economic issues, such as behavioral economics.

6. Enhancing economic sustainability: improvement of indicator approaches

Quantifying the sources of resources for the world and for specific regions can help explain the limitations to economic growth imposed by potential and real environmental constraints, especially the immediate availability of resources. This can build a situational scenario reflecting emerging demands and aspirations from different parts of the world.

Although economic sustainability is difficult to define and apply at some levels of assessment, many integrated indicators must be proposed, combining different economic, environmental, social, and institutional aspects that can replace or complement the coverage of development indicators, such as GDP, for example. A broader analysis of economic sustainability must reach and contribute directly to an assessment of *lato sensu* sustainability, which can be called Integrated Assessment, as it integrates all dimensions of SD.

In this approach to *lato sensu* sustainability, different indicators must be brought together with the objective of providing a transversal and comprehensive assessment of sustainable progress, distinguishing the role of each dimension in this scoreboard, including support and economic viability.

Indicators with direct links to economic activities have historically been explored with consensual and solidly established methods, while indicators in natural sciences, such as increased or decreased biodiversity, air quality, or water stress, have been revised and improved. Social and institutional indicators also need more attention from scientists, as they are still neglected in analyses for sustainability, as well as their theoretical and practical bridges and connections with other SD themes.

7. Sustainability indicators and economic indicators issue to SD management

Indicators are essential to analyze the progress of SD by measuring the most diverse earth systems (Griggs et al., 2013), whether they are linked to the economic, environmental, or social dimensions (Elkington, 1998). The scientific literature on indicators is vast, and several authors discuss the importance of this topic and the technical aspects of Sustainability Indicators (SIs) (Moldan et al., 2012).

SD is a complex concept (Innes and Booher, 2000; Rotmans and Loorbach, 2009; Loorbach, 2010) that is quite challenging in both practical and theoretical aspects. It has several dimensions in which a considerable number of actors need to be aware in order to make the earth systems redirect to its viable levels of sustainability. Therefore, there is a need for indicators capable of absorbing and reducing this complexity (Eustachio et al., 2019), transforming it in useful information for policy or decision-makers to take action. In this context, it is essential to discuss the main characteristics of the SIs present in the literature.

In this way, the literature provides a great amount of discussion on important characteristics these indicators should assume and how they should behave in practical aspects. Thus, much of the discussion on SIs orbit on the scientific rigor and the utility for public managers, focused on practical reporting and communication, based on specific goals (especially regarding the Sustainable Development Goals—SDGs). These elements are relevant aspects to assess sustainability and improve levels of sustainable development to universal application and have been discussing recurrent issues raised in the literature and desired in practical aspects (Hák et al., 2016; Hagan and Whitman, 2006).

It is common sense that SD is a complex phenomenon. For that reason, SIs are generally developed to absorb complexity by analyzing several variables alone or by connecting them to understand as a system of variables or indicators (Eustachio et al., 2019). In other words, the indicators should be capable of absorbing complexity, considering the whole is greater than the sum of its parts (Bertalanffy, 1968). Therefore, it is common to find in the literature indicators that are not only connected to a specific issue or dimension alone, but also considering several dimensions like the economic, environmental, and social. Through this point of view, it is not hard to find in the literature a combination of several variables or indexes, forming systems of indicators capable of dealing with this inherent complexity of SD (Meadows, 1998; Balkema et al., 2002; Bossel, 2002).

Besides, this aspect is present on the most worldwide known framework of SD—the SDGs and the several ways of measuring each one of the goals proposed by the United Nations (Colglazier, 2015; Lee et al., 2016). The correlation of these goals is evident, making possible to organize the measures connected to them in a system capable of dealing one or more aspects of SD (Eustachio et al., 2019). Example of that is the capacity of the economic system can affect environmental and social systems, and vice-versa. According to Hák et al. (2016), for example, the objectives of SD should not be related only to a set of objectives and goals, but, above all, to a system of indicators so that the responsible actors organize their agendas, leading the whole system toward sustainability.

In line with this, there is a distinction between two categories: indicators can be composed (aggregated) or detailed. This makes possible to understand issues in depth or relevant to its several kinds of actors that should use them, according to their levels and type of responsibility and which geographic scale is responsible (e.g., president or minister for countries and the mayor for cities). Because of that, an indicator can be used in several levels: detailed or comprehensive incorporating several indicators at the same time (Kilkiş, 2016), or by analyzing each one of them separately.

Also, SIs should be useful for those who need to analyze them to make smart decisions. Thus, it is important to consider the role of SIs of helping decision-makers in their activities to make the several systems returning to their normal level of sustainability (Griggs et al., 2013). In other words, they contribute to learning and adaptation of the system (Regmi et al., 2016) by showing gaps and leading decision-makers to be assertive and attempt sustainability issues. This means that indicators are powerful feedback tools (Hák et al., 2016), helping public managers, decision-makers, and civil society to improve their governance system and guide in the SD direction (Leal Filho et al., 2016).

Once they contribute to learning and adaptation, it is relevant to consider how SIs contribute to communication processes in two ways: the adoption of SIs by several actors and how they could be implemented. This leads to a couple of questions: Who are the people who will use such indicators? Who should have access to the indicators? How they should be used for decision-making? Who will be responsible for the development of such indicators? Furthermore, implementation and adoption end up being important to achieve the sustainability of the systems as communicating efficiently with their target audience, considering the context, the complexity of understanding them, and what could compromise the communication between the sender and receiver of what the indicator aims to communicate. Otherwise, the indicator may become useless, and plans, actions, and investments to achieve sustainability may not be effective. Communication can make the problem visible in a way that makes people aware of what needs to be done to achieve sustainability (Janoušková et al., 2018).

In line with the communication aspect, the indicators become essential since they contribute to feedback channels that can help policymakers to monitor and understand how some activity can be developed surgically for a more sustainable scenario. These feedback channels show the urgency and importance of a given issue to be tackled by someone (Barrera-Roldán and Saldívar-Valdés, 2002), and contributes to the adaptation processes discussed before.

In this context, communication-based indicators become important and useful. They are connected to the idea that information on SD must be communicated efficiently within and among the scientific, political, and practical spheres. Thus, this issue implies the need to bring together scientists, policymakers, and civil society to create a mutual understanding of the issue to be addressed and, consequently, to solve problems related to the sustainability of a region (Janoušková et al., 2018).

The communication-based characteristic of indicators shows that these indicators should go beyond instrumental uses. They have to call attention and, possibly, be useful by reflecting their data into attitudes and actions in order to direct resources and investments for change (Burford et al., 2016).

Another important aspect of SIs is the capability of dealing with recursion levels or different scale levels. Thus, SIs could be useful in dealing not only with small scales like neighborhoods, areas, or cities but also with higher geographic levels such as country or continent or, perhaps, understanding a given earth system such as the geosphere, hydrosphere, atmosphere, or biosphere. With this feature, SIs allow policymakers or decision-makers to access levels of recursion or any other geographic scale (Schmidt and Schwegler, 2008; Isah and Sodangi, 2013; Bossel, 2002).

It is worth considering that this aspect is hard to develop, since most of the new indicators discussed in the literature do not find available or required variables to compose them in the country or cities recursion levels. For example, most of the indicators are available in the country scale but are difficult to find in smaller scales such as cities. Also, when it comes to environmental dimension, there is a lot of information of the environmental dimension for countries, but a lack of them for cities (Valcárcel-Aguiar et al., 2019).

Thus, scale and the recursiveness are important features to develop SIs once they contribute to the process of controlling and monitoring whether the indicators are showing that society is moving toward SD through the assessment of goals concerning different levels of scale, such as regions, cities, states, countries, and continents (Burford et al., 2016; Hák et al., 2016).

The time dimension is also recurrent when analyzing SIs in a longitudinal way. The variable time allows SIs to assume dynamic characteristics, since they seek to understand some reality that changes repeatedly, measuring whether SD is progressing or regressing (Blancas et al., 2016). In addition to measuring the evolution or regression of the level of sustainability, the indicators and frameworks should also be reviewed over time, to stay as adherent as possible with the academic discussion and Agenda 2030 of SD (Janoušková et al., 2018).

Also, the SIs could serve for governance and decision-making processes by allowing integrated governance, designing review processes, searching for transparency and accountability mechanisms, and integrating several points of view in policy-making. These issues could generate adaptative and learning processes to pursue sustainability (Leal Filho et al., 2016; Pupphachai and Zuidema, 2017).

The role of SIs for policymakers is important, and they could contribute and give significant insights when considered the public policy cycle proposed by Howlett et al. (2009) in several aspects. First, they are essential for diagnosing a system, contributing for the problem identification stage. After, they become useful to support a proposed public policy (policy formulation and adoption stages). Finally, with its time feature, they could be useful in the stage of evaluating and monitoring a public policy.

In conclusion, it is relevant to consider that SIs is a developing area of study with several articles been published and an evolving literature. In this way, the number of sustainability indicators has grown, making their understanding, adoption, and usage more refined to understand and solve complex SD problems. Thus, there is a tendency to aggregate variables, use composed indicators and its recursion aspect not only to geographical areas, but also for economy sectors (e.g., fishing, agriculture, government, electric, mining) (see Rahdari and Rostamy, 2015; Roca and Searcy, 2012; Schianetz and Kavanagh, 2008; Abolina and Zilans, 2002).

Analyzing such complexity informed above, the next section has as main objective to show the reader how the SIs have evolved, also showing the complexity considering the most diverse indicators that have emerged from the literature, initiatives, and agencies.

7.1 Evolution of sustainability indicators

According to the previous section, SD is a complex phenomenon and SIs are a good way to absorb and reduce its level of complexity. Thus, to work with this characteristic, there are several perspectives and features SIs could assume in order to help policymakers or decision-makers to take action, improving the several systems connected to the three dimensions of sustainability (Elkington, 1998).

This complexity of SIs has evolved over time. The available SIs were not only produced in quantity, but also became more complex over the years. In this way, before Brundtland Commission, the discussion and proposition of indicators were linked mainly to the economic dimension such as GDP and economic growth. The problem of assuming only this perspective is that looking at purely economic indicators, all the other dimensions are disregarded, creating the possibility to generate negative externalities. Thus, the discussion expanded to understand the positive and negative impacts that the economic dimension could have on the social dimension (Vaghefi et al., 2015; Zhang, 2012; Ayres et al., 2001).

After that, mainly because of several growing initiatives and projects, the term sustainability gained strength and came to consider the environmental impacts. Thus, the sustainability indicators started to acquire holistic and transdisciplinary aspects, considering the economic, social, and environmental dimensions.

Table 9.1 provides this chronological perspective by showing the evolution of the main milestones according to its source and its reflection on how each one of the most important milestones impacted on the discussion of SIs. This chronological order shows the main historical facts that contributed to the discussion of sustainability and how this discussion reflected in the development of indicators over time.

Also, Table 9.1 indicates how the discussion on indicators has matured over time. During Meadows et al. (1972), World Charter for Nature (Wood, 1985) "Our Common Future" report (Brundtland, 1987), there was a consensus that quantitative information and indicators would play an important role on SD. Years later, from the Earth Summit, discussions on indicators, initiatives, and frameworks for sustainable development accelerated. After that, academics and scientists started the discussion and publication of the best indicators that could contribute to monitoring the evolution of SD. Finally, the United Nations through a working group process started to suggest the main issues that would eventually become the 17 goals and 169 targets (United Nations, 2015). After that, indicators were determined and set available for public consulting by both the IAEG U. N. (2016) and the World Bank Group (2020) at a national scale.

By this analysis, especially when it concerns the two main milestones (Millennium Development Goals and, subsequently, Sustainable Development Goals), several scientists and organizations started to publish ideas and trends, and suggest new indicators and indexes capable of measuring the level of sustainability in an aggregated way by a system of indicators or theoretical frameworks. Thus, to understand the evolution of SIs, it is necessary to understand the change that happened from the traditional world view toward the integrated view of Bertalanffy's systemic theory. Such change could make easy to understand the complexity of SD (Eustachio et al., 2019).

Table 9.1 Milestones and discussion on sustainability indicators.

Year	Source	Milestones	Scientific discussion on indicators
1972	Meadows	Meadows	Consensus that quantitative information and indicators would play an important role in sustainable development
1982	United Nations	World Charter for Nature	
1987	WCED	Our Common Future	
1992	United Nations	Earth Summit 1992	Chapter 40: "Indicators that shows us if we are creating a more sustainable world."
1995	United Nations	World Summit on Social Development in Copenhagen	Consistent and intense creation of several indicators, indexes, and survey of variables.
2002	United Nations	World Summit on Sustainable Development in Johannesburg	
2012	United Nations	Rio + 20 Millennium Development Goals	
2015	United Nations	Sustainable Development Goals	Incentive to measure the SDGs from national to local-scale measures: United Nations and The World Bank Group increase of micro indicators

Adapted from Eustachio, J.H.P.P., Caldana, A.C.F., Liboni, L.B., Martinelli, D.P., 2019. Systemic indicator of sustainable development: proposal and application of a framework. J. Clean. Prod. 241, 118383.

Thus, with the growing demand for indicators and the growing need to adopt more systemic measures to understand, evaluate, and make surgical decisions, several sources started to become committed to the process of developing relevant indicators. Table 9.2 brings this discussion by showing the main initiatives and indicators linked to sustainability, the source which produced this indicator, the main contribution and coverage in relation to each of the main dimensions according to Elkington (1998).

In addition to the initiatives described above, one of the biggest issues in evidence when analyzing chronologically is the vast number of indicators, indexes, and projects to measure sustainable development over time. Traditional indicators already have an inherent complexity, even more if they are considered to the analysis of recursion levels. However, with the evolution of the discussion of the indicators, especially after the Millennium development goals were announced in 2000 and, more recently, the sustainable development goals in 2015, there was an explosion and a growing need for additional new indicators so that it is possible to understand specific and particular issues.

This could be seen as a tendency as the literature explores very specific issues of SD and SIs like composite indicators to measure water, energy, and environmental systems (Kilkiş, 2016), sustainable urban livability (Valcárcel-Aguiar et al., 2019), development of countries of a specific continent (Bartniczak and Raszkowski, 2019), relationship between environmental pollution, economic development, and public health (Lu et al., 2017), vulnerability assessment indicator with variables

Table 9.2 Main initiatives for sustainable development indicators.

Indicator/frameworks/ initiatives	Source	Main contribution
PSR (pressure/state/ response)	OECD—Organization for Economic Cooperation and Development	"The PSR framework seems highly capable of showing information to end users in a causal way by differentiating between causes, effects and human responses to control the extent of anthropogenic impacts on nature." (Wolfslehner and Vacik, 2008, p. 2)
DSR (driving-force/ state/response)	UN/CSD—United Nations Commission on Sustainable Development	"DSR framework is developed by the United Nations Commission on SD (UNCSD) to provide a consistent set of indicators and to assess progress towards a sustainable energy future." (Meyar-Naimi and Vaez-Zadeh, 2012)
GPI (genuine progress indicator)	Cobb	"GPI are designed to more closely approximate the sustainable economic welfare or progress of a nation's citizens. The sustainable economic welfare implied here is the welfare a nation enjoys at a particular point in time given the impact of past and present activities. The notion of sustainable economic welfare being approximated is critical." (Lawn, 2003, p. 106)
HDI (human development index)	UNDP—United Nations Development Programme	"The UNDP's composite Human Development Index (HDI), based on life expectancy, educational attainment, and material living standards, is an attempt to reorient the assessment of development levels away from income to more broadly based measures." (McGillivray and White, 1993, p. 183)
MIPS (material input per service)	Wuppertal Institute — Alemanha	"MIPS means Material Input Per Service unit. In order to estimate the input orientated impact on the environment caused by the manufacture or services of a product, MIPS indicates the quantity of resources (known as 'material' in the MIPS concept) used for this product or service. Once one has the reciprocal, a statement can be made about resource productivity, i.e., it can be calculated how much use can be obtained from a certain amount of 'nature.'" (Ritthoff et al., 2002, p. 9)
DS (dashboard of sustainability)	International Institute for Sustainable Development—Canada	"The Dashboard of Sustainability (DS) is a mathematical and graphical tool designed to integrate the complex influences of sustainability and support the decision-making process by creating concise evaluations." (Scipioni et al., 2009, p. 364)
EFM (ecological footprint model)	Wackernagel and Rees	"The ecological footprint assesses people's use of natural capital by comparing their resource consumption and waste production to the regenerative capacity of the earth." (Wackernagel et al. 1999, p. 604)

Table 9.2 Main initiatives for sustainable development indicators.—cont'd

Indicator/frameworks/ initiatives	Source	Main contribution
BS (barometer of sustainability)	IUCN—Prescott-Allen	"The proposed barometer of progress toward a sustainable society overcomes this problem by providing: a simple qualitative classification of the conditions of the ecosystem and human system; and a means of combining assessment of the ecosystem and human system into a single reading (without trading one off against the other)." (Prescott-Allen, 1997)
SEEA (system of integrating environment and economic)	United Nations Statistical Division	"The concept SEEA was developed in response to international calls for sustainable development, notably by the World Commission on Environment and Development (WCED, 1987). The objective was to provide a quantifiable definition and database for the new paradigm by introducing environmental concerns into the national accounts." (Bartelmus, 2014, p. 886)
NRTEE (national round table on the environment and economy)	Human/Ecosystem Approach—Canadá	"The NRTEE has a strong track record working with—and balancing the interests of—a wide range of stakeholders. Its special expertise is in devising practical solutions to issues and problems with both environmental and economic implications." (Smith, 2002, p. 306)
EE—eco efficiency	WBCSD (World Business Council on Sustainable Development)	"Business strategy with regard to sustainability is currently dominated by an eco-efficiency approach that seeks to simultaneously reduce costs and environmental impacts using tactics such as waste minimization or reuse, pollution prevention or technological improvement." (Korhonen and Seager, 2008, p. 411)
SPI (sustainable process index)	Institute of Chemical Engineering—Graz University	"The Sustainable Process Index (SPI) is a measure developed to evaluate the viability of processes under sustainable economic conditions. Its advantages are its universal applicability, its scientific basis, the possibility of adoption in process analyses and syntheses, the high sensitivity for sustainable qualities, and the capability of aggregation to one measure." (Krotscheck, Narodoslawsky, 1996, p. 241)
ESI (environmental sustainability index)	World Economic Forum	The ESI is an index applied in the evaluation of nations' sustainability, being its main objective to establish a way for comparison of the sustainability of countries. To assist in the comparisons across countries with similar profiles, a cluster analysis is used. Cluster analysis provides a basis for identifying similarities among countries across multiple dimensions. The cluster analysis performed on the ESI-2005 data set reveal seven groups of countries that had distinctive patterns of results across the 20 indicators (Bab Siche et al., 2008, p. 630)

connected to soil, habitat, climate, and vegetation (Hong et al., 2016), soil indicators for decision-making processes (Jónsson et al., 2016).

Along with this growing number of indicators, many others were created to measure the level of SDs of specific economic sectors: the primary sector, with the extraction of raw materials such as mining, fishing, and agriculture; the manufacturing sector, which is concerned with the production of finished goods, such as factories, cars, clothing, and food, and; the tertiary sector by the production of intangible goods and services, such as the discussed tourism through the sustainable tourism indicator chart developed by Blancas et al. (2016).

This shift and growing complexity related in the last paragraphs could be better understood by analyzing Table 9.3. It shows the evolution, considering first the classical economic indicators and, then, showing how these indicators are expanded into more specific ones, either by organizations, initiatives, or by researchers who have found it important to develop specific indicators to understand, measure, and generate solutions for SD. In this way, Table 9.3 has the economic category, and there are other two classifications embedded (1) traditional indicators, commonly used and (2) indicators derived from the SDGs that were extracted from a research on the World Bank website.

To understand what the indicators can measure and contribute to the SD, it is necessary to build a hierarchical classification structure to categorize each indicator according to the methodological strategy of data collection and treatment, as well as its analysis reference or reference scenario. It is important to take the objective function to minimize environmental impacts and maximize economic benefits, identifying potential tradeoffs. The combination of traditional indicators with more sophisticated indicators can directly contribute to the construction of a discourse that seeks to reconcile the economic growth of the linear economy with new approaches and strategies for economic sustainability.

Table 9.3 Economic indicators to sustainability.

Economic dimension	
Traditional indicators	**Indicators—SDGs (World Bank)**
Gross domestic product (GDP) GDP per capita growth Employment Interest rates Consumer price index (inflation) Producer price index Current employment statistics Money supply Stock index Income and wages Consumer spending	Battle-related deaths | Bribery incidence | Completeness of birth registration | Firms expected to give gifts in meetings with tax officials | Intentional homicides | Primary government expenditures as a proportion of original approved budget | Debt service to exports | Exports of goods and services | Foreign direct investment, net inflows | GDP | GDP, PPP | GNI | GNI, PPP | Account ownership at a financial institution or with a mobile-money-service provider | Agriculture, value added per worker | Children in employment, total | Commercial bank branches | Employment in agriculture | Employment in industry | Employment in services | GDP per person employed | Industry, value added per worker | Informal employment | New business density | Services, value added per worker | Share of youth not in education, employment or training, total | Unemployment, total | Unemployment, youth total | Wage and salaried workers, total | Air transport, freight | Air transport, passengers carried | CO_2 emissions | Manufacturing, value added | Medium and high-tech industry | Railways, goods transported | Railways, passengers carried | Research and development expenditure | Researchers in R&D

The prominent discourse of SD faced on economic development has historically been confronted by different lines and schools of knowledge that make little effort to build a theoretically congruent and methodologically feasible approach. We must recognize that economic development needs to be intertwined with the other dimensions of SD, providing useful perspectives on the integration among them (direct and indirect relationship).

8. Final remarks and the way forward

The fundamental shock between aspects of sustainability and economics may be related to the transition from a conventional view of economic capital to a view that brings together capitals of different types, including natural or environmental, human capital, social capital, and capital from governance mechanisms. One of the main challenges for economists and approaches to economic indicators for sustainability is to maximize the return on investment while maintaining constant capital. This is similar to the management of investment portfolios, where the volume and proportion of capital invested is observed, considering a strategic vision to obtain future profits. More and more conventional means of measuring and managing the economic dimension within sustainability have failed to neglect the scarcity caused by the excessive use of natural resources, imprecisely reflecting environmental degradation and its consequences on human health.

Macroeconomic policies must guide development processes toward a sustainable standard, balancing economic growth and consumption and production patterns, following technical progress, the potential replacement of products and services, and the control of natural disasters. The command and control mechanisms for the environment are still inefficient for the management and conservation of natural resources, as well as for achieving satisfactory levels of sustainability. There is a growing perception among the general public that it is necessary to include nonmonetary, social, demographic, institutional, cultural, and geographical elements in the planning and political action processes. In general, we know that the economic dimension interacts with the other dimensions of sustainability; however, we do not know specifically the real impacts of these interactions.

Therefore, a sufficient and socially acceptable level of development of products and services offered by a use of resources in steady state or in decline, with stock patterns within ecological limits, is still an important research frontier. However, the theory and knowledge of the systems approach on the nexus between stocks-flows-services is still not sufficient to resolve this issue. The role of economic indicators and indices should not only simplify relevant information on complex problems in the sectors of the economy but also point out interdimensional connections, presenting complex relationships within the fundamental elements of sustainability in a highly communicative and logical format.

Analyzing visions of strong sustainability and weak sustainability, we realized that both concepts still remain limited when questioning economic growth. Directly comparing the strategic sustainability decisions offered by the different conceptual lines presented above, we do not mainly protect substitutability, but its practical clarification and theoretical-methodological consistency. In this sense, the SD requires a balanced and simultaneous action of the available economic mechanisms, promoting an interaction between the environmental, social, institutional, and technological aspects of an economy, highlighting the role of the production sectors and individual and autonomous industrial processes.

We must not ignore the fact that the existing sustainability standards increasingly demand concepts applicable in different contexts, and innovative since their birth. This implies the performance of innovative agents, who work with highly sophisticated tools, and have a high level of human inter-relationship, facilitating the management and implementation of innovative ideas for the economic field. Due to the complexity of the SD vision, in most cases its implementation needs to be supported by different stakeholders who help to make appropriate changes in practices, policies, and decision-making tools.

The continuity and stability of the economic normative structure oriented toward SD and the opportunities of the domestic and global market depend on more research, specifically in the field of SIs. Companies must trust in the need to maintain the economic profitability of their activities and investments in SIs, integrating in their portfolio market mechanisms that cover the visions of sustainability, extrapolating common propositions. In this sense, appropriate assessment tools and indicators that reward positive externalities and punish negative externalities need to be adapted to provide reliable tools for policymakers and decision-makers.

Finally, we must be aware of limitations of current economic approaches, considering the non-economics phenomena in a possible future downward stage for most economies worldwide. Economic efficiency and political prudence can be crucial factors in guiding transitory sustainability policies, moving toward new patterns of production and consumption, capable of preventing losses and damages in the other dimensions of the SD.

Acknowledgments

A special thanks to colleagues from Global Organization Learning and Developing Network (GOLDEN) and School of Economics, Business Administration and Accounting at Ribeirão Preto, University of São Paulo, Brazil, and co-authors who worked actively with their collaboration, valuable guidance, insights, and thoughts. We thank the Editorial team and anonymous reviewers for their comments and suggestions.

References

Abolina, K., Zilans, A., 2002. Evaluation of urban sustainability in specific sectors in Latvia. Environ. Dev. Sustain. 4 (3), 299−314.

Ayres, R., Van den Berrgh, J., Gowdy, J., 2001. Strong versus weak sustainability: economics, natural sciences, and consilience. Environ. Ethics 23 (2), 155−168.

Bab Siche, J.R., Agostinho, F., Ortega, E., Romeiro, A., 2008. Sustainability of nations by indices: comparative study between environmental sustainability index, ecological footprint and the emergy performance indices. Ecol. Econ. 66 (4), 628−637.

Balkema, A.J., Preisig, H.A., Otterpohl, R., Lambert, F.J., 2002. Indicators for the sustainability assessment of wastewater treatment systems. Urban Water 4 (2), 153−161.

Barrera-Roldán, A., Saldıvar-Valdés, A., 2002. Proposal and application of a sustainable development index. Ecol. Indicat. 2 (3), 251−256.

Bartelmus, P., 2014. Environmental−economic accounting: progress and digression in the SEEA revisions. Rev. Income Wealth 60 (4), 887−904.

Bertalaffy, L.V., 1968. General Systems Theory: Foundations, Development, Applications. Braziller, New York.

Blancas, F.J., Lozano-Oyola, M., González, M., Caballero, R., 2016. Sustainable tourism composite indicators: a dynamic evaluation to manage changes in sustainability. J. Sustain. Tour. 24 (10), 1403—1424.

Bossel, H., 2002. Assessing viability and sustainability: a systems-based approach for deriving comprehensive indicator sets. Conserv. Ecol. 5 (2).

Brundtland, G.H., 1987. Our common future—call for action. Environ. Conserv. 14 (4), 291—294.

Burford, G., Tamás, P., Harder, M.K., 2016. Can we improve indicator design for complex sustainable development goals? A comparison of a values-based and conventional approach. Sustainability 8 (9), 861.

Colglazier, W., 2015. Sustainable development agenda: 2030. Science 349 (6252), 1048—1050.

Costanza, R., Daly, H.E., 1992. Natural capital and sustainable development. Conserv. Biol. 6, 37—46.

Elkington, J., 1998. Partnerships from cannibals with forks: the triple bottom line of 21st-century business. Environ. Qual. Manag. 8 (1), 37—51.

Eustachio, J.H.P.P., Caldana, A.C.F., Liboni, L.B., Martinelli, D.P., 2019. Systemic indicator of sustainable development: proposal and application of a framework. J. Clean. Prod. 241, 118383.

Freeman, C., 1982. Economics of Industrial Innovation. MIT, Cambridge.

Friedman, B., 2005. The Moral Consequences of Economic Growth. Alfred A. Knopf, New York City.

Griggs, D., Stafford-Smith, M., Gaffney, O., Rockström, J., Öhman, M.C., Shyamsundar, P., et al., 2013. Policy: sustainable development goals for people and planet. Nature 495 (7441), 305.

Hagan, J.M., Whitman, A.A., 2006. Biodiversity indicators for sustainable forestry: simplifying complexity. J. For. 104 (4), 203—210.

Hák, T., Janoušková, S., Moldan, B., 2016. Sustainable Development Goals: a need for relevant indicators. Ecol. Indicat. 60, 565—573.

Hardi, P., Barg, S., 1997. Measuring Sustainable Development: Review of Current Practice. IISD, Winnipeg.

Hartwick, J.M., 1977. Intergenerational equity and the investing of rents from exhaustible resources. Am. Econ. Rev. 67 (5), 972—974.

Hong, W., Jiang, R., Yang, C., Zhang, F., Su, M., Liao, Q., 2016. Establishing an ecological vulnerability assessment indicator system for spatial recognition and management of ecologically vulnerable areas in highly urbanized regions: a case study of Shenzhen, China. Ecol. Indicat. 69, 540—547.

Hotelling, H., 1931. The economics of exhaustible resources. J. Polit. Econ. 39 (2), 137—175.

Howlett, M., Ramesh, M., Perl, A., 2009. Studying Public Policy: Policy Cycles and Policy Subsystems, vol. 3. Oxford University Press, Oxford.

IAEG, U.N., 2016. Final List of Proposed Sustainable Development Goal Indicators. Report of the Inter-Agency and Expert Group on Sustainable Development Goal Indicators (E/CN. 3/2016/2/Rev. 1).

Innes, J.E., Booher, D.E., 2000. Indicators for sustainable communities: a strategy building on complexity theory and distributed intelligence. Plann. Theor. Pract. 1 (2), 173—186.

Isah, S.S., Sodangi, L.S., 2013. Development of Sustainable Key Performance Indicator (KPI) Monitoring and Control System Using Viable System Model.

Janoušková, S., Hák, T., Moldan, B., 2018. Global SDGs assessments: helping or confusing indicators? Sustainability 10 (5), 1540.

Jespersen, K., Gallemore, C., 2018. The institutional work of payments for ecosystem services: why the mundane should matter. Ecol. Econ. 146, 507—519.

Jónsson, J.Ö.G., Davíðsdóttir, B., Jónsdóttir, E.M., Kristinsdóttir, S.M., Ragnarsdóttir, K.V., 2016. Soil indicators for sustainable development: a transdisciplinary approach for indicator development using expert stakeholders. Agric. Ecosyst. Environ. 232, 179—189.

Korhonen, J., Seager, T.P., 2008. Beyond eco-efficiency: a resilience perspective. Bus. Strat. Environ. 17 (7), 411—419.

Krotscheck, C., Narodoslawsky, M., 1996. The Sustainable Process Index a new dimension in ecological evaluation. Ecol. Eng. 6 (4), 241—258.

Kılkış, Ş., 2016. Sustainable development of energy, water and environment systems index for Southeast European cities. J. Clean. Prod. 130, 222−234.

Lawn, P.A., 2003. A theoretical foundation to support the index of sustainable economic welfare (ISEW), genuine progress indicator (GPI), and other related indexes. Ecol. Econ. 44 (1), 105−118.

Leal Filho, W., Platje, J., Gerstlberger, W., Ciegis, R., Kääriä, J., Klavins, M., Kliucininkas, L., 2016. The role of governance in realising the transition towards sustainable societies. J. Clean. Prod. 113, 755−766.

Lee, B.X., Kjaerulf, F., Turner, S., Cohen, L., Donnelly, P.D., Muggah, R., Waller, I., 2016. Transforming our world: implementing the 2030 agenda through sustainable development goal indicators. J. Publ. Health Pol. 37 (1), 13−31.

Loorbach, D., 2010. Transition management for sustainable development: a prescriptive, complexity-based governance framework. Governance 23 (1), 161−183.

Lu, Z.N., Chen, H., Hao, Y., Wang, J., Song, X., Mok, T.M., 2017. The dynamic relationship between environmental pollution, economic development and public health: evidence from China. J. Clean. Prod. 166, 134−147.

Lundvall, B., Borrás, S., 2009. Science, technology, and innovation policy. In: Oxford Handbook of Innovation. Oxford University Press.

MacNeil, J., Winsenius, P., Yakushiji, T., 1991. Beyond Interdependence. Oxford University Press, New York.

McGillivray, M., White, H., 1993. Measuring development? The UNDP's human development index. J. Int. Dev. 5 (2), 183−192.

Meadows, D.H., 1998. Indicators and Information Systems for Sustainable Development.

Meadows, D.H., Meadows, D.L., Randers, J., Behrens, W.W., 1972. The Limits to Growth.

Meckling, J., Allan, B.B., 2020. The evolution of ideas in global climate policy. Nat. Clim. Change 10 (5), 434−438.

Meyar-Naimi, H., Vaez-Zadeh, S., 2012. Sustainable development-based energy policy making frameworks, a critical review. Energy Pol. 43, 351−361.

Moldan, B., Janoušková, S., Hák, T., 2012. How to understand and measure environmental sustainability: indicators and targets. Ecol. Indicat. 17, 4−13.

Pearce, D.W., Turner, R.K., 1990. Economics of Natural Resources and the Environment. Harvester Wheatsheaf, London.

Pols, A.J.K., Romijn, H.A., 2017. Evaluating irreversible social harms. Pol. Sci. 50 (3), 495−518.

Prescott-Allen, R., 1997. Barometer of Sustainability: Measuring and Communicating Wellbeing and Sustainable Development. IUCN, Gland, CH.

Pupphachai, U., Zuidema, C., 2017. Sustainability indicators: a tool to generate learning and adaptation in sustainable urban development. Ecol. Indicat. 72, 784−793.

Rahdari, A.H., Rostamy, A.A.A., 2015. Designing a general set of sustainability indicators at the corporate level. J. Clean. Prod. 108, 757−771.

Raszkowski, A., Bartniczak, B., 2019. On the road to sustainability: implementation of the 2030 agenda sustainable development goals (SDG) in Poland. Sustainability 11 (2), 366.

Regmi, B.R., Star, C., Leal Filho, W., 2016. Effectiveness of the local adaptation plan of action to support climate change adaptation in Nepal. Mitig. Adapt. Strategies Glob. Change 21 (3), 461−478.

Ritthoff, M., Rohn, H., Liedtke, C., 2002. Calculating MIPS: Resource Productivity of Products and Services.

Roca, L.C., Searcy, C., 2012. An analysis of indicators disclosed in corporate sustainability reports. J. Clean. Prod. 20 (1), 103−118.

Ropke, I., 2004. The early history of modern ecological economics. Ecol. Econ. 50, 293−314.

Rotmans, J., Loorbach, D., 2009. Complexity and transition management. J. Ind. Ecol. 13 (2), 184−196.

Schianetz, K., Kavanagh, L., 2008. Sustainability indicators for tourism destinations: a complex adaptive systems approach using systemic indicator systems. J. Sustain. Tour. 16 (6), 601−628.

Schmidt, M., Schwegler, R., 2008. A recursive ecological indicator system for the supply chain of a company. J. Clean. Prod. 16 (15), 1658−1664.

Scipioni, A., Mazzi, A., Mason, M., Manzardo, A., 2009. The Dashboard of Sustainability to measure the local urban sustainable development: the case study of Padua Municipality. Ecol. Indicat. 9 (2), 364−380.

Smith, S.L., 2002. Devising environment and sustainable development indicators for Canada. Corp. Environ. Strat. 9 (3), 305−310.

Solow, R.M., 1974. The economics of resources or the resources of economics. Am. Econ. Rev. 64 (2), 1−14.

Stevens, C., 1993. Do environmental policies affect competitiveness? OECD Obs. 183, 22−26.

Turner, R.K., Pearce, D., Bateman, I., 1993. Environmental Economics: An Elementary Introduction. Johns Hopkins University Press, Baltimore.

United Nations, 2015. Transforming Our World: The 2030 Agenda for Sustainable Development. General Assembly 70 Session.

Vaghefi, N., Siwar, C., Aziz, S.A.A.G., 2015. Green GDP and sustainable development in Malaysia. Curr. World Environ. 10 (1), 1.

Valcárcel-Aguiar, B., Murias, P., 2019. Evaluation and management of urban liveability: a goal programming based composite indicator. Soc. Indicat. Res. 142 (2), 689−712.

Vogel, P., Maddalena, T., Catzeflis, F., 1986. A contribution to taxonomy and ecology of shrews (*Crocidura zimmermanni* and *C. suaveolens*) from Crete and Turkey. Acta Theriol. 31 (39), 537−545.

Wackernagel, M., Lewan, L., Hansson, C.B., 1999. Evaluating the use of natural capital with the ecological footprint: applications in Sweden and subregions. Ambio 604−612.

WCED, 1987. Our Common Future. World Commission on Environment and Development Oxford University Press, Oxford, pp. 1−87.

Wolfslehner, B., Vacik, H., 2008. Evaluating sustainable forest management strategies with the analytic network process in a pressure-state-response framework. J. Environ. Manag. 88 (1), 1−10.

Wood, H.W., 1985. The United Nations World Charter for Nature: the developing nations' initiative to establish protections for the environment. Ecol. Law Q. 12 (4), 977−996.

World Bank Group, 2020. International Development, Poverty, & Sustainability. Retrieved from: https://www.worldbank.org/.

Zhang, J., 2012. Delivering Environmentally Sustainable Economic Growth: The Case of China. Asia Society Report, pp. 2−25.

Modeling methods for the assessment of the ecological impacts of road maintenance sites

Behzad Bamdad Mehrabani[1], Luca Sgambi[1], Elsa Garavaglia[2], Negarsadat Madani[3]

[1]*Faculty of Architecture, Architectural Engineering and Urban Planning, Université Catholique de Louvain, Tournai, Belgium;* [2]*Department of Civil and Environmental Engineering, Politecnico di Milano, Milan, Italy;* [3]*Local Environment Management & Analysis (LEMA), Department UEE, Université de Liège, Liège, Belgium*

Chapter outline

Environmental Sustainability and Economy. https://doi.org/10.1016/B978-0-12-822188-4.00009-9

1. Introduction

Although air quality has slowly improved in Europe, more than 400,000 premature deaths are annually attributed to air pollution in the continent (EEA, 2020). Therefore, air quality assessment through emission modeling is of great importance. Emission models determine the amounts of various pollutants emitted by vehicles in transportation facilities. Dispersion models are then employed to determine the concentration of a particular pollutant at a specific location and time. Then, exposure models are used to estimate exposure to air pollution (the number of individuals who inhale pollutants). The effects of the pollutant on the population (health effects) can then be calculated (health impact assessment) using the above information. Finally, the total economic effects of pollution are calculated using the combination of the effects of air pollutants on humans and the environment.

The transportation sector is a major source of air pollution, with Road Maintenance Sites (RMSs) releasing a considerable amount of emissions. Road maintenance comprises a wide range of activities that keep road structures and assets as close to their as-constructed or renewed conditions as possible (Wisconsin Department of Transportation, 2017).

By causing delays in traffic and increasing fuel consumption, road maintenance activities result in multiple environmental and ecological impacts. However, a limited number of studies have directly investigated the impacts of RMSs on exhaust emissions on roads and highways. In other words, the majority of previous studies have neglected the application of traffic emission modeling in investigating exhaust emissions by maintenance-related traffic. Moreover, some of these studies have reported conflicting results. Therefore, there is a need to thoroughly examine the impact of RMSs on the amount of traffic emissions to identify the best modeling methods and the most important factors and variables affecting the amount of emissions. This research aims to address the above gap and assess the existing emission models to determine the environmental impact of RMSs. Thereby, this chapter offers recommendations on enhancing the accuracy of the existing models.

To investigate the environmental impact of RMSs on roads, the second section defines RMS activities and explores the different types of effects that RMS can have on the road network. The third section presents explanations regarding traffic simulation, since the first step in estimating the environmental impact of RMSs is the calculation of their traffic impacts. The fourth section of this chapter consists of an introduction to various types of pollutants and how they are exhausted. Section 5 explores various approaches to traffic emissions modeling in different traffic conditions, mainly in the presence of RMSs. A simple simulation is also presented in Section 6 to illustrate the negative environmental impacts of RMSs. Finally, suggestions are made to improve the performance of existing models (Section 7).

2. Road maintenance

Transport accounts for a quarter of greenhouse gas emissions, a figure that continues to rise as demand grows (European Commission, 2019). Demand for mobility will continue to grow over the next three decades. By 2030, annual passenger traffic will surpass 80 trillion passenger kilometers (50% increase); global freight volumes will grow by 70%; and an additional 1.2 billion cars will be on the road by 2050 (double today's total) (The World Bank, 2017). Higher demand means higher levels of pollution and an increased need for infrastructure maintenance. As a result, modern societies are seeking to develop a sustainable transport sector in which less pollution is caused. Under current trends, road transport and private cars remain dominant and have an ever-greater role to play in meeting this additional demand. One way to develop a sustainable transport sector is to reduce road transport pollutants since road transport is responsible for more than 70% of emissions (Alonso et al., 2019). RMS plays a major role in transport emission, which will be discussed below.

2.1 What is road maintenance and why is it important?

The Highway Capacity Manual (HCM) 2010 defines a work zone as "A segment of highway in which maintenance or construction operations reduce the number of lanes available to traffic or affect the operational characteristics of traffic flowing through the segment" (Transportation Research Board, 2010). Therefore, since an RMS is a subset of work zone segments, this chapter uses the word "RMS" instead of a "work zone."

Road maintenance activities are one of the main factors affecting the mobility of road networks. Although the cost of road maintenance activities is relatively high, such activities help avoid massive construction costs. Road maintenance activities are essential for a variety of reasons including safety and economic, environmental, and social well-being. In particular, such activities are important because they

- Promote road safety;
- Increase the availability and mobility of the transport network;
- Support the local and national economy by ensuring that freight and businesses can move efficiently and safely;
- Reduce vehicle operating costs;
- Improve travel time in the long-term; and
- Reduce vehicle emissions (by improving traffic conditions and vehicle speeds) in the long-term (O'flaherty, 2001; PIARC, 2014).

Therefore, road maintenance involves significant activities that lead to major advantages in the long run despite their short-term environmental impacts and costs. What follows presents the classification of road maintenance activities. Given the wide variety of such activities, it is beyond the scope of this chapter to provide an in-depth overview. The readers can refer to the relevant references cited below to find out more about each of these activities. It should be noted that given the purpose of the present chapter, which is to examine the environmental and ecological impacts of road maintenance activities, what matters is whether these maintenance activities result in lane closures. Previous surveys have shown that most pavement rehabilitation activities result in lane closures.

Road maintenance activities can be categorized based on (1) The road assets on which they are performed; (2) The type of activity being performed and (3) The time and period of the activity. The following presents a brief explanation of each category.

1. Road maintenance activities are classified by the road assets on which they are performed (Alberta Department of Transportation, 2000; Illinois, 2008; Texas Department of Transportation, 2018; Shahin, 2005) into the following categories: pavement, roadside, bridge, traffic control, right of way, drainage, signs, safety items, and illumination activities.

2. Road maintenance activities are divided by the time and period of the activity into the following groups (Robinson et al., 1998; CIHT, 2012; Wisconsin Department of Transportation, 2017; CORNWALL Council, 2018):
   - Routine Maintenance: Activities, such as vegetation control, drain clearing, and culvert clearing that are planned, scheduled, and performed (sometimes annually) to maintain or preserve the condition of the highway system to achieve an adequate level of service.
   - Programmed Maintenance (Preventive Maintenance): Activities that form part of a capital program and are usually carried out according to a planned schedule. These activities promote system service life by preventing deterioration. Examples include surface dressing, resurfacing, and the strengthening or reconstruction of roads and footways.
   - Reactive Maintenance (Corrective Maintenance): These activities are carried out as a response to inspections, complaints, or emergencies. Examples include filling potholes, clearing, and restoring safety measures following traffic accidents. These activities cannot be scheduled with certainty in advance.
   - Emergency: These activities are intended as a response to severe weather and other emergencies affecting highway networks. Examples include snow removal and the removal of debris or obstacles left by natural disasters, etc.
   - Restorative Maintenance: These activities restore the pavement section to an acceptable service level by removing or repairing existing distress. Besides, such activities retain the existing pavement while increasing width throughout the length of the section. Restorative maintenance activities are not the same as preventative activities, since they are performed on degraded pavements.

3. Road maintenance activities are classified into the following categories according to actions specific to seasonal weather conditions (Alberta Department of Transportation, 2000):
   - Winter maintenance (snow and ice control) activities such as snow plowing, sanding, snow removal, and snow fencing.
   - Spring clean-up activities such as sweeping, bridge washing, culvert/drainage clearing.
   - Summer surface maintenance activities such as dust control, line painting (center, shoulder, and lane lines), and message painting (stop bars, pedestrian crosswalks).

2.2 What are the impacts of road maintenance sites on the road network?

RMSs can be divided into short-term and long-term sites based on the duration of their activities. The longer the duration of the activities, the higher their environmental effects. The presence of RMSs on roads causes numerous changes in road conditions. The effects of RMS on the road network in the short term include (Robinson et al., 1998):

1. Higher Level of Service (LOS)
2. Higher road user and administration costs
3. Socioeconomic impacts
4. Lower road safety
5. Environmental degradation (pollutant emissions)

Since the objective of this chapter is to investigate the environmental and ecological impacts of RMSs, it only considers environmental-related impacts or impacts that lead to environmental and ecological issues (e.g., LOS, road safety, and pollutant emissions).

2.2.1 Level of service (LOS)

As stated by HCM, LOS is "a quantitative stratification of a performance measure or measures representing the quality of service." The existence of RMS causes changes in road LOS. In other words, RMSs may cause lane closure, road closure, and the reduction of lateral clearance between vehicles and roadside objects (Transportation Research Board, 2010). These circumstances influence capacity, speed, or both (which are indicators of LOS). How RMSs affect road capacity and vehicle speed depends on various factors such as heavy vehicle percentage (HVP), ramp, work zone speed, driver composition, weather conditions, work zone length, lane closure location, lane width, work time, work zone activity duration, number of closed lanes, number of opened lanes, road type, work zone intensity, and work zone grade (Weng and Meng, 2013). The impact of RMSs on capacity, speed, and road safety has been investigated by many studies. The most important of such studies are discussed below.

There are several approaches to the estimation of work zone capacity, which are categorized into parametric, nonparametric, and simulation groups. In parametric approaches, the predictor takes a predetermined form. Nonparametric approaches do not assume that the structure of a model is fixed (e.g., artificial neural network, etc.).

HCM is one of the most important authorities on how to calculate LOS for highway facilities. The manual states that work zones (RMSs) affect the capacity and speed of vehicles as well as the speed-flow relationship. The impact of RMSs on capacity is different from their impact on speed. The impact of RMSs on capacity and speed in HCM is estimated using the parametric approach. In what follows, the impacts of RMS on capacity and speed are discussed separately.

The following factors should be specified to determine the impact of RMSs on capacity: lane closure type (e.g., shoulder closure, three-to-two lane closure), barrier type, area type, lateral distance, and time of the day (daytime or nighttime). First, the Lane Closure Severity Index (LCSI) must be calculated by specifying the ratio of the number of open lanes during road work to the total (or normal) number of lanes (decimal) and the number of open lanes in the work zone. Then, barrier type (concrete and hard barrier for the long term; cone, plastic drum, or soft barrier for the short term), area type, lateral distance, time of the day (daytime or nighttime), and the amount of Queue Discharge Rate (QDR) are calculated using the LCSI value. Finally, the amount of capacity at RMS can be obtained by using QDR and the percentage of capacity reduction before breakdown. When using this methodology, the calculated capacity should not be greater than the capacity of the nonwork zone.

RMSs also affect the free-flow speed (FFS), which cannot be calculated using equations developed for nonwork zones. To obtain FFS at RMSs, we need an equation that takes the ratio of the nonwork zone speed limit to the work zone speed limit, the work zone speed limit, lane closure severity, barrier type, time, and total ramp density into account. The work zone speed limit has a direct relationship with FFS at RMSs while, LCSI, ramp density, soft barrier type, and time of the day (night time) have an indirect relationship with FFS at RMSs.

The above HCM methodology, as well as estimates by previous studies, can be employed to calculate the ratio of capacity and free-flow speed in the RMS condition to the non-RMS condition (the capacity reduction factor and the free-flow speed reduction factor). The results show that the capacity reduction factor for different work activities is from 0.68 to 0.95, while the free-flow speed reduction factor is from 0.78 to 1.0 (Edara et al., 2018).

A review of the effects of RMSs on capacity and speed reported by previous studies reveals that the presence of RMSs causes reductions in capacity and speed, which consequently leads to lower LOS and higher traffic congestion (Transportation Research Board, 2016).

2.2.2 Road safety

The presence of RMS affects road safety in two ways: (1) Through speed changes in the RMS sections (since speed is one of the most critical factors in road safety (Bamdad Mehrabani and Mirbaha, 2018)) (2) Through the threat it poses to workers. Although the data available for RMS crashes is limited and incomplete, previous studies have indicated that crash rates at RMSs are generally higher than non-RMS locations (Paolo and Sar, 2012). The majority of previous studies have concluded that the presence of a work zone (RMS) increases both crash severity and crash rate (Yang et al., 2015).

Some types of collisions are more likely to occur at RMSs. Previous research indicates that rear-end crashes are more common at RMSs, which is due to higher speeds at the beginning of RMSs compared to the temporary speed limit. Drivers reduce their speeds only when the lane width is reduced, resulting in high deceleration rates and a higher likelihood of rear-end crashes (Paolo and Sar, 2012). The main factors contributing to road crashes at RMSs are as follows:

- **Driver Expectations**: In many cases, drivers do not expect the presence of RMSs due to improper road signage.
- **Roadside Hazards**: Occasionally, the configuration of tools, signs, and maintenance vehicles leads to collisions.
- **Driver Behavior**: Driver behavior is one of the most important contributing factors in all types of crashes, including RMS collisions. For instance, whether a driver is a commuter or not might affect driving behavior at RMSs.
- **Unsuccessful Mitigation Strategies**: Although strategies have been developed to increase RMS safety, some measures might actually increase the risk of collisions.
- **Roadway Characteristics**: Road features such as lighting and pavement conditions are among the most critical factors in increasing crash risk at RMSs.
- **Environmental Conditions**: Weather conditions, driver visibility, etc., have a significant impact on crash risk.
- **Secondary Crashes Caused by Roadway Incidents**: Crashes occurring within the work zone may contribute to congestion and aggressive driving, leaving the upstream roadway susceptible to additional incidents. On the other hand, the crash rate in the congested traffic flow condition is nearly three times higher compared to the noncongested condition (at RMSs) (Waleczek et al., 2016).
- **Combined Effects**: Previous studies have shown that combined effects such as the combination of light conditions and weather conditions are among the most critical factors for RMS safety (Ghasemzadeh and Ahmed, 2019).

2.2.3 Pollutant emissions

The transport sector is responsible for around 26% of the overall Greenhouse Gas (GHG) emissions in Europe (Nocera et al., 2018), 70% of which is attributed to road transport. Emissions exhausted by

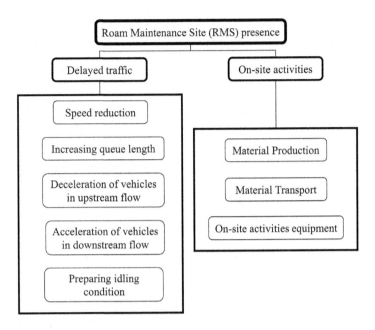

FIGURE 10.1

The causes of pollution in road maintenance sites.

road maintenance activities constitute the second most crucial factor in GHG emissions caused by roads (Deng, 2010). Therefore, it is important to examine the factors affecting such emissions caused by such activities.

RMSs usually lead to the closure of several lanes or the whole road, which severely impacts the transport network. Such impacts cause delays in traffic flow, which leads to additional emissions. On the other hand, road maintenance activities themselves produce emissions through fuel consumption by the maintenance equipment. Therefore, the existence of RMSs causes extra emissions in the road environment in two ways (Fig. 10.1): emissions caused by on-site activities and exhaust emissions by delayed traffic due to lane closure.

Numerous models have employed speed and queue length, which are significant traffic-related factors, as criteria for evaluating vehicle emissions. As mentioned in the previous section, vehicle speed decreases at maintenance sites. Previous studies indicate that higher traffic flow speeds (up to 60 km/h) lead to lower fuel consumption and vehicle emissions. Thus, the speed reduction in RMSs leads to extra emission. The relationship between speed and fuel consumption is shown in Fig. 10.2.

Another traffic-related factor affecting vehicle emissions at RMSs is queue length. Vehicle emissions are highest under long queue length and congested traffic flow conditions (Dong et al., 2019; Lizasoain-Arteaga et al., 2020; Pandian et al., 2009).

In addition, the vehicle speed is decreased (deceleration) when approaching an RMS and increased (acceleration) once the queue starts to discharge (outside the RMS). Acceleration increases pollutant emissions, especially at high speeds when the engine and emissions control systems are highly loaded.

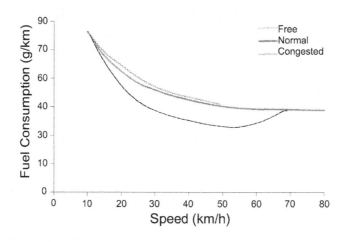

FIGURE 10.2

Fuel consumption factors for free and congested traffic (Samaras et al., 2019).

On the other hand, the emissions and fuel consumption rates of decelerating vehicles are independent of speed.

Another aspect of emissions caused by RMSs is the on-site activities themselves. On-site activities cause emissions in three ways: material production, material transport, and equipment and machines. To estimate material production emissions, an emissions factor, which is available for each material type, is multiplied by the amount of material (Liu et al., 2018, 2019). In addition, fuel consumption by construction equipment such as bulldozers and excavators is used to calculate the amount of emissions exhausted by the transport of materials and operations performed at RMSs. However, previous research (Huang et al., 2009) suggests that on-site activities constitute a small fraction (e.g., less than 10%) of the total energy consumption and pollutant emissions at RMSs. Therefore, this chapter does not consider the effects of on-site activities.

Among the impacts of RMSs on the road network described above, this chapter investigates the environmental effects of RMSs. Delayed traffic emissions are the most important factor in investigating the environmental impacts of RMSs. Delayed traffic caused by such sites has a variety of effects on the environment, which can be measured through traffic simulation. The following section provides a brief explanation of various approaches to traffic simulation.

3. Traffic simulation

3.1 What is traffic simulation modeling and why do we need it?

Researchers have long employed simulation as a powerful tool for simplifying reality. May (1990) defines simulation as a "numerical technique for conducting experiments on a digital computer." Simulations employ mathematical models and might include microscopic or macroscopic stochastic characteristics. Simulation is a viable alternative to analytical models, since such models have the highest degree of simplification in model representation.

On the other hand, traffic systems are very complex, nonlinear, and affected by multiple external and internal factors. Thus, there is a need to simplify reality to evaluate different options and scenarios. Simulation is capable of assessing the performance of several alternatives, supplied externally to the model by the decision-maker.

3.1.1 The definitions and assumptions of microscopic, mesoscopic, and macroscopic traffic simulation

Based on the intended application, there are three distinct approaches to simulation: microscopic, macroscopic, and mesoscopic. The main differences between these approaches are the level of aggregation and the tradeoff between optimizing simulation speeds and estimating traffic states (or traffic phenomena) as accurately as possible. These approaches are briefly described in what follows.

- **Microscopic Traffic Simulation Modeling**: This approach considers the dynamics of individual vehicles (or the response of individual drivers). Microscopic traffic simulation modeling is based on the description of the motion of individual vehicles in the traffic stream. In other words, microscopic models consider the interactions of *individual* vehicles. Information considered by these models include acceleration, deceleration, speed, and the lane-changing behavior of individual vehicles. Microscopic models describe vehicles (as well as their drivers) individually in the form of varying characteristics and multiple classes. This type of simulation takes highly detailed information from road sections into consideration and is usually used for modeling on the street or intersection level. Microscopic parameters for describing traffic flow are headway, gap, and occupancy.
- **Macroscopic Traffic Simulation Modeling**: These models are the aggregated versions of microscopic models. Macroscopic models are based on the aggregate behavior of the traffic flow and do not consider individual vehicles. Instead of tracking vehicles on an individual level, macroscopic simulations take place on a section-by-section basis, which is why they are employed by studies that address the behavior of the entire transport network. The variables employed by macroscopic models are general descriptors representing traffic conditions as flows at a high level of aggregation without distinguishing their parts. Macroscopic parameters, which characterize the traffic stream as a whole, are volume (rate of flow), speed, and density.
- **Mesoscopic Traffic Simulation Modeling**: These models combine some of the characteristics of macroscopic and microscopic models. Mesoscopic models consider most parameters at a high level of detail but describe the activities and the interactions between vehicles at a much lower level of detail compared to microscopic models. Mesoscopic models assume that vehicles move together as packets or platoons. The simulation of the flow of buses on a Bus Rapid Transit (BRT) lane is an example of mesoscopic simulation.

4. Pollution

Although the transportation sector creates economic value by increasing mobility, it leads to environmental issues such as pollution. Water, soil, and air pollution are three main types of pollution resulting from the activities of the transportation sector. Transport has long been a significant source of air pollution and, consequently, there is substantial concern over the effects of transport emissions on human health.

RMSs usually lead to air pollution, rather than soil or water pollution. To explore the environmental effects of RMSs more accurately, various types of contaminants, especially those emitted by RMS, are discussed below.

4.1 What are the different types of air pollution?

Air pollution can be defined as "the presence of any liquid, solid, or gas compound in the atmosphere at such concentration values that can directly or indirectly affect humans, animals, and/or plants" (Hussain and Keçili, 2020). There are two types of air pollutants: primary pollutants and secondary pollutants. Primary pollutants are compounds that are emitted as such (i.e., they are directly emitted in the atmosphere). Secondary pollutants are formed through chemical reactions in the atmosphere (i.e., they are formed indirectly). Most of the pollution caused by the transportation sector results from primary pollutants.

Four major types of sources that contribute to air pollution are stationary sources (those that are fixed in location), mobile sources (primarily transportation), fires, and biogenesis (naturally occurring emissions) . Each of these sources produces different types of pollutants. Nevertheless, the National Ambient Air Quality Standards (NAAQS) recommends that the following criteria air pollutants be taken into account by air pollution studies (USEPA, 2016): lead (Pb), carbon monoxide (CO), oxides of nitrogen (NO_x), volatile organic compounds (VOCs), sulfur dioxide (SO_2), and particular Matter (PM). The following is a brief description of each of the six criteria pollutants.

- **Lead (Pb)**: Lead is a soft, blue-gray metal, usually found in the form of lead compounds.
- **Carbon Monoxide (CO)**: CO is a toxic, colorless, and odorless gas, which is generated through the incomplete combustion of hydrocarbons. The amount of produced CO depends on the air-to-fuel ratio (A/F) in the combustion process.
- **Oxides of Nitrogen (NO_X)**: Mixtures of nitrogen monoxide and nitrogen dioxide—$\Sigma(NO, NO_2)$—are generally described as nitrogen oxides (NO_x). NO_x is produced during combustion.
- **Volatile Organic Compounds (VOCs)**: VOCs react with NOx to form ozone. Hydrocarbon (HC) emissions, which are classified as Volatile Organic Compounds (VOCs), are a result of incomplete combustion and fuel evaporation.
- **Sulfur dioxide (SO_2)**: The compound is emitted by sulfur-containing fuels such as coal and oil. It is colorless and has a sharp nasty smell.
- **Particulate Matter (PM)**: Particulates are fine particles of material suspended in the air. PM is produced by industries, natural sources, motor vehicles, agricultural activities, mining and quarrying, and wind erosion. Primary particles are emitted directly from sources such as construction sites, unpaved roads, fields, smokestacks, and fires. Secondary particles are generated through reactions between sulfur dioxides, nitrogen oxides, and other compounds in the atmosphere. Most fine PMs are secondary particles. Criteria matter are categorized into two groups: PM with a size of 10 microns or less (PM10) and particulate matter with a size of 2.5 microns or less (PM2.5).

4.2 What types of air pollution are caused by road maintenance?

Pollution emitted by the transportation sector is commonly referred to as mobile source air pollution, which includes pollution emitted by aircraft, commercial marine vessels, locomotives, nonroad equipment, and on-road vehicles (USEPA, 2016). On-road vehicles, which are investigated by the present research, emit pollutants in three ways (Tsanakas, 2019): (1) exhaust emissions (CO_2, CO, NO_x, HC, PM, etc.): emissions produced primarily from the combustion of petroleum products such as petrol, diesel, natural gas, and liquefied petroleum gas; (2) evaporative emissions (HC, VOC, etc.): vapors that are emitted by fuel and engine systems; and (3) tire and clutch abrasion (PM, etc.): the mechanical abrasion and corrosion of vehicle parts (tires, brakes, and clutch), the road surface wear, and the corrosion of the chassis.

The role of RMSs in causing pollution is similar to that of other subsets of the transportation sector. As mentioned in Section 2, RMSs can emit pollution in two ways: through delayed traffic and through on-site activities. Delayed traffic (mobile sources) exhausts significant amounts of only four of the six criteria pollutants; VOC, CO, NO_x, and PMs (Kutz, 2004; Meyer and Elrahman, 2019). As a result, the majority of studies on vehicular emissions (Lizasoain-Arteaga et al., 2020) have evaluated the above four pollutants to measure air pollution. On the other hand, the primary source of pollution emitted by on-site activities is fuel consumption by equipment and machines employed at the site. Therefore, the type of emissions caused by on-site activities is similar to the type of emissions produced by delayed traffic. However, previous studies have recommended that the following six pollutants be examined to enhance the accuracy of life cycle assessment studies (Lizasoain-Arteaga et al., 2020): CO_2, CO, NO_x, CH_4, C_6H_6, NH_3, VOC, and PM2.5.

5. Modeling approaches

Emissions modeling, which is described below, is the first step in evaluating the health and economic effects of air pollutants. The amount of exhaust emissions caused by RMSs can be calculated using traffic emission models. To more precisely examine various approaches to modeling, Section 5.1 introduces traffic emission modeling methods and Section 5.2 presents the models that analyze the effects of RMSs on exhaust emissions.

5.1 Traffic emission modeling

There are various approaches to emission modeling. In a few studies, regression modeling has been employed to determine the relationship between exhaust emissions and various road features such as slope and elevation. However, the majority of past studies have employed models based on emission factors. These models employ emission factors, traffic-related data, vehicle fleet composition, and other local characteristics to estimate the exhaust emissions in grams per vehicle and grams per kilometer. The emission factor is a critical component in emission modeling. An emissions factor is "a representative value that attempts to relate the quantity of a pollutant released to the atmosphere with an activity associated with the release of that pollutant." (Cheremisinoff, 2011). A dynamometer test usually provides the emission factor for each vehicle type in gram per kilometer, gram per second, or gram per fuel burned.

Table 10.1 Parameters affecting traffic emissions.

Characteristics	Parameters
Traffic-related data	Speed, traffic flow, queue length and delay, traffic density, driving behavior, driving mode (accelerating, decelerating, idling, stop-and-go), driving style (e.g., aggressive driving), vehicular composition, startup mode
Local characteristics	Road type, road geometry (e.g., curvature and longitudinal grade), ambient temperature, altitude, relative humidity, weather conditions, atmospheric pressure, pavement quality
Vehicle characteristics:	Vehicle type, vehicle age, mileage, emissions control equipment, engine load, vehicle weight and size, maintenance frequency, engine size, engine type, engine dynamics (engine speed, power demand, etc.), air-to-fuel mass ratio
Fuel	Fuel type, oxygen content, sulfur content, volatility, density

Many parameters affect the emission factor and, consequently, the total traffic emissions. In general, these parameters can be categorized into four groups: traffic-related data, local characteristics, vehicle characteristics, and fuel. Each of these parameters is listed in Table 10.1.

Like traffic simulation models (Section 3), emissions models are divided into three general categories: microscopic, macroscopic, and mesoscopic (Quaassdorff et al., 2016; So et al., 2018; Samaras et al., 2019). Microscopic emission models are as detailed as traffic flow microscopic models. These models typically require a wide range of detailed geometric and driving behavior data (the exact speed profile of the vehicles as input). However, some macroscopic models estimate fleet emissions on a regional or country-wide scale. Most macroscopic emission models utilize average speed as input. Some mesoscopic emission models combine the capabilities of the two models mentioned above. These models can be used in traffic-link-based urban emission modeling. In mesoscopic emission models, both average speed and driving dynamics are incorporated into emission calculations.

Below is a brief explanation of how these models work, the software used for modeling, and the most important models in each category.

5.1.1 Microscopic emission models

Microscopic emission models estimate emissions by evaluating the driving speed and acceleration characteristics/profiles on a vehicle-by-vehicle and second-by-second basis. These models are more suitable for the assessment of interventions on single roads, single junctions, or short sections of highways. The data for this type of modeling is usually imported from microscopic traffic models. These models provide the instantaneous emission factors of each vehicle from all vehicle categories as output.

Microscopic emission models are categorized into two groups: cycle variable models and modal models. Cycle variable models do not use instantaneous information. Several driving cycle variables (e.g., idle time, average speed, and acceleration) determine the emission factors in the cycle variable model. On the other hand, modal emission models use instantaneous information directly from the trajectory information. These models depend on specific engine or vehicle operating modes. There are two types of modal models: speed-based models and engine-based models. The instantaneous speeds and accelerations of each vehicle are essential data for speed-based models. Speed-based microscopic emission models are based on empirical measurements relating vehicle emissions to the type, instantaneous speed, and acceleration of the vehicle (Panis et al., 2006). The mathematical formula of these models is as follows (Ma et al., 2012):

$$E_{i,j}(t) = f\left(C_j, a, v, \ldots, t, \Phi_{i,j}\right) \tag{10.1}$$

In which:

$E_{i,j}(t) =$ the emissions or fuel consumption of gas type i for the vehicle category j

C_j and $\Phi_{i,j} =$ Model parameters

$a =$ acceleration

$v =$ speed

Moreover, in some speed-based models, instantaneous speed and acceleration data are not directly imported. Instead, such models use the Vehicle Specific Power (VSP) and Engine Stress (ES) to describe the relationship between transient operating conditions and vehicle emissions. VSP and ES are functions of instantaneous speed, instantaneous acceleration, road grade, and vehicle mass.

The most crucial disadvantage of speed-based models is that although such models can account for driving dynamics, they cannot explain the operating characteristics of the engine. In engine-based emissions models, engine functions such as power and speed are modeled based on speed profile and road grade input. Fuel consumption and pollutant emissions are calculated based on such engine functions (So et al., 2018).

Many microscopic emission models have been developed by previous research. Some examples are as follows (Ma et al., 2012; Samaras, 2019):

- AVL CRUISE
- Comprehensive Modal Emissions Model (CMEM)
- Emissions from Traffic (EMIT)
- International Vehicle Emissions (IVE)
- Motor Vehicle Emissions Simulator (MOVES)
- Panis emissions model
- Passenger Car and Heavy Duty Emissions Model (PHEM)
- POLY
- Virginia Tech Microscopic model (VT-Micro)

5.1.2 Macroscopic emission models

Macroscopic traffic emission models use average aggregated network parameters to estimate traffic emissions. In principle, such models are used together with macroscopic traffic flow models. Macroscopic traffic emission models are usually used for modeling emissions at the network level or large road sections and mostly utilize average speed as input. In other words, in macroscopic emission models, the analysis of single-vehicle emissions is not necessary. Total emissions are calculated by multiplying vehicle activity with the corresponding emission factor expressed in gram per kilometer. The conceptual formula for calculating traffic emission is as follow:

$$\text{Traffic emissions} = \text{Vehicle Activity} * \text{emissions Factor} \tag{10.2}$$

Vehicle activity is usually expressed as vehicle kilometers traveled. Moreover, vehicle fleet composition is another crucial parameter since emission factors are provided based on various types of vehicles. Ease of use and the limited amount of required data have made such models widely popular.

Macroscopic emissions models can be categorized into three groups (Kanagaraj and Treiber, 2019): area-wide models, average-speed models, and traffic-variable models. For further detail about the above models, please refer to Kanagaraj and Treiber (2019).

Here are some of the most important models in this category:

- Emission Factors (EMFAC)
- Assessment and Reliability of Transport Emissions Models and Inventory Systems (ARTEMIS)
- UK's National Atmospheric Emissions Inventory (NAEI)
- Computer Program to calculate Emissions from Road Transport (COPERT)
- Handbook Emissions Factors for Road Transport (HBEFA)
- Mobile-Source emissions (MOBILE)

5.1.3 Mesoscopic emission models

Mesoscopic and macroscopic emission models employ similar processes to estimate emissions. The only difference is the level at which they are employed. Macroscopic emission models are usually employed at the regional or county-wide levels while mesoscopic models are used at the link level. In other words, mesoscopic models are traffic-link based. It should be noted that many studies have not differentiated between macroscopic and mesoscopic emissions models and have referred to both models as macroscopic emission models (Kanagaraj and Treiber, 2019; Tsanakas, 2019). Besides, many of the presented macroscopic emission models are also applicable to mesoscopic emissions modeling.

The most important category of mesoscopic emission models is traffic situation models. In these models, the provided emission factors correspond to different vehicle classes, road categories, and traffic situations. Traffic situations are divided into four classes: free flow, heavy, saturated, and stop-and-go. Roads are categorized by their environment, speed limit, and type. Each traffic situation is qualitatively defined based on the characterization of the traffic situation. Please refer to Quaassdorff et al. (2016), Samaras et al. (2019), and Liu et al. (2018), for more information.

The most widely used models referred to by previous studies as both mesoscopic and macroscopic models are:

- The Handbook of Emission Factors for Road Transport (HBEFA)
- Computer Program to Calculate Emissions from Road Transport (COPERT)

5.2 Traffic emission modeling in the presence of RMSs

Few studies have directly assessed the impact of work zones or RMSs on vehicle emissions. Majority of previous research on the effects of RMSs on emissions have employed Life Cycle Assessment (LCA). Section 5.2.1 presents studies that have examined the effects of delayed traffic caused by RMS using LCA. Then, Section 5.2.1 explores the studies that have directly investigated the effects of different lane closure scenarios (different RMS configuration) on exhaust emissions (without using LCA).

5.2.1 Evaluating the effects of RMSs in life cycle assessment

LCA is a methodology that comprehensively evaluates the total environmental burden of roads. Five stages are commonly identified during the life cycle of roads: material production, construction, use (which includes leaching, rolling resistance, albedo, and lighting), maintenance (which includes delayed traffic emissions in addition to the replacement of the layers), and end of life. In most of these studies, it has been concluded that the emissions exhausted during the maintenance phase should be considered in the LCA (Lizasoain-Arteaga et al., 2020; Liu et al., 2019).

One of the studies that have evaluated the effects of RMSs in LCA has been conducted by Huang et al. (2009). This study concludes that the existence of RMSs increases the amount of emissions. It recommends the application of microscopic emission models instead of macroscopic emission models, since the latter usually overestimate the amount of exhaust emissions. This research recommends that the duration of maintenance operations be limited to reduce exhaust emissions.

Galatioto et al. (2015) concluded that increases in traffic levels result in an exponential increase in emissions during road activities due to the oversaturation and delay caused by reduced capacity after lane closure. Although the microscopic approach performs well in emissions modeling, the study recommends that the effects of RMSs on exhaust emissions be examined at the network level as well, since RMS operations are usually performed at the network level, rather than the link level. The factors affecting exhaust emissions by RMSs are identified by the researchers as the type, duration, and timing of road activities.

Another study examining the effects of RMSs on vehicle emissions using LCA was conducted by Hanson and Noland (2015). The study used a model called "completed Greenhouse Gas Assessment Spreadsheet for Transportation Capital Projects (GASCAP)" for calculating emissions during the whole life cycle of a road. To estimate the exhaust emissions, the model employs the HCM methodology to obtain road capacity and queue length in work zones and MOVES to obtain the emission factors. The study, which investigated the effects of the presence of an alternative route through macroscopic modeling, found that the amount of exhaust emissions increases significantly in full closure scenarios.

Inti et al. (2016) also concluded that the delayed traffic caused by RMSs should be considered in LCA. Using macroscopic emission modeling in the MOVES environment, they found that the longer the duration of lane closure, the higher the amount of exhaust emissions. The following parameters are incorporated in the model developed by the study: projected AADT at maintenance time, hourly demand distributions, vehicle classification, work zone speed, vehicle speeds during queues, lane closure duration, timing, work-zone length, and lane capacity during maintenance.

Lizasoain-Arteaga et al. (2020) investigated the effects of congestion during RMS operations using two approaches: macroscopic and microscopic emission modeling. A model developed by Panis et al. (2006), which is integrated into AIMSUN, is applied for microscopic emission modeling. The results of this study indicate that congestion levels play a significant role in the amount of exhaust emissions by RMSs. The study also found that the accuracy of microscopic modeling is higher than that of macroscopic modeling, as macroscopic models may overestimate the amount of emissions. According to the authors, the existence of alternative routes is one of the most critical factors that should be taken into account.

In short, it can be concluded that LCA studies indicate that RMSs use fossil fuels (and consequently exhaust emissions) in two ways: gasoline and diesel used by delayed traffic, and the fuel and electricity used for machinery and equipment. These studies have found that fuel used by delayed traffic cause more exhaust emissions than the machinery and equipment employed by RMS activities (Liu et al., 2019).

5.2.2 Evaluating the effects of RMSs in different lane closure scenarios

Studies that directly examine the effects of RMSs in different lane closure scenarios have employed both microscopic and macroscopic emission models. Zhang et al. (2011), who used the microscopic emission modeling (CMEM), compared the amount of exhaust emissions in the presence of RMSs and lane closure conditions with the amount of emissions in the free-flow condition. According to this study, the effects of RMSs on exhaust emissions is not the same for different vehicle types (light-duty and heavy-duty vehicles) and pollutants. For instance, the amount of CO, HC, NOx, and CO_2 exhaust emissions by heavy vehicles is higher compared to the free-flow condition, which is not true for light vehicles. The study also found that the amount of exhaust emissions by both light-duty and heavy-duty vehicles in transition zones was higher compared to the other parts of RMSs. This indicates that the acceleration and deceleration behavior of vehicles have a major influence on the amount of emissions.

Gu et al. (2018) have recently conducted a study on the impact of lane closure caused by RMSs by utilizing the macroscopic emission modeling approach (using average speed as input) and the microscopic emission modeling approach (using vehicle operating model as input). Traffic data was obtained using VISSIM software and used as input for the MOVES model. The study investigates the effect of RMS schedule (daytime or nighttime) on the amount of emissions and concludes that road maintenance operations at nighttime exhaust less emissions due to less traffic congestion. It should be noted that Alvanchi et al. (2020), who used VISSIM and the ad-on emissions model EnViVer (which is integrated into the VISSIM), found similar results and stated that maintenance schedule has a significant effect on the amount of vehicle emissions.

The above studies have used the microscopic emission modeling approach. However, few studies have employed the macroscopic modeling approach. For instance, a study by Kim et al. (2018), which employed the macroscopic emission modeling approach, indicates that mitigating traffic congestion from heavy (average speed 5 mph) to medium congestion (average speed 15—25 mph) in a work zone would reduce fuel consumption and GHG emissions by 40% on a freeway and by 32% on a multilane road. This study, which incorporates Average Annual Daily Traffic (AADT) data into its emission model, evaluates different levels of congestion at RMSs. It is worth mentioning that this research does not compare the congested traffic condition in the presence and absence of RMSs. The comparison carried out by the authors is limited to the congested traffic condition at RMSs and the free-flow conditions in the absence of RMSs.

Although majority of past studies have concluded that the existence of RMSs increases the exhaust emissions by vehicles, a limited number of studies have concluded that RMSs can actually decrease pollutant emissions. For instance, research by Avetisyan et al. (2014) used the On-road Simulation Emissions Estimation Model (ORSEEM) (microscopic emission modeling approach) to investigate the impact of the work zone and traffic incidents on exhaust emissions. The authors concluded that the existence of a work zone reduces emissions. The reason behind this result can be stated as follows: the researchers have only used vehicle speed for emission modeling, which means that the modeling is not accurate enough, and the congested conditions are not evaluated (the evaluation is limited to scenarios without queues).

Another study that reached similar results was conducted by Wang et al. (2014). The authors first estimated the road capacity and the queue length in the presence of RMSs using the HCM work zone methodology. Then, these data were presented as input to the MOVES model (macroscopic approach). The study concluded that in the absence of congestion, the amount of delayed traffic emissions caused by RMS is less significant in comparison with other road construction and maintenance phases. They recommended that the delayed traffic emissions caused by RMSs be evaluated in the presence of congestion.

In summary, it can be concluded that the presence of RMSs increases exhaust emissions. A few studies have identified the delayed traffic emissions caused by RMSs as insignificant because the scenarios evaluated by such studies do not examine the congestion condition. As a result, congestion caused by lane closure (RMS) can be identified as the most critical factor in exhaust emissions.

A review of the methodologies employed by previous studies reveals that there are two distinct approaches to the evaluation of the effects of RMSs on emissions: macroscopic emission modeling and microscopic emission modeling. The majority of studies use microscopic emission modeling because of its superior performance compared to macroscopic emission modeling, which is because vehicles experience different operating modes at RMSs. Thus, aggregated data may lead to inaccurate estimations. The microscopic emission modeling approach first simulates traffic flow using traffic simulator software (such as AIMSUN, VISSIM, etc.). Second-by-second and vehicle-by-vehicle traffic-related data (e.g., instantaneous speed, instantaneous acceleration, etc.) are then imported into emission models such as MOVES, CMEM, etc. Another group of studies have employed the emissions models integrated into traffic simulation software such as the Panis model in AIMSUN or the EnViVer model in VISSIM. On the other hand, macroscopic emission model studies typically obtain aggregated traffic data by HCM methodology and then incorporate these traffic data (such as average queue length or average speed) into emission models. In other words, it can be stated that previous studies have only used existing emission modeling approaches, and to the best of the authors' knowledge, researchers have not yet developed a specific methodology for delayed traffic emissions caused by RMSs.

The following simulates a very simple case study in the AIMSUN environment using the Panis microscopic emissions model to demonstrate that the delayed traffic (congestion condition) caused by RMSs increases emissions.

6. Application

Given that some studies have reported conflicting results, a simple simulation is presented in order to thoroughly examine emission modeling in RMS and to determine whether delayed traffic caused by RMS increases or decreases the amount of emission in road sections. AIMSUN was selected because of its ability to model road network geometry, the behavior of individual vehicles in response to traffic, and its easy-to-use graphical interface. In particular, the software is capable of modeling traffic emissions and provides three pollution emissions models: the QUARTET Pollution Emissions Model, the Panis Pollution Emissions Model, and the London Emissions Model. As reported by past research, microscopic emission models exhibit superior performance compared to macroscopic emission models. The Panis model, which is a microscopic emission model first released in 2006 and widely employed by previous studies, is used for modeling in this chapter. The model provides the exhaust emissions of four pollutants (CO_2, NOx, VOC, and PM) based on the type, the instantaneous speed, and the acceleration of the vehicle. Please refer to Panis et al. (2006) for more information on this model.

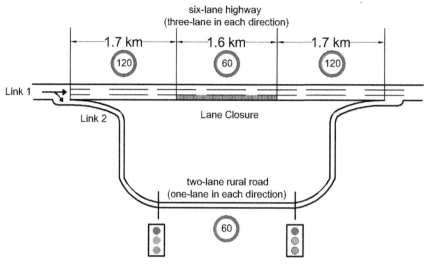

FIGURE 10.3

The simulation case study.

Fig. 10.3 illustrates the case study for the simulation. As can be seen in this figure, two links are simulated in this chapter. The first link is a six-lane highway (three lanes in each direction) with a total length of 5 km, and a 1.6 km long lane closure in midsegment. The second link is also considered as an alternative road, which is a two-lane (one lane in each direction) rural road. For the sake of simplification, traffic flow is only simulated in one direction. Traffic flow is set to 6000 personal cars per hour (pc/hr), which leads to LOSE.

In reality, alternate routes usually have a higher length compared to the main road. In addition, they have an intersection with other routes. The alternate route length is therefore set to 7 km in this simulation to promote accurate results. There are two signalized intersections on link 2. The signalized intersections cycle is set to 80 s with a green time of 50 s for link 2.

It should be noted that since the purpose of this study is to investigate the extra exhaust emission in the presence of RMSs, the cycle length of intersections in the alternative road was assumed to be a hypothetical value. Moreover, the volume of other approaches in these intersections was not considered.

Different scenarios intended for simulation are presented in Table.10.2. As shown in this table, four scenarios are simulated. There is no lane closure in the first scenario. In the second scenario, one lane is closed in the midsegment of link 1 due to RMS operations. To consider the effects of the existence of an alternative route, half of the traffic flow is assigned to the alternative route (scenarios 3 and 4). Fig. 10.4 shows the exhaust emissions by vehicles during a 1-hour simulation.

As shown in Fig. 10.4, the amount of all pollutants in the presence of lane closures (scenario 2) is higher than the no-lane-closure scenario (scenario 1). The presence of alternative routes has also been

Table 10.2 Scenarios for RMS lane closure.

Scenario	Operating type		Link flow (Pc/h)		Signalized intersection	
	Link 1	Link 2	Link 1	Link 2	Link 1	Link 2
1	No lane closure	No lane closure	6000	0	No signalized intersection	—
2	Lane closure		6000	0		—
3	No lane closure		3000	3000		Yes
4	Lane closure		3000	3000		Yes
Simulation duration: 1 h						

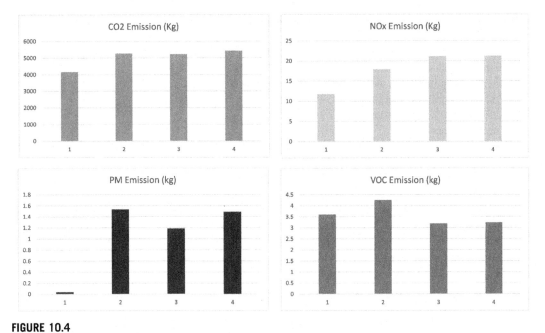

FIGURE 10.4

Exhaust emissions for different scenarios.

investigated in scenarios 3 and 4. An examination of the exhaust emissions by vehicles in these scenarios (in lane closure conditions) shows that when there is an alternative route (scenario 4), the emission values of CO_2 and NOx are higher compared to the no-alternative-route condition (scenario 2). However, this is not the case for PM and VOC emissions. On the other hand, when there is an alternative route, the amount of exhaust emissions for all pollutants for the lane closure condition is higher than the no-lane-closure condition. Therefore, the results reported by the majority of previous studies, which indicate that exhaust emissions are higher in the lane closure condition compared to the no-lane-closure condition, are confirmed.

In this study, the alternate route was considered as a two-lane highway (one lane in each direction). It was also assumed that when there is an alternative route, the existing demand is split 50/50 between the two available routes. It is therefore suggested that future studies consider different traffic assignment scenarios and functional classes for alternative routes, as the amount of exhaust emissions in the alternative route depends on these two parameters.

7. Conclusions

The following conclusions can be drawn regarding the ecological effects of RMSs:

- In addition to vehicle and local characteristics (as mentioned in Table 10.1), the following factors, which are specific to RMSs, affect the amount of exhaust emissions:
  o Congestion formation
  o Closure schedule
  o The length of the transition zone
  o RMS length
  o Acceleration, deceleration, idling, and stop-and-go behavior at RMSs
  o The exitance of an alternative route
  o The type, duration, and timing of road work
  o Lane closure configuration
  o The RMS speed limit
  o Vehicle speeds in case of queue formation
  o Number and capacity of lanes during maintenance
- The amount of exhaust emissions at RMSs can be various for different types of pollutants.
- The amount of RMS emissions can vary for light-duty and heavy-duty vehicles. Therefore, future research should thoroughly investigate the behavior of each vehicle type at RMSs.
- Previous studies have only used the existing approaches to emissions modeling and, to the best of the authors' knowledge, researchers have not developed a specific methodology to model delayed traffic emissions caused by RMSs to date. Therefore, there is a need to calibrate the existing emission models since the acceleration, deceleration, stop-and-go, and idling behavior of drivers at RMS congestion may differ from ordinary traffic congestion.
- In general, microscopic emission models have superior performance in modeling RMS emissions. It can be stated that the MOVES model and the Panis model are most widely used due to their higher accuracy.
- In future studies, it is advisable to investigate the impact of RMSs at the network level, where such activities are usually performed.
- In some cases, when there is an RMS in one direction, traffic from the opposite direction is affected as well, which should also be investigated by future research.
- In future studies, the impact of different lane closure configuration scenarios (e.g., different transition zone lengths) should be considered because the configuration can affect the acceleration and deceleration behavior of vehicles (which are the most critical factors in exhaust emissions).

- The presence or absence of traffic congestion in RMSs is the most critical factor in the amount of exhaust emissions at these zones. Therefore, it is necessary to perform a sensitivity analysis for the amount of exhaust emission in different LOS scenarios.
- In reality, usually, when a maintenance operation is performed on a highway, an alternative route is specified, or some drivers experimentally use alternative routes. The traffic assignment is very important in these cases. Therefore, the drivers' route choice in these cases should be investigated in future studies.

References

Alberta Ministry of Transportation, June 2000. Highway Maintenance Guidelines and Level of Service Manual, Section 1 − Highway Maintenance Guidelines.

Alonso Raposo, M., Ciuffo, B. (Eds.), Alves Dies, P., Ardente, F., Aurambout, J-P., Baldini, G., Baranzelli, C., Blagoeva, D., Bobba, S., Braun, R., Cassio, L., Chawdhry, P., Christidis, P., Christodoulou, A., Corrado, S., Duboz, A., Duch Brown, N., Felici, S., Fernández Macías, E., Ferragut, J., Fulli, G., Galassi, M-C., Georgakaki, A., Gkoumas, K., Grosso, M., Gómez Vilchez, J., Hajdu, M., Iglesias, M., Julea, A., Krause, J., Kriston, A., Lavalle, C., Lonza, L., Lucas, A., Makridis, M., Marinopoulos, A., Marmier, A. Marques dos-Santos, F., Martens, B., Mattas, K., Mathieux, F., Menzel, G., Minarini, F., Mondello, S., Moretto, P., Mortara, B., Navajas Cawood, E., Paffumi, E., Pasimeni, F., Pavel, C., Pekár, F., Pisoni, E., Raileanu, I-C., Sala, S., Saveyn, B., Scholz, H., Serra, N., Tamba, M., Thiel, C., Trentadue, G., Tecchio, P., Tsakalidis, A., Uihlein, A., van Balen, M., Vandecasteele, I., 2019. The Future of Road Transport − Implications of Automated, Connected, Low-Carbon and Shared Mobility. EUR 29748 EN, Publications Office of the European Union, Luxembourg. ISBN 978-92-76-14318-5, doi: 10.2760/668964. JRC116644.

Alvanchi, A., Rahimi, M., Mousavi, M., Alikhani, H., 2020. Construction schedule, an influential factor on air pollution in urban infrastructure projects. J. Clean. Prod. 255, 120222.

Avetisyan, H.G., Miller-Hooks, E., Melanta, S., Qi, B., 2014. Effects of vehicle technologies, traffic volume changes, incidents and work zones on greenhouse gas emissions production. Transport. Res. Transport Environ. 26, 10−19.

Bamdad Mehrabani, B., Mirbaha, B., 2018. Evaluating the relationship between operating speed and collision frequency of rural multilane highways based on geometric and roadside features. Civil Eng. J. 4 (3), 609−619.

Cheremisinoff, N.P., 2011. Chapter 31 − Pollution Management and Responsible Care. Academic Press, Waste, ISBN 9780123814753, pp. 487−502. https://doi.org/10.1016/B978-0-12-381475-3.10031-2.

CORNWALL Council, 2018. Highway Maintenance Manual.

Deng, F., 2010. Greenhouse Gas Emissions Mitigation in Road Construction and Rehabilitation: A Toolkit for Developing Countries. World Bank Group.

Dong, Y., Xu, J., Liu, X., Gao, C., Ru, H., Duan, Z., 2019. Carbon emissions and expressway traffic flow patterns in China. Sustainability 11 (10), 2824.

Edara, P., Bharadwaj, N., Sun, C., Chang, Y., Brown, H., 2018. Understanding the Impacts of Work Zone Activities on Traffic Flow Characteristics.

EEA, 2020. The European Environment - State and Outlook 2020-Knowledge for Transition to a Sustainable Europe. European Environment Agency, ISBN 978-92-9480-090-9. https://doi.org/10.2800/96749. TH-04-19-541-EN-N.

European Commission, 2019. Sustainable Mobility, the European Green Deal. EUGreenDeal, ISBN 978-92-76-13901-0. https://doi.org/10.2775/395792. NA-02-19-958-EN-N.

Galatioto, F., Huang, Y., Parry, T., Bird, R., Bell, M., 2015. Traffic modeling in system boundary expansion of road pavement life cycle assessment. Transport. Res. Transp. Environ. 36, 65−75.

Ghasemzadeh, A., Ahmed, M.M., 2019. Exploring factors contributing to injury severity at work zones considering adverse weather conditions. IATSS Res. 43 (3), 131−138.

Gu, C., Farzaneh, R., Pesti, G., Valdez, G., Birt, A., 2018. Estimating vehicular emission impact of nighttime construction with VISSIM and different MOVES emission estimation approaches. Transport. Res. Rec. 2672 (25), 174−186.

Hanson, C.S., Noland, R.B., 2015. Greenhouse gas emissions from road construction: an assessment of alternative staging approaches. Transport. Res. Transport Environ. 40, 97−103.

Highway Capacity Manual, 2010. HCM2010. Transportation Research Board. National Research Council, Washington, DC, p. 1207.

Highway Capacity Manual, 2016. HCM2016. Transportation Research Board. National Research Council, Washington, DC.

Huang, Y., Bird, R., Bell, M., 2009. A comparative study of the emissions by road maintenance works and the disrupted traffic using life cycle assessment and micro-simulation. Transport. Res. Transport Environ. 14 (3), 197−204.

Hussain, C.M., Keçili, R., 2020. Chapter 1 − Environmental Pollution and Environmental Analysis, Modern Environmental Analysis Techniques for Pollutants. Elsevier, ISBN 9780128169346, pp. 1−36. https://doi.org/10.1016/B978-0-12-816934-6.00001-110.1016/B978-0-12-816934-6.00001-1.

Illinois, D.O.T., 2008. Bureau of Local Roads and Streets Manual.

Inti, S., Martin, S.A., Tandon, V., 2016. The necessity of including maintenance traffic delay emissions in life cycle assessment of pavements. Proc. Eng. 145, 972−979.

Kanagaraj, V., Treiber, M., 2019. Fuel-consumption and CO_2 emissions modells for traffic. In: International Climate Protection. Springer, Cham, pp. 155−160.

Kim, C., Ostovar, M., Butt, A.A., Harvey, J.T., 2018. Fuel consumption and greenhouse gas emissions from on-road vehicles on highway construction work zones. In: International Conference on Transportation and Development 2018.

Kutz, M., 2004. Handbook of Transportation Engineering, vol. 768. McGraw-Hill, New York, NY, USA.

Liu, Y., Wang, Y., An, D., 2018. Life-cycle CO_2 emissions and influential factors for asphalt highway construction and maintenance activities in China. Int. J. Sustain. Transp. 12 (7), 497−509.

Liu, Y., Wang, Y., Li, D., Feng, F., Yu, Q., Xue, S., 2019. Identification of the potential for carbon dioxide emissions reduction from highway maintenance projects using life cycle assessment: a case in China. J. Clean. Prod. 219, 743−752.

Lizasoain-Arteaga, E., Indacoechea-Vega, I., Alonso, B., Castro-Fresno, D., 2020. Influence of traffic delay produced during maintenance activities on the Life Cycle Assessment of a road. J. Clean. Prod. 120050.

Ma, X., Lei, W., Andréasson, I., Chen, H., 2012. An evaluation of microscopic emission models for traffic pollution simulation using on-board measurement. Environ. Model. Assess. 17 (4), 375−387.

May, A., 1990. Traffic Flow Fundamentals. Prentice Hall, Englewood Cliffs, New Jersey.

Meyer, M.D., Elrahman, O.A., 2019. Chapter 4 − Air and Water Pollution: An Important Nexus of Transportation and Health, Transportation and Public Health. Elsevier, ISBN 9780128167748, pp. 65−106. https://doi.org/10.1016/B978-0-12-816774-8.00004-9.

Nocera, I., Dianin, A., Cavallaro, F., 2018. Chapter Nine − Greenhouse Gas Emissions and Transport Planning: Toward a New Era?, Advances in Transport Policy and Planning, vol. 1. Academic Press, ISBN 9780128152942, pp. 245−280. https://doi.org/10.1016/bs.atpp.2018.07.009. ISSN 2543-0009.

O'flaherty, C.A., 2001. Soils for road work. In: O'flaherty, C.A. (Ed.), Highways the Location, Design, Construction and Maintenance of Pavements, p. 133.

Pandian, S., Gokhale, S., Ghoshal, A.K., 2009. Evaluating the effects of traffic and vehicle characteristics on vehicular emissions near traffic intersections. Transp. Res. Transport Environ. 14 (3), 180−196.

Panis, L.I., Broekx, S., Liu, R.H., 2006. Modeling instantaneous traffic emission and the influence of traffic speed limits. Sci. Total Environ. 425, 69−78.

Paolo, P., Sar, D., 2012. Driving speed behavior approaching road work zones on two-lane rural roads. Proc. Soc. Behav. Sci. 53, 672−681.

Quaassdorff, C., Borge, R., Pérez, J., Lumbreras, J., de la Paz, D., de Andrés, J.M., 2016. Microscale traffic simulation and emission estimation in a heavily trafficked roundabout in Madrid (Spain). Sci. Total Environ. 566, 416−427.

Robinson, R., Danielson, U., Snaith, M.S., 1998. Road Maintenance Management: Concepts and Systems. Macmillan International Higher Education.

Samaras, C., Tsokolis, D., Toffolo, S., Magra, G., Ntziachristos, L., Samaras, Z., 2019. Enhancing average speed emission models to account for congestion impacts in traffic network link-based simulations. Transp. Res. Transport Environ. 75, 197−210.

Shahin, M.Y., 2005. Pavement Management for Airports, Roads, and Parking Lots, vol. 501. Springer, New York.

So, J., Motamedidehkordi, N., Wu, Y., Busch, F., Choi, K., 2018. Estimating emissions based on the integration of microscopic traffic simulation and vehicle dynamics model. Int. J. Sustain. Transp. 12 (4), 286−298.

Texas Department of Transportation, September 2018. Maintenance Operations Manual, (512), pp. 463−8630 (All rights reserved).

The Chartered Institution of Highways & Transportation (CIHT), 2012. Sixty Years of Highways Maintenance.

The World Bank, 2017. Global Mobility Report 2017: Tracking Sector Performance, ISBN 978-0-692-95670-0.

The World Road Association (PIARC), 2014. The Importance of Road Maintenance, 2014R02EN.

Tsanakas, N., 2019. Emission Estimation Based on Traffic Models and Measurements, vol. 1835. Linköping University Electronic Press.

USEPA, 2016. U.S. Environmental Protection Agency, Office of Air Quality Planning and Standards, 2016, 2014 National Emissions Inventory, Version 1 Technical Support Document. Air Quality Assessment Division, Emissions Inventory and Analysis Group. Research Triangle Park, North Carolina.

Waleczek, H., Geistefeldt, J., Cindric-Middendorf, D., Riegelhuth, G., 2016. Traffic flow at a freeway work zone with reversible median lane. Transp. Res. Proc. 15, 257−266.

Wang, T., Kim, C., Harvey, J., 2014. Energy consumption and Greenhouse Gas Emission from highway work zone traffic in pavement life cycle assessment. In: International Symposium on Pavement Life Cycle Assessment, Davis, California, USA.

Weng, J., Meng, Q., 2013. Estimating capacity and traffic delay in work zones: an overview. Transport. Res. C Emerg. Technol. 35, 34−45.

Wisconsin Department of Transportation, 2017. Highway Maintenance Manual, Bureau of Highway Maintenance.

Yang, H., Ozbay, K., Ozturk, O., Xie, K., 2015. Work zone safety analysis and modeling: a state-of-the-art review. Traffic Injury Prev. 16 (4), 387−396.

Zhang, K., Batterman, S., Dion, F., 2011. Vehicle emissions in congestion: comparison of work zone, rush hour and free-flow conditions. Atmos. Environ. 45 (11), 1929−1939.

Circular economy and urban metabolism

Earth, wood, and coffee: empirical evidence on value creation in the circular economy

11

Stephan Kampelmann[1], Emmanuel Raufflet[2], Giulia Scialpi[3]

[1]*Laboratory for Landscape, Urbanism, Infrastructure and Ecology, Faculty of Architecture La Cambre-Horta, Université Libre de Bruxelles, Bruxelles, Belgium;* [2]*Département de Management, HEC Montréal, Montréal, QC, Canada;* [3]*Faculty of Architecture, Architectural Engineering and Urban Planning, University of Louvain UCLouvain, Ottignies-Louvain-la-Neuve, Belgium*

Chapter outline

Environmental Sustainability and Economy. https://doi.org/10.1016/B978-0-12-822188-4.00014-2

1. Introduction

How to build circular value chains that reduce the economic pressure on ecosystems to provide raw materials and resources? How to coordinate diverse actors from different sectors of society when setting up loops around materials? While circular economy (CE) research has paid considerable attention to new circular business models for companies, insufficient attention has been paid to new models of value creation that go beyond the micro-scale and individual companies (Arnsperger and Bourg, 2016; Babbitt et al., 2018).

Tate et al. (2019) consider the emergence of circular value systems as a process in which actors play different roles in establishing and managing novel information and material flows to identify more sustainable uses around available resources. We follow and extend this approach by applying a system-level conceptual framework to the empirical cases of three circular value chains around material flows (excavated earth, urban wood, and coffee grounds) that we observed through interviews and document analysis in three different countries (France, Canada, and Belgium).

Our overall objective in this chapter is to examine how circular value creation emerges in the form of value chains/systems. To address this objective, we consider two sets of research questions:

1. How can circular value systems be conceptualized? What are the main components of such systems?
2. How do circular value systems emerge? What are the actors, roles, and processes at work? What holds these systems together?

This chapter is organized as follows. We first review the literature on sustainable value creation and assess the contributions from value chain management and system analysis. Second, we sketch the theoretical contours of a framework for the analysis of circular value systems. Third, we apply this framework to the empirical cases describing emerging circular value systems in Canada (on urban wood), France (on excavated earth), and Belgium (on coffee grounds). We then discuss the empirical results from a comparative perspective. The final section concludes the chapter.

2. Literature review
2.1 Sustainable and circular value chains

In logistics and management research a value chain is defined as a chain that "encompasses all activities associated with the flow and transformation of goods from raw materials stage (extraction), through to the end user, as well as the associated information flows. Material and information flow both up and down the supply chain. Supply chain management (SCM) is the integration of these activities through improved supply chain relationships to achieve a sustainable competitive advantage" (Handfield and Nichols, 1999). Moreover, researchers have defined *sustainable* value chains as the "management of material, information and capital flows as well as cooperation among companies along the supply chain while taking goals from all three dimensions of sustainable development, i.e., economic, environmental and social, into account" (Seuring and Muller, 2008).

Several literature reviews on sustainable value chains have been published in the past decade (Koberg and Longoni, 2019; Pagell and Shevchenko, 2014; Panigrahi et al., 2019; Touboulic and Walker, 2015). Seuring and Meuller (2008) highlight that the implementation of sustainability in an existing value chain is achieved by a focal company which decides, designs, and sells the products and services, as it connects and influences both suppliers and clients within a firm-centric chain. In this view, the focal company decides to implement sustainability in the value chain based on corporate decisions, which themselves may result from pressures and incentives from government, customers, and other stakeholder groups.

Research on sustainable value chains has contributed to improving the understanding of the process of transformation of an existing value chain to a sustainable value chain. Yet, the literature falls short of explaining the emergence of circular value chains for several reasons. First, this research area assumes the preexistence of a value chain to begin with. However, circular economy initiatives can face situations in which the recycling or reuse of existing material resources do not exist. In other words, the underlying value chain needs to be created, not transformed. The emergence of circular value chains is thus not well explained by this research area.

Second, there is a large diversity of sustainability initiatives, including environmental performance measures, environmental management systems, environmental certifications and ecolabels, environmental innovation, environmental partnerships and collaborations, sustainable supply chain management, and sustainable human resource management (DeBoer et al., 2020). A similar range of initiatives can arguably be found in moves in the context of the circular economy of materials that we are concerned with in this paper. Such diversity calls for empirical research on specific and comparable types of initiatives. In this paper, we focus on initiatives that aim at the establishment of circular value chains around specific material flows.

Third, the literature on sustainable value chains assumes that a focal organization has the capacity and power to influence other actors in the value chain. Yet, emergent circular value chains are often characterized by high levels of power distribution in a value chain in which diverse actors from different sectors may not be coordinated and oriented by the decisions or corporate policies of a single organization.

2.2 Toward a systemic perspective on value creation in the circular economy

A recent strand of literature on supply chain management has pointed out that research on circular economy stands to gain from adopting more comprehensive conceptual frameworks. According to Tate et al. (2019), system analysis could be useful in this regard. They invite researchers to shift from "linear supply chain thinking" to "interconnected, circular, ecosystem thinking." They argue that this new focus could contribute to addressing global ecological and social challenges. Unfortunately, current research at the intersection of circular economy and supply chain management still lacks a systemic perspective. Bals et al. (2020) argue that literature in this area "frames the world in dyads instead of 'networks' or 'systems'. In order to move beyond short-sighted recycling solutions that still result in waste, a systemic perspective is needed that embraces cross-industry flows and more actors (e.g., taking care of the reverse logistics) than in traditional supply chains, establishing circular value cycles."

Different avenues for developing such a systemic perspective on value creation are possible. Tate et al. (2019) have opted for exploring direct analogies with natural ecosystems and formulating "biomimetic insights" that could be helpful for transitioning toward more circular business practices. Looking at natural ecosystems for inspiration is indeed a popular way of conceptualizing circular economy (Athanassiadis and Kampelmann, 2021).

Moving from simple models of value *chains* to the analysis of value *systems* implies being able to get a grip on complexity in the form of a larger number of interrelated variables. Several academic traditions have successfully explored and analyzed complex systems. While theological worldviews have long crowded out scientific forms of system analysis, the study of systems has reappeared in Western philosophy in the wake of Enlightenment. The foundations of ecology as a scientific discipline were laid in the 1930s; ecology pioneers such as Eugene P. and Howard T. Odum built on them to develop the ecosystem concept and apply it to forest and ocean systems in the 1960s (Odum and Bartlett, 1971). In the 1970s, system analysis started to develop formalized tools for describing complex biotopes, notably the notion of interconnected biogeochemical cycles within ecosystems (Duvigneaud, 1974). Almost simultaneously, a group of MIT scientists applied system analysis to the evolution of a matrix of biophysical and economic variables to estimate the "limits to growth" beyond which global production systems would become unstable. Historically and intellectually rooted in engineering science, today system analysis is a full-fledged academic discipline. In parallel, other disciplines have developed specific theories on monetary, solar, scholar, or legal systems.

As stated by Kampelmann (2017), one of the most relevant theories for applying system thinking to circular economy research can be extracted from the interdisciplinary literature concerned with the analysis of social-ecological systems and their metabolism (Fischer-Kowalski and Haberl, 2007). While most authors study social-ecological systems at the macroscopic scale, a rapidly expanding literature looks at territorial metabolism, and more specifically at the metabolism of urban agglomerations (Duvigneaud and Denayer-De Smet, 1975; Barles, 2010; Kampelmann and De Muynck, 2019). These advances could build on earlier research including the so-called Brussels School under the influence of Paul Duvigneaud, who initiated a first wave of progress in the understanding of social-ecological systems in the 1970s. Indeed, the book *L'écosystème Belgique* [Ecosystem Belgium] by Billen et al. (1983) is an early attempt at applying methods of scientific ecology to the analysis of the industrial system of Belgium. Its subtitle "A study in industrial ecology" is a hint that the book already anticipated the distinction between linear and circular economic systems.

In terms of methods, the system analysis of Duvigneaud and colleagues was mostly based on elaborate flow diagrams informed by biogeochemical data and analyses. This echoed the kind of representations used in scientific ecology to describe phenomena like the nitrogen or the carbon cycle. Unsurprisingly, the complexity of a social-ecological object such as the Belgian industrial system was framed by Billen and his coauthors in analogy to Duvigneaud's seminal representation of a forest, in which the interconnections of numerous cycles give rise to an overarching ecosystem. By contrast, frameworks developed by social scientists such as Elinor Ostrom and others put the emphasis on the roles of agency, as well as on the rules and institutions that govern social-ecological systems (Ostrom, 2009a). Tate et al. (2019) draw on network theory and complex adaptive systems (CAS) to formulate biomimetic insights on networked business systems. Such a strategy has helped to stress the need for interorganizational balance in value systems, for instance in the form of a balance between producers, consumers, scavengers, and decomposers (Babbitt et al., 2018).

We argue that the analysis of circular value systems implies accounting for both types of variables: on one hand, material flows that have formed the backbone of the analysis of economic systems in the tradition of industrial ecology based on Duvigneaud and, on the other hand, agencies and institutions that have received more attention by social scientists like Ostrom. To achieve such a combination, Kampelmann (2017) has proposed a framework for analyzing systemic relationships in the context of circular economy and applied it to the case of organic waste management. In the next section, we summarize the structure of this framework and submit our case that it can be a valuable tool for organizing and analyzing case study material on emerging circular value systems.

3. Analytical framework

Our framework for analyzing circular value systems encompasses heterogeneous concepts drawn from different disciplinary traditions and organized in three distinct sets.

The first set concerns agency: it includes the actors of the systems and the interactions between these actors. It aims to answer questions about the different types of actors in value systems; their roles as entrepreneurs, producers, scavengers (such as reverse logistics companies and overstock/salvage retailers), decomposers (recyclers and waste treatment companies), end users, or policymakers. In the case of emerging circular value systems, we are interested in the coordination mechanisms that bring about new value systems, the leading and secondary actors in these mechanisms, as well as the strategies they deploy.

The second set of variables is concerned with the biophysical elements and characteristics of the geographical spaces of the system (e.g., a valley, a neighborhood, a center, a periphery, etc.), its artifacts (a factory, different types of products), and the material and energy flows passing through the system. For the study of circular value chains, we are not only interested in the quantitative aspects of these flows but also how certain qualitative features influence the emergence of circular value chains.

Finally, the third set refers to framing elements such as the domain of the system, its rules, and its geospatial scale. Institutionalists such as Ostrom have emphasized the significance of these elements for understanding the underlying conditions of both agency and biophysical phenomena. The term "domain"—in the sense of "sphere of action"—aims to capture the limits of the system at hand. Depending on the research objective, the domain can be defined in terms of a natural ecosystem (such as a forest, i.e., the limits of the system are defined by the frontiers of the forest) or an anthropogenic system (such as the organic flows of a city, i.e., other material flows or organic flows in other cities are outside of the domain). Defining the limits of a system is to some extent arbitrary; researchers are often confronted with difficult tradeoffs between a more comprehensive representation of systemic interactions (which calls for increasing the domain to include more variables or a larger scope) and the need for keeping the amount of information manageable. An example of addressing this tradeoff is to formulate the system's domain as a "nexus," such as the energy-water-food nexus. Indeed, restricting the limits of the system to the interactions between energy, water, and food at the scale of a country can be justified by the particularly strong causal relationships between these three elements (Bazilian et al., 2011). Our use of the term "rules" refers to the sociological understanding of the institutions that "constrain, or from some points of view determine, the behavior of a special social group" (Scott and Marshall, 2005, p. 311).

These generic elements of a social-ecological system thus combine the perspectives of Paul Duvigneaud and Elinor Ostrom: the material flows as well as the scale and spaces that organize these flows are directly borrowed from the analysis of ecosystems as described by the Ecological Synthesis (Duvigneaud 1974); the domain of the system, its actors, and their interactions plus the rules that govern the system are inspired by the Institutional Analysis and Development Framework of Ostrom (2009b) (Table 11.1).

Table 11.1 Overview of the categories included in the framework for the analysis of value systems.

Agency	Biophysical elements	Framing elements
Key actors	Material and energy flows	Domain
Interactions between these actors	Spaces and artifacts	Scale
		Rules

Based on Kampelmann, S., 2017. Circular economy in a territorial context: the case of biowaste management in Brussels. In: Cassiers, I., Maréchal, K., Méda, D. (Eds.), Post-growth Economics and Society. Exploring the Paths of a Social and Ecological Transition, Series in Ecological Economics, Routledge, New York.

4. Case studies

In order to analyze the emergence of circular value systems, we describe and compare three recent in-depth case studies. The case studies are comparable due to their focus on the circularity of a particular material flow (wood, earth, and coffee grounds, respectively), but also provide substantial institutional variations to reflect how the involvement of different actors affects the emergence of circular value chains in different contexts.

4.1 Case 1: urban wood in Quebec, Canada

This case is based on three interviews with the director and the coordinator of Bois Public (BP) carried out in May and June 2019. We also interviewed a board member of Montreal's Environmental Council and a representative of the Department of Forestry of Greater Montreal. Moreover, we visited production sites and a carpentry school in May 2019.

Montreal, much like other cities in North America, experiences a massive occurrence of the emerald ash borer (*Agrilus planipennis*). This insect lays eggs on the bark of ash trees (*Fraxinus*). After hatching, the larvae eat into the outer strata of the tree and can cause their death within 3–5 years. City authorities all over North America proceed to preemptive cutting of affected ash trees in order to prevent accidents in city streets, parks, or gardens. In many places, the massive production of ash wood that results from these cuts has sparked place-based initiatives that strive to find local uses for the trees. Such initiatives have been particularly vibrant and organized in the United States, where they are federated by an "Urban Wood Network."

4.1.1 Agency

The idea to convert and use ash trees in Montreal was sparked by informal discussions between key *actors*: the mayor of the Municipality of Rosemont-La Petite Patrie (one of the *arrondissements* of Montreal's metropolitan area) and a social entrepreneur who later became the founder and first director of BP. These conversations took place in 2014 and established a joint interest for using ash trees locally. From the perspective of the municipality, cutting down ash trees in large numbers was undesirable: the local population cherished these neighborhood trees as valuable amenities and the removal of the ash trees represented a financial cost. Moreover, it became clear that the default approach for handling the wood from preemptive felling was not compelling in neither environmental nor economic terms. Most of the wood was cut into small chips that can be used as mulch or ground cover in parks, or as biomass for energy creation. Observing that the inner wood of ash trees was not affected by attacks of the insects, the social entrepreneur proposed a pilot project in which some trees were converted into higher value-added uses such as benches and other types of park furniture. The project aimed for creating higher economic and social value compared to ground cover or energy, while at the same time lowering environmental impacts. This strategy boils down to "wood cascading," which is a key principle of forest-based circular bioeconomy (Jarre et al., 2020).

The pilot project in Rosemont took place in 2015/2016 and gave rise to the creation of BP as an organization with the mission to coordinate and organize the necessary operations. Influenced by the background of its founder in social economy initiatives and given the public interest for local wood projects, BP was created as a social enterprise.

The way that BP set up the local value chain—from trees to wooden objects—focused on partnerships with other organizations that offered operational experience, especially sawmills and carpentry firms. BP basically acts as a broker who organizes the interventions of all other actors along the chain in bilateral negotiations and agreements: the company established how the municipality would collect the felled trunks, contracted a sawmill and kiln to convert and dry the wood, and worked with local carpentry firms for the secondary transformation of the dried boards.

BP also benefited from collaborating with Montreal's carpentry schools and the dynamic urban woodworking scene. In its first years, BP worked particularly closely with two social enterprise carpentry firms and a carpentry school whose incubator for new woodworking businesses is located in the same building as BP's offices. On our visit to this building, we were able to observe a bustling cluster of numerous small carpentry firms, but also other connected activities such as design companies and fab labs.

4.1.2 Biophysical elements

According to our interviewee from the Forestry Department, there are about 200,000 ash trees on the island of Montreal, many of which are street trees or occur in large parks such as the Mont-Royal. The exact proportion of trees that have been or will be affected by the ash borer epidemic is yet unknown. However, in its first year of operations (2016), BP converted 260 trunks (*billots*) of ash trees, followed by 457 in 2017, 739 in 2018, and 1800 in 2019. The latter increase is due to an agreement with the Forestry Management Department—which oversees all large parks on the island of Montreal—to work with BP for the local use of these trees.

The ash tree produces hardwood with many relevant applications in woodworking. By contrast, wood chips made from the bark or trunk of ash trees are less valued than chips from other tree species. A particularity of ash trees felled in urban agglomerations is the presence of nails or other pieces of metal in the trunks. These nails destroy the blades of sawmills and increase the costs and risks of sawing city trees.

The value chain that has been assembled by BP focuses on trunks that could be converted into timber and used in carpentry. The other parts of the trees—notably the branches of the canopy—continue to be chopped into chips and used as mulch or biomass. The bark and outer layers of the tree, i.e., the part where the ash borer larvae are active, must be destroyed to prevent spreading the disease.

Neighborhood parks and public spaces have played a key role in the development of the local value system for ash trees in Montreal. Since the ash trees grow in public spaces, their visibility and accessibility meant that the local population—including public officials like the mayor of Rosemont-Petite Patrie and local carpentry firms—had a vivid, sometimes very emotional, relationship with individual trees or tree stands. The fact that the trees grew in urban spaces and within residential areas distinguished them from the vast forest with millions of anonymous trees so abundant in Canada.

Once felled, the conversion of the ash trees was to be executed as locally as possible. However, there are no conventional sawmills within the administrative borders of Montreal. Indeed, commercial sawmills tend to be large operations converting thousands of cubic meters of wood per year; they are typically located in industrial or rural areas outside of urban agglomerations. Transporting small loads to these sawmills can be economically challenging. In order to carry out the pilot project in 2015/2016, BP therefore made use of a machine that is relatively common in Quebec: a mobile sawmill. This technology is used in rural parts of Quebec to build cabins or wooden houses from trees growing on the same plot. Given the availability of mobile sawmills around Montreal, BP was able to covert the trees within the urban agglomeration.

In order to deal with the larger volumes of trees that result from the successful establishment of a local value chain/system, BP has first shipped larger batches of trees by trailers to a sawmill in Mascouche, outside of Montreal. Following the increase in the number of trees converted by BP in 2019, a discussion has started whether it would be economically viable to set up a new permanent sawmill on the island of Montreal. This sawmill could be run by BP, the public authorities, or other partners.

4.1.3 Framing elements

The *domain* of the system set in place by BP has extended over time. Initially focused on ash trees in a local park of Rosemont-La Patrie, BP later started projects in other neighborhoods and municipalities. It now operates at the scale of the entire Island of Montreal, an area of c.500 km^2. Moreover, the focus on ash trees has given way to a larger attempt to create value from other tree species, although ash still represent the vast majority of BP's turnover.

While the operations of BP focus on Montreal, it is interesting to note that the company's activities are not restricted to this city. In fact, it has started consulting other municipalities in Quebec to support the creation of other local value chains.

The most important *rules* that have affected the set-up of the local value system around BP and its external partners are phytosanitary rules related to city trees. Indeed, the opportunity of a short-circuit value chain was not sparked by the presence of trees alone, but by the decision of public authorities to preemptively cut ash trees before they die and rot, which would make their timber less valuable. BP's operations are also affected by the special care that must be taken to avoid a further spread of the emerald ash borer, which means that handling ill ash trees is more complicated and costly compared to healthy trees.

Finally, the rules regarding the governance of tree stands in public parks have been consequential. If the local mayor of Rosemont-Petite Patrie did not have the authority to use ash trees from municipal park, the initial pilot project and indeed the entire value system might have failed to emerge. The multi-level system of Montreal, in which *arrondissements* (boroughs) manage local-level parks and street trees and the authorities of greater Montreal oversee larger parks, has proved to be conducive for allowing small-scale experimentation followed by upscaling to the larger volumes of regional-level parks.

4.2 Case 2: raw earth construction in Paris, France

This case is based on the analysis of documentation provided by the project coordinator of the Earth Cycle project and an interview with a staff member of the International Center for Earthen Construction (CRAterre) conducted in March 2020.

The use of excavated soil as building material has a long history; different premodern civilizations developed applications such as mud cladding or adobe bricks. While materials based on excavated earth are still widespread in traditional and vernacular architectures around the world, their use in the building industry received renewed interest in the context of efforts to curb the global increase in the consumption of other materials derived from fossil, nonrenewable resources, especially steel and concrete (Tangtinthai et al., 2019). Moreover, many building projects involve excavations to create space for underground structures such as garages, which means that excavated earth is available in vast quantities. Building with raw earth could therefore make a twofold contribution to circular economy: decreasing the use of nonrenewable building materials and turning an abundant waste flow into a resource for sustainable construction.

4.2.1 Agency

The Earth Cycle project started in March 2018 and is financially supported by the European Fund for Urban Innovative Actions (UIA). The estimated date for the launch of physical operations is February 2021.

The initiative aligned the interests of multiple *actors* to create pilots for the production of earth-based building material made with the excavated soil from major construction sites in Greater Paris. The emergence of the Earth Cycle project benefited from strong political support and is demand-based, i.e., the production is oriented toward a specific market created by the stakeholders themselves.

The project team is composed of a mix of private and public actors, which ensures that it includes different steps of the future value system. Stakeholders in the Earth Cycle project include the City of Sevran, a growing city that coordinates the project and which will facilitate the construction of a buildings materials factory; Greater Paris Developer (GPA), a public urban planning agency in Ile de France, which plays the role of a project developer; Greater Paris Transportation (SGP), a company overseeing the construction of a new underground metro line in the suburb of Paris and which will supply the raw material for the project; Quartus, an urban developer specialized in the construction of innovative and ecological neighborhoods and that will provide expertise on building with bio- and geo-sourced materials; Antea Group, a private company involved in urban planning and circular economy and included as experts in soil management and material data systems; Joly & Loiret, an architecture agency experienced with raw earth as building material and that will be in charge of product development; and the nonprofit organization Competence Emploi that will carry out training modules.

Moreover, the consortium includes five different research organizations (IFSTTAR, Sciences-Po Paris, AE&CC/ENSAG, Amàco, and CRAterre) to provide expertise on specific aspects of the value chain for raw earth building materials such as knowledge on earth-based architecture; the assessment of soil transport and processing; the analysis of opportunities and barriers for the replicability of pilot projects; the development of training programs; and the establishment of standards to certify earth construction techniques (ATEX) and training (ECVET).

Prior to the Earth Cycle initiative, the raw earth value chain in France has followed a polynuclear development driven by relatively uncoordinated action and geographically limited scope. Only a handful actions have been carried out on a larger scale such as the European PIRATE project (2012—15) for testing and certifying the construction skills related to earth building across Europe and a strategic guideline for six construction techniques (rammed earth, adobe, cob, wattle and daub, plasters, and light earth) conceived in 2012 by a working group including the Directorate of Habitat, Urbanism and Landscape (DHUP), the Minister for Ecology, Sustainable Development, and Energy (MEDDE), laboratories, companies, producers, builders, and professional trade unions.

Compared to past initiatives promoting raw-earth construction in France, Earth Cycle is the first pilot project characterized by a "demand-pull" rather than a "product-push" orientation (Habert, 2019). This is not uncommon for innovations developed in the context of the circular bioeconomy in the construction industry, where "science and technology push" plays an important role in "competing against the incumbent concrete frame construction system" (Lazarevic et al., 2020). By contrast, the strategy of Earth Cycle is to exploit the convergence of interests and complementarities between functions of heterogeneous partners within Greater Paris, including actors with a demand to find downstream uses for excavated earth and others with demand for building projects incorporating raw earth materials.

The use of excavation soils for building purposes is still rare in urban environments, but a handful of other initiatives exist in Europe. Terrabloc produces compressed earth blocks in Geneva, Switzerland. The earth used by Terrabloc is typically sourced on-site, except when the need for larger volumes requires importing soil from other urban locations. BC Materials produces bricks, plasters, and a mix for rammed earth walls in Brussels, Belgium. Here the earth is sourced from building sites or rural quarries. Both companies combine the services of architecture and materials manufacturing, i.e., they design and build projects with their own materials. Earth Cycle is quite different from such an in-house value chain, as it strives to develop a more systemic approach that aims for cooperation across multiple entities and at a much larger scale.

The project interacts in different ways with the existing supply chain of excavated soil management. The latter includes the following phases: diagnosis; extraction; transport and preparation of the material for reuse; collection and sorting; valorization or landfill. Earth Cycle closely collaborates with the actors in charge of the first four phases in an attempt to convince them of the benefits of a more laborious and time-consuming process of transforming excavated earth into building materials. Thanks to this broad setup, Earth Cycle includes actors representing both the incumbent "linear" and the envisaged "circular" value chain for excavated earth. The variety of involved actors arguably reflects the wide range of expertise and competences that must be mobilized in the transition to a circular value chain, from urban planning and large-scale public works over architecture and product development to industrial process management and training activities.

The need to engage with heterogeneous interests and perspectives is also borne out by the legal form of the soil processing facility that Earth Cycle plans to set up. This facility will operate as a public interest cooperative (SCIC) pursuing the goals of both economic efficiency and public interest. The membership of the cooperative is expected to include producers of goods and services, employees, local authorities, and public institutions. Any benefits will be reinjected into the cooperative as investments and other development activities.

4.2.2 Biophysical elements

Even though excavated material is not always included in the amount of construction and demolition waste, available statistics for France suggest that it can represent up to 80% of the total waste mass (European Commission, 2011). Greater Paris is a hotspot for the production of excavated earth. It includes the City of Paris and its suburbs, an area whose construction sites generate an estimated 20 million tons of excavated earth per year. Additional 40 million tons will be extracted in the next 10 years for works related to the Greater Paris Express featuring 200 km of tracks, 68 stations, and 7 technical centers (Gasnier, 2019).

The excavated earth that can be used for construction projects is the inorganic part of the soil, consisting of gravel, sand, silt, and clays in different proportions. The aim of the Earth Cycle project is to use these soils to produce earth-based building materials directly on-site. This would reduce material transports, the amount of waste sent to landfill, pressure on natural resources such as sand and gravel used to make concrete, and CO_2 emissions of the building projects (Benhelal et al., 2013).

The pilot factory will be located in Sevran, at around 20 km of the construction sites of two planned train stations. The choice was motivated by a desire to reduce transportation but also by a strategy to relocalize blue-collar jobs that have disappeared in the area. The facility will include two covered areas (one for production and one for administration) and two storage areas (one for earths and one for products). Three production lines will be developed for operations related to extrusion (earth panels used to make wall linings fixed to wood or metal frame studs); compression (compressed earth blocks, produced with a press and used for masonry); and plasters (earth-based coatings without artificial colorings or pigments and mortars to build the earthen masonry).

The plan to use excavated soils from the two construction sites—around 50,000 tons—immediately raised logistical problems: since the new facility is not expected to absorb and convert materials with the same speed as the excavations, additional storage space for stocking and drying the material will be needed. To limit the volumes to be stored, a scan of extracted soil will eliminate samples that are not suitable as raw materials for construction (e.g., due to pollution).

4.2.3 Framing elements

Contrary to many other efforts to render the construction sector more sustainable, the *domain* of the value system envisaged by the Earth Cycle project is restricted to an exclusive focus on soil and its potential journey from excavated sites to reuse in construction projects. Moreover, in a reaction to the rapid urbanization and infrastructure building that currently occurs in peripheral spaces around the French capital, the *scale* of the project is limited to the area of Greater Paris. Earth Cycle is thus an urban soil project. Being set in an urban context allows the project to benefit not only from the considerable economic and political clout of Paris-based institutions; it can also tap into the attention that the urban population pays to the benefits and nuisances of large-scale construction projects.

The scale of the production facilities will ultimately depend on factors like the possibility for mechanization and the anticipated production volume. The strategy of Earth Cycle is to increase the market share of raw earth construction by multiplying production facilities throughout the territory. In other words, the project wants to replicate its prototype processing factory on other sites, which is why replicability in other contexts is considered a priority for the research on the project. Challenges in this regard include the specificity, complexity, and size of Parisian construction sites, as well as the diversity of construction and soil management companies.

As regards the relevant *rules* of the emerging value system, compared to conventional building materials raw earth still lacks suitable normative references. This is a main barrier for the development of the value chain. As a consequence, Earth Cycle is working on three Atex type A, a procedure to evaluate the technical performances of a product that will be used on multiple building sites. It will allow commercializing the products from the pilot factory.

The policy landscape in Paris is certainly conducive for the emergence of a value system around the use of excavated earth as building material. The French capital has adopted a circular economy strategy and identified public procurement and large urban projects as opportunities for transitioning the urban economy (Athanassiadis and Kampelmann, 2021). What is more, the French law for energy transition (TEPCV) and several EU Directives foster reduction of waste and emissions and therefore create favorable conditions for Earth Cycle. This being said, the regulations in the market for building materials do not impose the internalization of environmental externalities into prices. This means that competing materials—foremost concrete and steel —remain the default choice in most nonresidential building projects. Compared to the inroads that wood has made in recent years as alternative building material (Panwar et al., 2015), raw earth construction has not seen significant increases in demand. The strategy of Earth Cycle follows the French tradition of voluntarist policies employing public instruments to coordinate and influence market forces.

4.3 Case 3: coffee grounds in Belgium

This section is based on empirical material collected in 2019 and 2020 during meetings and workshops organized within the context of a circular economy initiative on coffee ground in Belgium. Additionally, an in-depth interview with two staff members of the King Baudouin Foundation was conducted in March 2020. We also relied on the project documentation (including internal notes and minutes of meeting) that were provided to us by the project coordinator.

4.3.1 Agency

In Belgium, the first applications of circular economy thinking to coffee grounds focused on using the material as substrate in the production of mushrooms. Over the last decade, several start-ups ("Permafungi," "Champignon de Bruxelles") have launched operations to cultivate varieties of shitake or pleurotus using coffee grounds. The sourcing and markets of these companies are inherently urban.

In 2018, one of the major philanthropic organizations in Belgium—the King Baudouin Foundation (KBF)—started to develop activities related to circular economy. This choice was guided by KBF's objective to seek high societal impact and a desire to identify an area where the foundation's specificities would have the highest possible added value. KBF wanted to take advantage of its unique position as a federating entity and its privileged access to prominent figures from politics, administration, and the business world. KBF then took stock of existing circular economy initiatives in Belgium and Europe and decided to narrow down the scope of its intervention to the food sector before settling its focus on coffee grounds in late 2018. Coffee grounds were seen as having economic and environmental potential beyond what had already been achieved by pioneering start-ups; in line with its position as "independent assembler," KBF did not impose any specific vision, economic model, or technological hypothesis regarding how coffee residues could become more circular in the future.

While KBF was the main driving force behind the initiative, the foundation worked in close collaboration with two external partners: the Belgian federation of the agro-food industry (FEVIA) and the Royal Union of Coffee Roasters "Koffie-Café" (KC). These sectoral partners were interested in finding new approaches to coffee grounds, which they perceived as a waste product that could benefit from interesting new technologies or business models. They decided to form a consortium with KBF to push circular economy for coffee grounds. The three organizations shared the perception that (a) more value can be created from coffee grounds in Belgium and that (b) emerging technological, organizational, and social innovations could foster such value creation. The consortium worked with the private consultancy firm Möbius to manage the interactions with both the coffee grounds producers and the companies or research institutions proposing novel solutions for coffee recycling. KBF provided the financial resources for carrying out the first phases of the project.

The trio KBF-FEVIA-KC recognized that finding solutions to coffee grounds ending up as low-value waste would require a systemic approach that would integrate a wide range of different actors. However, working with all types of coffee brewers—from households to restaurants to larger institutions—was quickly discarded as impractical. Given KBF's access to big economic and administrative players, the trio decided to target large producers of coffee grounds, such as institutional kitchens in big administrations, hotels, prisons, or airports. In a relatively brief period in 2019, the consortium managed to spark vivid interest for coffee grounds recycling in large organizations such as Belgacom, the airport of Zaventem, the Federal Ministry of Finance, the municipal administration of Genk, and the port of Antwerp. These organizations were brought in contact with innovators experimenting with high value-added solutions for the use of coffee grounds. In a second phase scheduled to start in late 2020, some of these organizations will design and experiment with new value systems for coffee grounds.

4.3.2 Biophysical flows

The current state of the system in which KBF and its partners intervened is expansive, global, and linear. Coffee production is generally carried out in the Global South through large coffee monoculture plantations owned by international capital. The largest producer countries are Brazil, Indonesia, and Ethiopia, who respectively accounted for 43.7%, 9.6%, and 7.5% of global exports in 2019. According to the World Coffee Organization, the global consumption of coffee beans in 2019/2020 was 10.2 million tons. Europe is with 3.3 million tons by far the largest importer and consumer of coffee beans in the world. Belgium consumes annually c.50,000 tons of coffee, or c.5 kg per inhabitant per year. Coffee typically reaches European markets in the form of green coffee beans. The roasting—or torrefaction—is then carried out in Europe. In Belgium, several large roasting companies run industrial-scale operations. The coffee roasters sell and ship roasted coffee—either as powder or roasted beans—to wholesalers, retailers, or restaurants. The actual brewing of coffee happens at the place of consumption.

Coffee grounds are the leftover material after the brewing process. Due to the extremely scattered spatial distribution of coffee grounds, it is rarely collected selectively. In Belgian cities with a separate organic waste collection, some coffee grounds are collected with other biowaste and will then be used in composting or anaerobic digestion processes. Otherwise it will be collected with the unsorted waste fraction and incinerated. Coffee grounds are rich in potassium, magnesium, and phosphorus; these minerals were extracted from the soil of plantations in the Global South and metabolized by the coffee plant. The constant extraction of minerals from the soil in the growing regions must inevitably deplete its mineral content and reduce its productivity over time, unless this process is counterbalanced by artificial mineral fertilization.

KBF and its partners have evaluated the first phase of the circular economy project on coffee grounds as a success, particularly as regards the mobilization of large producers and innovators. In retrospect, they have identified several success factors that are specific to coffee grounds. First, the fact that coffee grounds represent a sizable flow of around 75,000 tons per year (after brewing, the coffee ground is wet and heavier) and thus a significant supply of materials. Second, the absence of tensions and political or economic sensitivities around coffee grounds which could hamper cooperation between different actors in new value system. And third, the fact that coffee grounds constitute a very tangible and well-known flow. As one interviewee put it: "Whereas circular economy can be abstract, coffee grounds allowed us to tell a story that almost everyone can personally relate to."

If we consider the global value chain of coffee production, most of the *spaces* it involves lie outside of the scope of KBF's circular economy project on coffee grounds: the growing regions in the southern hemisphere, the infrastructure facilities of ports, road networks, and storage facilities that allow to move enormous quantities of coffee around the world; even the Port of Antwerp, which is a partner of the project, entered the project from the perspective of a local producer of postconsumer coffee grounds rather than its role as hub in the global coffee trade.

The spaces that the project focuses on are the locations of coffee brewing and subsequent collection of coffee grounds in institutional kitchens and large organizations. The project mapped these spaces and the underlying processes as organizational ecosystems, including facility management, suppliers, waste operators, etc. Given the current state of the project, the most important *artifacts* are therefore those that are instrumental to waste management such as separate collection bins, storage containers, or transportation vehicles. Future phases of the project, however, could introduce a much wider range of objects and artifacts. Indeed, the envisaged experimentation with innovative recycling processes for coffee grounds could bring new forms of plastic substitutes, coffee-based paper, cardboard, or different forms of substrates and their horticultural production more firmly within the remit of the circular economy system that the project seeks to put in place.

4.3.3 Framing elements

The *domain* of the circular economy system underlying KBF's project was subject to a lengthy and deliberate thought process. In retrospect, many alternative foci could have been possible, such as other flows in the food sector or the flow of coffee in general (including growing regions) rather than the narrow focus on coffee grounds. According to the interviewees and project documents, narrowing the scope of the domain was essential to the success of this circular economy initiative, making the remit and objectives more palpable and realistic to business people or public administrations outside of the circular economy milieu. Another way of restricting the domain of the circular economy system was to focus on large producers of coffee grounds.

Apart from these two system boundaries—i.e., a specific flow and a specific type of producer—many other boundaries have been deliberately left unspecified. Importantly, the type of technology and recycling output have not yet been defined.

The *scale* of the envisaged system is currently limited to Belgium; global value chains having been deliberately left out of the conversation. Given the low economic value of coffee grounds, it is arguably not realistic to close the loop with growing regions. The only way of reducing the linear economy of coffee would be to reduce its consumption in importing countries such as Belgium—arguably an objective that the Belgian federation of coffee roasters would hardly support.

The scale of operations in the envisaged value system around coffee grounds was left open by KBF and its partners. During the exploration phase, they worked with large producers from all three Belgian regions and emphasized regional and institutional diversity. What emerged from this phase is thus a national conversation on the circular economy of coffee grounds with a polycentric pattern of subregional nodes around the large producers.

According to our interviews, the *rules* underpinning flows of coffee grounds in Belgium have so far not yet been the object of much attention. Coffee grounds are classified as post-consumer waste and are thus regulated by waste management laws. Unlike animal byproducts, coffee grounds are not associated with high sanitary risks. There are also no specific obligations for coffee producers to deal with end-of-pipe waste of their product. The absence of any specific regulation such as Extended Producer Responsibility on coffee grounds has been perceived as a facilitating factor that allows experimentation and new forms of alliances and interactions between different actors.

By the end of 2021, a European directive with higher targets for separate organic waste management will have to be transposed into Belgian law. This will mean that the separate collection of organic waste will become a legal obligation for the organizations that are part of the project (currently the separate collection of biowaste is not obligatory in many parts of Belgium). This might create an additional momentum for the project; however, the new regulations could also undermine the initiative if companies and other organizations perceive the separate collection of coffee grounds and other organic fractions as redundant and too costly.

5. Discussion of results

In this section we adopt comparative perspective to address our two sets of research questions. A summary of all three case studies is presented in Table 11.2.

Table 11.2 Summary of factors that have been potentially conducive for the emergence of a value system in each of the three case studies.

Case study	Urban wood/Montreal	Raw earth/Paris	Coffee grounds/Belgium
Agency—*coordination*	Individual entrepreneur and coordinator convening diverse interventions of other actors in the system	Consortium of complementary actors	Neutral philanthropic foundation as federating agent
Agency—*secondary actors*	Interested municipal authorities Presence of necessary skills and organizations in local economy	Strong role of public actors Interest from research organizations and planning professionals	Focus on large players Access to political and private decision-makers
Agency—*motivation*	Supply-push-driven motivation	Demand-pull motivation	Supply-push motivation
Agency—*context*	Influence of social enterprise philosophy	Influence of public administration and industrial policy thinking	Influence of general interest and societal ambitions
Material flows—*input*	Material with strong emotional value for local populations (neighborhood trees)	Large supply of soils makes upscaling and substantial investments realistic	Coffee grounds are easy to communicate and very relatable (straightforward narrative)

Continued

Table 11.2 Summary of factors that have been potentially conducive for the emergence of a value system in each of the three case studies.—cont'd

Case study	Urban wood/Montreal	Raw earth/Paris	Coffee grounds/Belgium
Material flows—*output*	Interesting applications in woodworking	Attractive alternative to more polluting construction materials that are in high demand (concrete)	Technological innovations and startup initiatives sparked interest for material
Spaces and artifacts	Preexistence of necessary technology and infrastructure at the regional scale	Public subsidies for setting up new pilot infrastructure (soil factory) that is to be replicated in other places	Focus on large producers of coffee grounds; technological choices left deliberately open
Domain and scale	Successive increase of domain (albeit with a continuous focus on ash trees) and scale (from neighborhood to metropolitan scale)	Domain fixed at beginning of project (soil processing); scale to be increased beyond pilot project through replication	Coffee grounds as starting point, but technological trajectory and scale of applications intentionally left open at the beginning of the project
Rules	Phytosanitary rules created massive supply of ash wood	Potential scale of value chain could make changes in norms worthwhile (Atex type A norms)	No important legal or normative obstacles for reusing coffee grounds

5.1 Conceptualization of circular value systems

A central proposition of this chapter is that a two-tiered framework combining concepts related to agency, biophysical and framing elements can be helpful for conceptualizing circular value systems. Drawing on the case study descriptions we can highlight strengths and weaknesses of the proposed framework.

The most relevant strength of our framework is that it provides a common grid against which heterogeneous real-world cases can be assessed. Without such as tool, the diversity of the underlying initiatives might lead to incommensurate and idiosyncratic material that cannot be used for comparative work. This is particularly relevant for research applying inductive methods such as "thick descriptions" that have received renewed interest in literature on economic sustainability (Geertz, 2008; Gibson-Graham, 2014). While the latter approaches are workable in isolated cases, understanding patterns across cases requires a framework that is sufficiently malleable to capture a wide variety of empirical observations while also imposing a set of ex ante variables. We experienced that the concepts included in our framework turned out to be sufficiently broad to capture the richness of different empirical contexts. The framework forced us to think through all relevant dimensions in each case, rather than focusing on the most intriguing or innovative aspect of each initiative. This point is particularly relevant in research carried out in a multidisciplinary team, as it ensures that the different disciplinary angles are applied consistently and systematically. All three case studies confirm that circular value chains are meandering sociotechnical objects whose description warrants the mobilization of a wide range of concepts drawn from different disciplinary traditions.

A weakness of the framework is its limited propensity to allow for formalization, i.e., formal modeling and quantitative testing of hypotheses. This weakness is, of course, the flipside of the framework's openness for qualitative input and conceptual heterogeneity. Indeed, the framework provides a panorama of the value system at hand but fails to specify a sequence or causal relationship between conceptually different aspects of it. This weakness might be overcome through repeated application and a progressive formalization of variables and their relationships—much in the way that Ostrom's generic framework for analyzing the sustainability of social-ecological system has become increasingly formalized (Ostrom, 2006b). In light of our case study material, however, the prominence of many open-ended and context-specific processes would probably require the definition of a third and fourth tier of variables to allow for comparable quantitative measurement. For example, one could define a series of common metrics to characterize the biophysical flows of wood, earth, and coffee (e.g., tonnage, distribution in space), but we have seen that certain flow characteristics can be extremely relevant in some cases and completely irrelevant in others. The presence of nails in the material flow, for example, is highly salient in the Montreal case but hardly worth noting for the case of raw earth excavations. Defining common variables with consistent relevance across all cases—such as "frequent presence of impurities in the material flow"—only defers the problem that the *value* of wood, earth, and coffee is to some extent incommensurate on a very fundamental level. We will get back to this issue below.

Another weakness we observed is that some objects could not be classified straightforwardly under a single conceptual category of our framework. The importance of carpentry schools for the value system around urban wood is a case in point. Such schools are associated with agency and interactions with other actors of the system; but they also turned out to be important physical spaces for the storage of small quantities of wood, the testing of new woodworking machines, the fabrication of prototypes, and the organization of exhibitions. The polyvalence of certain objects creates confusion, for instance when the same object is presented from different conceptual angles. We have tried to address this issue by placing each object under what we perceived as the dominant concept; for the case of carpentry schools, for instance, we have chosen to focus on their agency rather than their quality as physical spaces.

5.2 Emergence and dynamics of circular value systems

All three selected case studies have in common that they are concerned with initiatives applying circular principles to material flows. Based on our empirical material summarized in Table 11.2, we submit that coordination issues are central for understanding how emerging circular value systems are "held together."

First, a common element across all three initiatives is the pluralistic forms of coordination and management, and that power in its diverse forms is distributed among heterogeneous players from different sectors of society. This situation of distributed power contrasts with the firm-centric view proposed by literature on sustainable value chains (Seuring and Meuller, 2008). The pluralistic nature of these emergent circular value chains/systems appears to affect their coordination. A first element of their coordination concerns the main actor who convenes the circular value chain. Each of the cases is convened by a different actor (cf. Table 11.2). While the Montreal value chain is assembled by an individual entrepreneur from the social economy who calls on municipal authorities to protect a local public good, the Paris case around excavated earth is coordinated by a consortium of organizations. The Brussels coffee grounds value chain is facilitated by a foundation in the form close to a social incubator. The three initiatives thus surpass sustainability initiatives of an individual company, which remains a focus in research on circular bioeconomy (DeBoer et al., 2020): they are attempts to assemble novel processes of value creation in which a network of actors starts to align interests and competences, albeit in very diverse ways.

A second element of coordination concerns narratives and storylines, which is another aspect of circular bioeconomy that has received interest from scholars. Giurca (2020), for instance, has suggested that storylines regarding the transition toward circular economy of sustainable materials "reflect different interests and discursive struggles over policy outcomes." Narratives represent powerful forms of mobilization and vectors of coordination in all three emerging value chains we observed. The dominant narrative in the Montreal experience comes in the form of a powerful call to retain materials within the urban landscape, tapping into the value associated with provenience and emotional bonds with local trees. Relatively small volumes of material, as well as costly small-scale and customized processes are thus justified by the symbolic value of the Montreal ash timber. The dominant narrative in Paris mobilizes a less emotional approach to sustainability and instead develops a broader and more evidence-based business case formulated in terms of extrinsic volume rather than intrinsic meaning. The storyline in the Belgian experience is framed in terms of daily refuse produced by each household, rendering the flow familiar, relatable, and easy to communicate—but also less charged with intrinsic value compared to ash wood. Narratives matter because they motivate participants with other more significant professional, organizational, or administrative affiliations; they appear to have been especially helpful for actors to remain in the open-ended and time-consuming conversations and processes that are involved in the setting up of new value systems.

A third element of coordination concerns the material flow per se. Creating material loops is arguably the leitmotif of circular value chains. Yet, the materials presented in the three case studies are extremely diverse. Two of them concern relatively small quantities of organic material, while the third one focuses large flows of inorganic materials. The quality and quantity of these materials influence the need as well as the type of investments needed to set up the infrastructure for the circularization of the value chain. Although all cases struggled to varying degrees with characterizing and sorting of heterogeneous material, the coordination of the value chain was always greatly facilitated by the clear focus on a specific flow. This suggests that more general circular economy policies that do not focus on a specific flow—such as the city-level policies in London, Paris, and Brussels reviewed by Athanassiadis and Kampelmann (2021)—lack a key converging force that facilitates successful coordination.

A fourth element of coordination concerns the scaling-up of a circular value system. The Montreal experience suggests that scale increased over time, from small neighborhood-level pilot to regional infrastructure investments. The Paris case started upfront with regional ambitions of metropolitan actors to make system-level investments in production capacity. The project enabled access to long and expensive procedures for technical certification (ATEX), essential to validate products and enable their use in construction. The analysis of actors, however, suggests gaps regarding the engagement of industrial partners. The participation of the public sector is crucial for the development of appropriate policies and demand from public procurement, but the involvement of industrial partners is key to achieve operational upscaling so that projects surpass anecdotal volumes. Although market forces pushing for a value chain based on raw earth materials are still weak, the Earth Cycle project could become emblematic and increase the use of excavation soils in the construction sector in France and elsewhere. Finally, the Belgian case suggests that coordination by an independent broker is helpful for attracting initial interest from larger industrial players, but the project is still too immature to gauge its potential for upscaling.

6. Conclusion

The comparative analysis of three initiatives around sustainable materials offers insights into the relative success of circular value creation. All three value chains are better understood as complex value systems and display considerable diversity in their agency, biophysical, and framing elements. The conceptual grid proposed by our framework proved to be relevant for characterizing the different configurations of these elements.

The case studies suggest that circular value systems are live and open-ended experiments initiated by diverse actors with equally diverse motivations. The initiators manage to mobilize individuals and organizations around material flows as powerful leitmotif in a federating narrative or storyline. These processes lead to the reorganization of social, material, symbolic, and administrative interactions; they also generate social learning and mobilizations, improved material flows, new technological and social solutions, as well as potential environmental benefits to be measured in greenhouse gas emission reductions or tons of materials diverted from landfill.

While quantitative evaluation of these value systems is important, we invite researchers to be cautious about the type of metrics relevant to appreciate emergent circular value systems. Circular value creation is diverse; both their analysis and evaluation require interdisciplinary work. And although it is often possible to measure in quantitative terms—for instance in monetary units—*how much value* a given value system generates, the cases in this chapter suggest that an equally interesting question is to ask *which kind of value* the circular economy initiatives harnesses. The flow-based storylines in our case studies mobilized very different kinds of value such as extrinsic volume (raw earth), intrinsic meaning (urban timber), and relatability (coffee grounds). We think that a better understanding of the different kinds of value associated with material flows opens new avenues for research and business development around circular value systems.

Further research on other empirical cases is necessary to validate both the choice of concepts included in the proposed framework (our first research question) and our tentative observations regarding the emergence of circular value chains/systems (second research question). Future applications should notably assess whether all included variables are relevant across a wide range of cases and experiment with different strategies to address the polyvalence of certain objects. A stabilized framework allowing to characterize and compare a growing number of empirical cases on value creation in the circular economy would be a strong basis for evidence-based learning from the numerous experiments that are currently underway.

References

Arnsperger, C., Bourg, D., 2016. Vers une économie authentiquement circulaire. Rev. OFCE 1 (145), 91–125.

Athanassiadis, A., Kampelmann, S., 2021. Opportunities and limits of circular economy as policy framework for urban metabolism. In: Barles, S., Marty, P. (Eds.), A Research Agenda for Urban Metabolism (forthcoming).

Babbitt, C.W., Gaustad, G., Fisher, A., Chen, W.Q., Liu, G., 2018. Closing the loop on circular economy research: from theory to practice and back again. Resour. Conserv. Recycl. 135, 1–2.

Barles, S., 2010. Society, energy and materials: what are the contributions of industrial ecology, territorial ecology and urban metabolism to sustainable urban development issues? J. Environ. Plann. Manag. 53 (4), 439–455.

Bals, L., Tate, W.L., Ellram, L.M., 2020. Circular Economy Supply Chains – from Chains to Systems. Call for Papers for Contributions to an Edited Collection.

Bazilian, M., Rogner, H., Howells, M., Hermann, S., Arent, D., Gielen, D., Steduto, P., Mueller, A., Komor, P., Tol, R.S.J., Yumkella, K.K., 2011. Considering the energy, water and food nexus: towards an integrated modelling approach. Energy Pol. 39, 7896–7906.

Benhelal, E., Zahedi, G., Shamsaei, E., Bahadori, A., 2013. Global strategies and potentials to curb CO2 emissions in cement industry. J. Clean. Prod. 51, 142–161. https://doi.org/10.1016/j.jclepro.2012.10.049.

Billen, G., Toussaint, F., Peeters, P., Sapir, M., Steenhout, A., Vanderborght, J.P., 1983. L'ecosysteme Belgique. Essai D'écologie Industrielle. CRISP, Brussels.

DeBoer, J., Panwar, R., Kozak, R., Cashore, B., 2020. Squaring the circle: refining the competitiveness logic for the circular bioeconomy. For. Pol. Econ. 110, 101858.

Duvigneaud, P., 1974. La Synthèse Écologique: Populations, Communautés, Écosystèmes, Biosphère, Noosphère. Doin, Paris.

Duvigneaud, P., Denayer-De Smet, S., 1975. L'ecosysteme Urbs. L'ecosysteme urbain Bruxellois. In: Duvigneaud, P., Kestemont, P. (Eds.), Productivite biologique en Belgique. Editions Duculot, Gembloux, pp. 581–597.

European Commission, 2011. Service Contract on Management of CDW. Available at: https://ec.europa.eu/environment/waste/pdf/2011_CDW_Report.pdf. (Accessed 11 March 2020).

Fischer-Kowalski, M., Haberl, H. (Eds.), 2007. Socioecological Transitions and Global Change: Trajectories of Social Metabolism and Land Use. Edward Elgar, Cheltenham.

Gasnier, H., 2019. Construire en terres d'excavation, un enjeu pour la ville durable. Art et histoire de l'art. Université Grenoble Alpes.

Geertz, C., 2008. Thick description: toward an interpretive theory of culture. In: The Cultural Geography Reader. Routledge, pp. 41–51.

Gibson-Graham, J.K., 2014. Rethinking the economy with thick description and weak theory. Curr. Anthropol. 55 (S9), S147–S153.

Giurca, A., 2020. Unpacking the network discourse: actors and storylines in Germany's wood-based bioeconomy. For. Pol. Econ. 110, 101754.

Habert, G., 2019. The Earth Cycle Project Journal No 1. Urban Innovative Actions (UIA).

Handfield, R.B., Nichols Jr., E.L., 1999. Introduction to. Supply Chain Management. Prentice Hall, Englewood Cliffs, NJ.

Jarre, M., Petit-Boix, A., Priefer, C., Meyer, R., Leipold, S., 2020. Transforming the bio-based sector towards a circular economy-What can we learn from wood cascading? For. Pol. Econ. 110, 101872.

Kampelmann, S., De Muynck, S., 2019. Les implications d'une circularisation des métabolismes territoriaux – une revue de la littérature. Pour, 236.

Kampelmann, S., 2017. Circular economy in a territorial context: the case of biowaste management in Brussels. In: Cassiers, I., Maréchal, K., Méda, D. (Eds.), Post-growth Economics and Society. Exploring the Paths of a Social and Ecological Transition, Series in Ecological Economics. Routledge, New York.

Koberg, E., Longoni, A., 2019. A systematic review of sustainable supply chain management in global supply chains. J. Clean. Prod. 207, 1084–1098.

Lazarevic, D., Kautto, P., Antikainen, R., 2020. Finland's wood-frame multi-storey construction innovation system: analysing motors of creative destruction. For. Pol. Econ. 110, 101861.

Odum, E.P., Barrett, G.W., 1971. Fundamentals of Ecology, vol. 3. Saunders, Philadelphia, p. 5.

Ostrom, E., 2009a. A general framework for analyzing sustainability of social-ecological systems. Science 325 (5939), 419–422.

Ostrom, E., 2009b. The institutional analysis and development framework and the commons. Cornell Law Rev. 95, 807.

Pagell, M., Shevchenko, A., 2014. Why research in sustainable supply chain management should have no future. J. Supply Chain Manag. 50 (1), 44–55.

Panigrahi, S.S., Bahinipati, B., Jain, V., 2019. Sustainable supply chain management: a review of literature and implications for future research. Manag. Environ. Q. 30 (5), 1001–1049.

Panwar, R., Kozak, R., Hansen, E. (Eds.), 2015. Forests, Business and Sustainability. Routledge.

Scott, J., Marshall, G., 2005. A Dictionary of Sociology. Oxford University Press.

Seuring, S., Muller, M., 2008. From a literature review to a conceptual framework for sustainable supply chain management. J. Clean. Prod. 16 (15), 1699–1710.

Tangtinthai, N., Heidrich, O., Manning, D.A.C., 2019. Role of policy in managing mined resources for construction in Europe and emerging economies. J. Environ. Manag. 236, 613–621. https://doi.org/10.1016/j.jenvman.2018.11.141.

Tate, W.L., Bals, L., Bals, C., Foerstl, K., 2019. Seeing the forest and not the trees: learning from nature's circular economy. Resour. Conserv. Recycl. 149, 115–129.

Touboulic, A., Walker, H., 2015. Theories in sustainable supply chain management: a structured literature review. Int. J. Phys. Distrib. Logist. Manag. 45 (1–2), 16–42.

Further reading

Anulewicz, A.C., McCullough, D.G., Cappaert, D.L., 2007. Emerald ash borer (*Agrilus planipennis*) density and canopy dieback in three North American ash species. AUF 5 (33), 338–349.

Eurostat, 2016. Waste Generation. Available at: https://ec.europa.eu/eurostat/statistics-explained/index.php/Waste_statistics. (Accessed 11 March 2020).

van Kote, G., 2013. Le Grand Paris Face à Une Montagne de Déblais. Available at:https://www.lemonde.fr/economie/article/2013/03/22/le-grand-paris-facea-une-montagne-de-deblais_1852567_3234.html. (Accessed 11 March 2020).

Leilavergne, E., 2012. La filière terre crue en France. Mémoire DSA Terre, École Supérieure d'Architecture de Grenoble.

CHAPTER

A story of resilience and local materials: sourcing bio-based materials in Norman wetlands, France

12

Giulia Scialpi

Faculty of Architecture, Architectural Engineering and Urban Planning, University of Louvain UCLouvain, Ottignies-Louvain-la-Neuve, Belgium

Chapter outline

Environmental Sustainability and Economy. https://doi.org/10.1016/B978-0-12-822188-4.00010-5

This chapter presents the case study of *Marais du Cotentin et du Bessin* Natural Park in Normandy (France), which hosts a vast wetland area of international ecological interest. Historically, these wetlands have been used for farming and their vegetation has been incorporated in traditional architecture. Due to changes in both farming practices and architectural features, these lands need constant maintenance to avoid reforestation, protect biodiversity, and maintain the provision of a broad range of ecosystem services. Protection, restoration, and maintenance activities reveal new uses for the biomass produced by wetland vegetation.

The aim of this study is to understand how a selection of Norman wetland plants could be used for the production of building materials, thereby increasing the value of the local species and ensuring the site's maintenance.

The research revolves around two main workflows. One involves a territorial analysis which identifies territorial resources, agents, and dynamics in order to assess the possibility of creating a value chain. The second involves experimental research to provide a first characterization of the mechanical, physical, and hygrothermal properties of the selected wetland plants (purple moor, rush, sedge, and reed). The tests performed have revealed that wetland plants have characteristics comparable with other bio-based materials that are used for insulation or in earth-based composites. Among the plants selected, only reed was previously investigated for its suitability as a building material. Our results can provide new insights into the use of wetland plants in the construction sector.

1. Introduction

Resilience is a polysemic concept with interpretative flexibility depending on the context in which it is used, the purpose it serves, and the disciplinary lens through which it is observed (Baggio, 2015).

In this transdisciplinary context, the concept of resilience has been applied to studies on complex socioecological systems (Sanchez-Zamora et al., 2014a, 2014b; Perrotti et al., 2020), whose framework is an important methodological and conceptual reference model used to analyze resilience from a territorial perspective (Sanchez-Zamora et al., 2019). In this research, a territory/region is considered a complex system in continuous evolution defined by its specific resources (territorial resources) with which local stakeholders (territorial agents) interact, whether they establish institutional agreements or not (territorial dynamics) (Sánchez-Zamora et al., 2016). The ability to anticipate, prepare, respond, and adapt to new and ongoing changes produced by territorial dynamics is defined as "territorial resilience" (Sanchez-Zamora et al., 2019). In this chapter, the concept of territorial resilience is used as a framework for analysis in the case study of *Marais du Cotentin et du Bessin* Regional Natural Park (PnrMCB), which hosts a vast wetland area of international ecological interest.

In recent years, the appeal of wetlands has been enhanced by the recognition of their provision of a broad range of ecosystem services (e.g., supply of biomass as raw material, water purification, regulation of the water cycle, areas for recreation, and wildlife habitat) (Millennium Ecosystem Assessment, 2005).

In many countries, wetlands have been protected by legislation and degraded wetlands have been restored (Kusler and Kentula, 1990; Hodge and Mcnally, 2000; Cui et al., 2009).

When protection and restoration measures are applied, it is often necessary to find uses for the biomass produced by the wetland vegetation and new sources of income for local communities (Köbbing et al., 2013). The rising demand for food, energy biomass, and renewable raw materials has increased interest in wetland plants, because they are highly productive and do not compete for land that is used for food production (Kroon, 2013).

Current possible uses for wetland plant can be divided into industrial, energy, agricultural, and water treatment (Köbbing et al., 2013). In this chapter, industrial uses such as the manufacture of construction materials are investigated.

The state-of-research on the use of wetland plants for construction purposes is mainly limited to *Phragmites australis* (reed) traditionally used for thatch roofing (Köbbing et al., 2013). In this chapter, besides *P. australis*, other plants typical of the Norman wetlands are studied, such as *Cladium mariscus* (sedge), *Juncus effusus* (common rush), and *Molinia caerulea* (purple moor). No published works dealing with the use of these wetland plants as building materials, such as earth-based composites or insulation materials, were found.

The process of giving economic value to wetland plants is an incentive to preserve them and their many functions that are beneficial for nature and humanity. This can only be achieved if the exploitation is recognized as being environmentally acceptable, commercially feasible, and a source of economic gain for all stakeholders (Kroon, 2013). For this reason, the project was planned around two main workflows: territorial analysis and experimental research. Through territorial analysis resources, agents and dynamics were identified in order to assess the creation of a value chain, which is intended as a set of interlinked activities that deliver products and services by adding value to bulk materials (Lokesh et al., 2018). At the same time, the mechanical, physical, and hygrothermal properties of the selected wetland plants were identified through experimental research.

The aim of this research was to test wetland plants from PnrMCB for their use as building materials and analyze how this value chain would improve territorial resilience, so as to address the following research questions: (1) *How can wetlands plants from PnrMCB be employed as building materials?* (2) *How can the valorization of wetland plants for the production of building materials improve the territorial resilience of PnrMCB?*

The chapter is organized as follows: first, the presentation of the case study of PnrMCB; second, the methodology applied to both territorial analysis and experimental research; third, a presentation of results; fourth, a discussion of the empirical results, followed by the conclusion.

1.1 Case study: Marais du Cotentin et du Bessin Natural Regional Park, France

PnrMCB is located in Normandy (France) and consists of 148,000 ha, 119 municipalities, and 74,000 inhabitants. Its landscape consists of bocage, moorland, and coastline which include a vast wetland area of international ecological interest. The wetlands in the Park are divided between the "lowlands" with almost no altitude and characterized by large valleys and rivers, and the "highlands" with altitudes up to 50 m. The lowlands consist mainly of swamps and prairies, as opposed to the agricultural areas of the highlands.

The wetlands occupy 30,000 ha (accounting for 20% of the Park), a remarkable area from a biodiversity perspective, considering that 50% of French wetlands were lost during the 20th century (ONB, 2016). The loss increased considerably the interest in conserving these biodiversity-rich areas.

While completely submerged by water during the winter, the wetlands are maintained by local communities through mowing and grazing. Without these activities, the wetlands risk disappearing due to reforestation. Over time, changes to agricultural, craft and domestic practices have resulted in a gradual decrease in the use of wetlands as well as in the importance of their role in the local economy.

Wetland maintenance is made difficult owing to adverse weather conditions. Farmers, sub-contracted by town halls, are the main maintenance operators, followed by nature reserve managers or private individuals. They receive European subsidies and sell the collected plants as litter or mulch for growing carrots. This short supply chain seems to be well established, but it is highly dependent on subsidies and a small number of economic outlets.

This study was requested by the Landscape Planning Department and the Biodiversity and Water Resources Department as a first approach to understanding the link between safeguarding and conserving the wetlands on the one hand and the evolution of farms and habitat on the other. The project involved a multidisciplinary approach, namely maintaining the wetlands in order to preserve their biodiversity and develop their plants for use as eco-materials in the local construction sector.

The research is part of two projects carried out by the PnrMCB: the TEPCV (Territoires à Energie Positive pour la Croissance Verte) and the CobBauge (Interreg France Channel—England). The former is a program to finance local implementation of energy transition objectives such as the valorization of local resources in ecoconstruction (PnrMCB, 2016). The latter is a European cross-border cooperation program to test and establish a novel prefabricated cob technology and to meet current building standards (CobBauge). Cob is a traditional construction technique that uses local soil and agricultural fibers; the latter could be replaced with wetland plants.

Cob-technique earth architecture is an important part of Park heritage. Since its creation, the Park has safeguarded cob architecture, with the renovation of more than 250 buildings over the past 8 years (PnrMCB, 2016). Through improved use of organic fibers, which have been integrated in earth architecture for centuries, the thermal performance of heritage renovation and contemporary cob is evolving (Laborel-Préneron et al., 2016). The creation of a value chain based on wetland plants could be connected to cob improvement.

2. Materials and methods
2.1 Territorial analysis

The case study was analyzed between November 2017 and April 2018, using the lens of territorial resilience. A territorial approach to resilience is intended to establish a conscious and adaptive relationship between a given region and its resources, which are associated with different forms of territorial capitals (Sanchez-Zamora et al., 2014a, 2014b). In this research, six of them (natural, human, social, cultural, economic, and manufactured) were considered (Sanchez-Zamora et al., 2019). A methodology specific to each form of capital is presented below.

2.2 Natural capital: wetland plants and stocks localization

The studied bio-based materials are part of the nonagricultural biomass that grows naturally in PnrMCB wetlands. Four plants were selected by the Biodiversity and Water Resources Department and the Landscape Planning Department and supplied by farmers and companies in charge of the maintenance of several natural sites: *M. caerulea* (purple moor), *C. mariscus* (sedge), *P. australis* (reed), and *J. effusus* (common rush). Selection criteria were based on an estimate of fiber quantities and qualities (e.g., presence in nature reserves; similitudes with agro-based materials already in use; sourcing facilities).

The exploration of natural capital began with the identification of plants. An ID card for each plant was completed. The card contained information about nomenclature, botanical aspects, harvesting methods, and yield provided by nature reserve managers and farmers (see surveys below); possible uses in construction were found through a Google Scholar search using the keywords "construction materials" followed by the scientific/popular name of the resource.

Stocks were difficult to quantify due to the fact that plants are not monospecific and grow spontaneously.

Two kinds of tools were used for this purpose: Natura2000 Maps and Maps of the National Botanical Conservatory of Brest. The former provided quantitative data and, in tandem with QGIS software, mapped land occupied by each species. However, at the time of this study, the maps were still incomplete and the only information available was on the Douve valley. Data on the Western Coast and Taut, Aure, and Vire valleys were completed in 2019. The latter, by means of the eCalluna application (http://www.cbnbrest.fr/), shows the geographical distribution and development of flowering plants and ferns over the entire Channel area. Since no quantitative data were available, only a qualitative comparison of their presence was possible.

In addition to map analysis, nature reserve stock was assessed. This was easier to calculate, since managers possessed data concerning areas, plant species, and harvesting practices. As for plants, an "ID card" for each reserve, containing information about plants, their maintenance, and land-use management, was completed using data from the most recent nature reserve activity reports and manager surveys (see human capital). To quantify the stock, all Park nature reserves were visited: Nature Reserve of Mathon, Nature Reserve Domaine de Beauguillot, Espace Naturel Sensible in Saint Côme du Mont, Nature Reserve of the Taute, The Hunting and Wildlife Reserve of the Bohons, National Nature Reserve of the Sangsurière and Adriennerie.

2.3 Human capital: craftsmen, farmers, managers of reserves, and social enterprises

In analyzing human capital, three target groups were identified—farmers, managers of reserves, and social enterprises—for their involvement in the maintenance of wetlands and plant extraction, and craftsmen—as potential customers for bio-based materials in construction. The surveys involved six craftsmen, four farmers, six reserve managers, and two social enterprises. This provided data was not available in official reports on barriers, timing, machinery, etc., and that was used as a template for project follow-up (see appendices).

The only farmers who work in the nature reserves are the four surveyed, as well as the two social enterprises. Six craftsmen belonging to ARPE (Norman Cluster of Eco-construction), including its former director, were surveyed, to test a questionnaire that ARPE intended to launch through its online platform for a bigger audience in the second phase of the project. Since the reduced number of craftsmen surveyed was insufficient to produce information on market trends, the "Survey and Analysis reports on perceptions, practices and expectations about bio-based materials of craftsmen and enterprises in Normandy" (Nomadeis, 2014) was consulted in order to have more information about local bio-based materials, in terms of cost, supply, and use.

All the surveys were conducted on-site, except for those concerning craftsmen, who were surveyed by phone, and the results were analyzed by a single researcher and recorded whenever possible. The same core questions were asked to everyone surveyed. The questions for nature reserve managers, farmers, and social enterprises revolved around three main topics: resources (to collect data about plants and stocks in terms of amount and localisation), maintenance (to collect data about the work required to preserve wetlands in terms of cost, frequency, resources, etc.), and management (to collect data about current outlets, storage and treatments after harvesting). Finally, the questions for craftsmen revolved around three main topics: works (services provided and professional qualification), materials (to collect data about the materials in use in terms of origin, price and selection criteria), and construction (to collect data about barrier and drivers for the use of bio-based materials). The full list of questions is placed in the appendix to the chapter.

2.4 Economic and manufactured capital: current market and available infrastructure

Data on market trends were obtained from official reports on bio-based materials, such as "Bio-based building materials. Survey on perceptions, practices and expectations of craftsmen and enterprises" (Nomadéis, 2014) and "Census of bio-based products available on the market and identification of targeted public" (Ademe, 2016).

The aim was to identify the current market for bio-based materials in the region and thus guide future product development.

Based on the above reports, "ID cards" were drafted to classify the most popular bio-based materials, such as *Cannabis sativa* (hemp) and its derivatives, *Linus usitatissimum* (flax) and its derivatives, *Triticum aestivum* (straw), and *Miscanthus*. The ID cards included the following information: origin; stock and yield; products and uses; physical characteristics; prices; strengths and weaknesses of their technical performance.

PnrMCB activity reports provided more information about European aid for wetland maintenance and other twinning projects (e.g., TEPCV and CobBauge), thus painting a clearer picture of the current market for wetland plants.

Finally, manufactured capital was explored mainly through the surveys described above, in particular the "maintenance" set of questions concerning tools and machinery, and the "management" set of questions concerning treatments after harvesting, storage, and transport.

2.5 Social capital: the role of the park, the wetland administration, and potential partners

The role of the PnrMCB and its interaction with public and private stakeholders were analyzed through two kinds of documents: activity reports and the Park's *Atlas cartographique*, both of which provided an overview of trends, opportunities, and challenges to the Park over the medium and long term.

When the first results of territorial analysis and experimental research were available, a working group analyzed potential value chain barriers and drivers. Everyone involved in the study, whether through surveys or site visits, was invited to take part in the working group. The group was composed of four PnrMCB agents from Biodiversity and Water Resources and the Landscape Planning Departments; two craftsmen; the director of one of the social enterprises; an agent from Biodiversity and Water Resources Department of the Boucles de la Seine Natural Regional Park (180 km from PnrMCB); a project manager from ARPE (Norman Cluster for Eco-construction); two project managers from Enerterre Association (an association for the ecological and sustainable renovation of houses through participative building sites).

Working group members were invited to identify value chain drivers and barriers through a SWOT (strengths, weaknesses, opportunities, and threats) analysis. SWOT analysis is a commonly used tool for examining simultaneously both internal factors (i.e., strengths and weaknesses) and external factors (i.e., opportunities and threats), with the aim of supporting decision-making and intervention strategies. It can help identify consensus and divergences in the stakeholders' vision and be used as a starting point for developing an action plan and improving project teamwork. For the purpose of the analysis, the value chain was structured into four phases: extraction (organic fibers from wetland maintenance); production (composite materials of organic fibers and local soil used as binder); distribution (transport, sales, etc.); implementation (professional training; prototype construction, etc.).

2.6 Cultural capital: wetland plants and traditional architecture

Cultural capital refers to the ability of human societies to adapt, deal with, and modify the natural environment (Berkes and Folke, 1994). The historical development of wetlands, the know-how, and the local building cultures are considered part of the cultural capital and have been investigated to understand whether the plants, the objects of this study, have been part of local customs and traditions, architectural or otherwise.

Cultural capital was explored by analyzing Park database publications to understand traditional applications for wetland plants and their performance and links to landscape and traditional architecture. Targeted topics included heritage with a special focus on cob (to learn whether wetland plant fibers were traditionally mixed with earth to produce composite materials); wetland customs and traditions (to learn about plant uses and wetland maintenance methods); and roof thatch (to learn whether fibers were used in traditional roofing methods).

2.7 Experimental research

The physical, mechanical, and hygrothermal characteristics of the selected wetland plants were tested through experimental research to ascertain their viability as insulation materials or earth-based composites.

Plants (*M. caerulea, C. mariscus, P. australis,* and *J. effusus*) were sourced from two sites, the Nature Reserve of Mathon and the National Nature Reserve of the Sangsurière and Adriennerie. They were then stored and dried naturally in a warehouse.

Based on existing literature concerning the role of plant aggregates and fibers in construction (Laborel-Préneron et al., 2016), the following tests were performed on bulk fibers: microscope observation; absorption coefficient; absolute and apparent density; tensile strength; thermal conductivity.

Scanning electron microscope (SEM) observation revealed the internal structure of the fibers and estimated their adhesion in the earth matrix (the rougher the fiber texture, the greater the adhesion with the matrix) (Aruan Efendy and Pickering, 2014). The SEM was also used to measure the transverse section of the smaller fibers (such as *M. caerulea*), which is required to estimate their tensile strength.

Absorption coefficient measures the ability of fibers to absorb water, which may affect the water content of composite material mixtures (e.g., raw earth and fibers) and produce variations in dry volume, thus modifying the fiber/matrix interface (Ghavami et al., 1999; Segetin et al., 2007).

Absolute and apparent density can be correlated with thermal and mechanical characteristics of the composite materials. The increase in porosity, and consequently a lower density, decreases thermal conductivity and compressive strength (Bouguerra et al., 1998).

Tensile strength is one of the most studied mechanical properties in the literature concerning plant aggregates (Laborel-Préneron et al., 2016) so as to test the flexural reinforcement provided by the fibers.

Finally, thermal conductivity was tested to understand the thermal insulation that could be provided by bulk fibers in a prefabricated system (such as loose-fill insulation systems). The compressive strength and thermal conductivity of composite materials (insulation panels and light cob blocks) were tested to study the effects of plant particle additions.

In order to produce the composite materials, two types of soils were used as a binder, both coming from a quarry in Lieusaint, 30–40 km from Carentan Les Marais, the Park Administration Headquarters. Test descriptions are in the appendix to the chapter.

3. Results
3.1 Territorial analysis

The analysis of natural capital revealed a stock often mono-specific and with potential harvest yields between 20 t/ha (*Cladium*), 5 t/ha (*Phragmites*), 3–7 t/ha (*Juncus*), and 2 t/ha (*Molinia*). Unfortunately, *Juncus effusus* is not present in Natura2000 Maps because it grows too sporadically to characterize an area; thus, it remained the most difficult resource to quantify in this research. However, images from Brest conservatory indicated a large presence of the plant throughout the entire region.

Through the analysis of Nature Reserve reports some extra stocks were added (see stocks quantification table in the appendix), but their amount was still small compared to other bio-based materials in the region: wheat straw (*T. aestivum*) was the most widespread (200,000 ha), followed by flax (*L. usitatissimum*) (5000 ha) and hemp (*C. sativa*) (500 ha) (Nomadéis, 2014). However, analysis of these areas should take into account that the surface required for wetland production is notably lower than that required for harvest residues (straw), but 1.5–2 times the area required for cultivated biomass (*Miscanthus*) (Kroon, 2013).

Plant identification revealed a lack of use in contemporary construction, except for *Phragmites australis*, which is used to produce panels, roof thatch, thermal blocks, etc. (COFREEN, 2013) (Köbbing et al., 2013) and *C. mariscus*, used only in the renovation of roof thatch.

The use of plants in construction was further investigated through the analysis of Ademe's "Reports on perceptions, practices and expectations about bio-based materials of craftsmen and enterprises in Normandy" (see human capital), which provided information about local bio-based materials in terms of cost, supply, and use. Of the surveyed enterprises, 38% reported using bio-based materials (note that wood was excluded from the study). The bio-based materials used most in the construction sector are: wood derivates (50%), *C. sativa* (hemp) (46%), cellulose (39%), *L. usitatissimum* (flax) (10%), and *T. aestivum* (straw) (6%) (Nomadéis, 2014). Surveyed enterprises also pointed out barriers such as the lack of product knowledge (cited by 32%), a higher purchase price (31%), and the lack of customer awareness (21%). More than half of the enterprises (58%) reported interest in receiving more information about the technical characteristics and specificities of bio-based materials. Craftsmen surveyed by Ademe agreed that there is a lack of communication with professionals about the characteristics of these materials; a lack of training to address resource variability and nonstandardization; and a lack of visibility for the existing bio-based materials. The only wetland plant used by surveyed craftsmen was *P. australis*, which is used in clay plaster support panels.

While craftsmen would be potential buyers and users of new building products, farmers would be the key players in sourcing the value chain. They have extensive, unique knowledge of these lands and their strengths and weaknesses, as can be assumed by the fact that they have always overseen the maintenance of the wetlands. They carry out annual mechanical maintenance, for which they may apply for European subsidies (see economic capital). Currently, because farming is disappearing from these areas, plants are sold mainly as mulch for growing only carrots and litter for animals (see economic capital). Hence, European subsidies are pivotal to wetland maintenance.

Moreover, older generations represent the vast majority of farmers: 45% are over 50 years old, compared to only 20% under 40. Farmers over 50 were surveyed in 2000 and 2010 concerning their business recovery prospects, which proved to be weak (Legrain, 2017). All four farmers surveyed expressed interest in alternative market for wetland plants.

Wetland maintenance is not only managed by farmers but also by two social enterprises which were also surveyed. Two social enterprises work in the Park and offer professional training to the unemployed, who represent a substantial workforce. The companies are in charge of maintaining specific spaces that cannot be reached by automated maintenance provided by farmers. An alternative market, such as the production of building materials, could support their activity in winter, when outdoor maintenance activity is reduced.

Nature reserve managers hire both farmers and social enterprises to ensure the maintenance of these protected spaces. They are aware of biodiversity issues, logistics, and difficulties. The combination of these three factors has a significant impact on the maintenance period. Maintenance is organized by geographic area; some species need to be cut annually (*J. effusus* and *P. australis*), others every 3 to 5 years (*Cladium mariscus*) or biannually (*Molinia caerulea*). In most cases, cutting takes place after birds nest (September, March, mid-November, February), when the ground is frozen or dry and therefore capable of bearing the weight of equipment. Costs vary depending on whether the plants are exported or left on site, and on-site accessibility. The maintenance of 1 hectare costs from €3000/ ha (automated) to €15,000/ha (manual). After mowing, the plants are exported, left on site or (in some special situations) burnt. In the case of exportation, farmers often charge less for maintenance provided they can keep use the plants as litter for animals or mulch for carrots.

The analysis of social capital revealed the role of the Park, which identifies priority sectors, supports research activities, and leads and coordinates deliberations and exchange between stakeholders. This study is in line with the Park's policy of securing technical and financial support for making communal wetlands more attractive to farmers. The wetlands cover an area of approximately 30,000 ha, of which less than 2000 ha were under traditional collective management in 2015. Farmers' growing lack of interest in collective management led in the long term to the privatization of these areas through agricultural lease (Parc des Marais du Cotentin et du Bessin, 2007). The reduced number of small farms reduced interest in communal wetlands, which may affect their future maintenance.

Moreover, the Park could broaden project participation to ensure maintenance of the raw materials stock. A series of visits and meetings established collaboration with Boucles de la Seine Natural Regional Park (about 180 km from PnrMCB), which has 110−130 ha covered with reeds which are harvested every year.

In recent years, the Park has also participated in a program to support the renovation of traditional earthen construction (cob) and thatched roofs. In addition to heritage conservation (for which it provides subsidies), the Park has been involved in the European Interreg project CobBauge. The project's aim is to improve the thermal performance of traditional cob by increasing the organic fiber content and facilitating its use in contemporary architecture through prefabricated construction systems. The link between the study and the CobBauge project is particularly important for this study because wetland plants have been integrated into studies of cob to test its thermal properties (see experimental research).

Cultural capital analysis, in accordance with the wetlands' historical development, established a link between traditional architecture and the plants under study. *Phragmites, Juncus,* and *Cladium* were used for thatched roofs, which were probably mixed with cob. Traditional uses revealed plant properties: *Cladium* was known for its resilience and thus used for thatched roof ridges; *Juncus* for its sponginess, which accelerated the drying of earth mixes; *Molinia* for its flexibility, which made it ideal for mattresses, etc.

Analysis of economic and manufactured capital led to drawing a picture of current market for wetland plants, current applications for bio-based materials, and available infrastructures for the development of the value chain.

After studying farmer surveys and analyzing Normandy Chamber of Agriculture reports, two main economic outlets for wetland plants were identified.

The first is the European agricultural subsidy, an economic measure to ensure the maintenance of these protected areas. Farmers and governments sign a contract based on the biodiversity of wetlands. The stricter the guidelines, the greater the subsidy farmers receive. In the Park, the granting of subsidies began in 1992 and have been in place since then. The measure is particularly important for lower quality and/or difficult-to-access land. Today, the subsidy is from €94 to €300/ha, depending on the specifications; for example, a harvest later in the year is subsidized at a higher rate than an earlier harvest is. Farmers can apply for subsidies up to a maximum of €16,000/year. The second outlet is when the wetland plants are used as mulch for carrots, litter, and hay. Mulch for carrots is one of the most important outlets, because wetland plants perform better than straw. Today, less litter and hay are sold because there are fewer animals and farms. The material is sold at a cost of €40−50/t with a maximum radius of 40 km.

The price of bio-based materials for construction purposes is variable: €1000/t for flax, €750/t for hemp, €450/t for *Miscanthus,* and €150/t for straw. They are used in private construction sites (97%) and mainly in renovation projects (57%) (Nomadéis, 2014). The most popular application is in insulation products: loose-fill insulation, panels/rolls, plaster, blocks, green concrete, prefabricated wall systems, etc. The biomass content of the products varies between 80% and 100%, for an average value of 92% (Ademe, 2016).

When considering the development of a value chain and related costs, manufactured capital is key to production support. Stakeholders such as farmers or social enterprises can be involved at many stages (e.g., harvest, storage and production) and provide useful infrastructure (e.g., machines, storage, dryers). Farms and storage facilities were inspected to determine whether they could be used for multiple purposes, such as housing production activities. This is particularly important during winter when outdoor activity is reduced. Since the stock of plants varies from one year to the next, multiple resources could be stored in the same place so as to optimize the use of infrastructure.

3.2 Recommended resilience strategies

Using SWOT analysis as a starting point, a working group discussed the territorial analysis findings (see social capital) to identify potential value chain barriers and drivers.

Factors and indicators were highlighted for each phase of the value chain. Factors were used to investigate a system's assets and vulnerabilities, which can be classified according to space (local/global), time (slow/fast), and their nature (human/economic/cultural/environmental/social). These factors were quantified by indicators that were measurable over time and defined sociologically (knowledge of history, culture, identity). Indicators were used to assess risk exposure, highlight regional strengths and vulnerabilities, orientate action strategies, and evaluate their impact. Finally, intervention strategies based on the principles defined by the Stockholm Resilience Center were proposed, namely, maintaining diversity and redundancy; managing connectivity; managing slow variables and feedbacks; fostering complex adaptive system thinking; encouraging learning; broadening participation; promoting a polycentric governance system. These principles use to provide guidance on key opportunities for improving the resilience of socio-ecological systems. They are context-dependent, highly interconnected, and interdependent. Applying any of these principles in isolation, as well as keeping an inappropriate combination of principles, will rarely lead to enhanced resilience (Biggs, 2012).

The main findings are summarized in Table 12.1.

Table 12.1 Resilience strategies for wetland plants value chain.

Value chain phases	Factors		Indicators	Strategies
Extraction phase	Natural capital	Variability in stocks caused by the irregular evolution of the Marais	- Naturalistic reports and monitoring	*Maintain diversity, redundancy and multifunctionality*: project outputs must be complementary to those already existing (mulch, litter, European subsidies).
		Water level variations that affect time of harvesting and transport	- Natura2000 maps - Maps of the national Botanical conservatory of brest	*Broaden participation*: involving multiple partners means having a diversified source of supply.
		Climatic conditions that may drastically reduce the harvest period	- Surveys	*Manage slow variables*: taking into consideration changes in agricultural practices (e.g.,
		Scattered location that could affect production and complicate the quantification of the resources		
		Loss of biodiversity		

Continued

Table 12.1 Resilience strategies for wetland plants value chain.—cont'd

Value chain phases	Factors		Indicators	Strategies
	Cultural	Loss of know-how related to the Marais	- Surveys	generational turnover).
	Human capital	Farmers will no longer be interested in wetland maintenance in the future	- Census - Report on trends in agriculture	
		Aging of farmers		
Production phase	Manufacture	Identification of processing locations, storage, and stakeholders	- Surveys - Reports	*Maintain diversity, redundancy and multifunctionality:* infrastructure must have multiple uses; harvesting systems and products must be compatible; products should be composites to accommodate resource heterogeneity (e.g., mix of fibers or oil and fibers).
	Economic	Adjustment of production to the demand, quantity, and quality of the raw material available		
	Natural	Heterogeneity of resources		
Distribution phase	Economic capital	Existing bio-based materials value chains	- Surveys - Reports on bio-based materials market trends	*Manage connectivity:* connectivity represents the quality and quantity of the interrelationships between elements. Therefore, the appointment of a facilitator/ intermediary who could coordinate the stakeholders is fundamental.
	Economic and human	Lack of demand and awareness about these products		*Broaden participation:* the creation of a multidisciplinary working group can help refine the demand and earn the confidence of the stakeholders involved.

Table 12.1 Resilience strategies for wetland plants value chain.—cont'd

Value chain phases	Factors		Indicators	Strategies
Implementation phase	Human capital	Lack of knowledge about product technical performance Unawareness of craftsmen	- Surveys - Reports on bio-based materials market trends	*Encourage learning:* the use of these fibers in the construction sector means embracing variability in their performances, by taking advantage of professionals' knowledge and understanding of the resource. Experimental research provides data on technical properties and allows comparing their characteristics with traditional materials. *Accept complex adaptive systems:* the greatest limit to the use of organic fibers is trying to use them for the same purposes of mineral products, assuming that they would have the same characteristics.
	Natural	Variability in product properties		
	Social	Weak political will Insufficient funding	- Activity reports - Working group reports	
	Cultural	Loss of traditional architecture	- Inventory of local heritage	

3.3 Experimental research on wetland plants

Prototypes of composite materials and light cob were produced in the Bas-Quesnay site (20 km from the Park Headquarters) in January and February 2018.

Composite material samples (different proportions of fibers and earth) in the form of 26 blocks (300 × 300 × 70 mm) and 26 cubes (150 × 150 × 150 mm) were produced for thermal conductivity testing. For the CobBauge project 54 cubes (150 × 150 × 150 mm) of light cob (earth and fibers) were prepared for compressive strength testing.

Sample dimensions were established according to test procedures; they were dried in special ovens and tested together with bulk fibers in the laboratory of the School of Building Engineering of Caen (ESITC).

An analysis of results revealed that wetland plants have characteristics comparable with the bio-based materials that are used most for earth construction, such as *T. aestivum* (straw*), L. usitatissimum* (flax), and *C. sativa* (hemp).

According to water absorption test results, after 24 h *J. effusus* (475%) presented the higher value, *M. caerulea* (300%) was comparable to both *T. aestivum* (280%−350%) (Bouasker et al., 2014) and *C. sativa* hurds (280%), and *P. australis* (175%) and *C. mariscus* (190%) were comparable to *L. usitatissimum* (185%) (Tuan Anh Phung, 2018; Pacheco-Torgal and Jalali 2011).

In composite materials water absorption is an important factor in the regulation of humidity inside buildings. However, a high absorption rate can lead to a loss of the interface with the matrix and a loss of dimensional stability (Laborel-Préneron et al., 2016), and for this reason *J. effusus* may be considered the least suitable for earth-based products (Fig. 12.1).

Absolute density measurement highlights the porosity of materials: wetland plants results were within a range of values similar to those of *T. aestivum, L. usitatissimum,* and *C. sativa* (1300−−1500 kg/m³). In earth-based composites, an increase of aggregate or fiber content leads to an increment of porosity, as well as a decrease of density and thus of thermal conductivity. These results need to be referenced with water absorption coefficients because thermal insulation decreases when water content increases. This can be explained by the fact that water, which replaces the air contained in the pores, is a better heat conductor than the air it replaces (Meukam et al., 2003).

Apparent density is more difficult to compare with other bio-based materials, since measurement procedures are not always provided in the literature (Laborel-Préneron et al., 2016). Among the plants tested using the same procedure, *M. caerulea* presented the lowest density (20 kg/m³), followed by *T. aestivum* (33 kg/m³), *J. effusus* (46 kg/m³), *C. mariscus* (68.5 kg/m³), and *P. australis* (86.9 kg/m³),

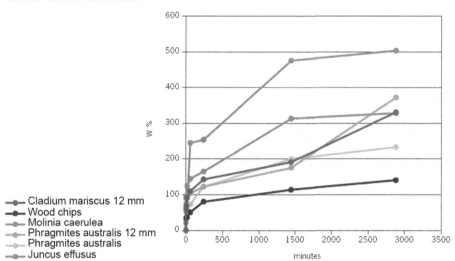

ABSORPTION COEFFICIENT

FIGURE 12.1

Absorption coefficient.

and *L. usitatissimum* (97.8 kg/m³). Apparent density can be useful in predicting the thermal properties of a loose-fill insulation system but is less useful in predicting the same for composite materials, since the bulk arrangement of plant aggregate or fibers will be modified when they are introduced into the mixture (Laborel-Préneron et al., 2016) (Fig. 12.2).

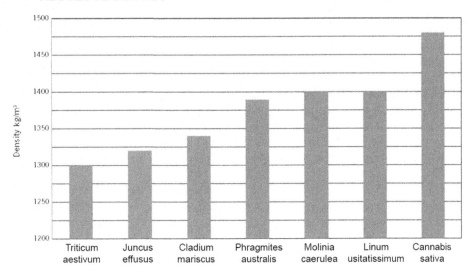

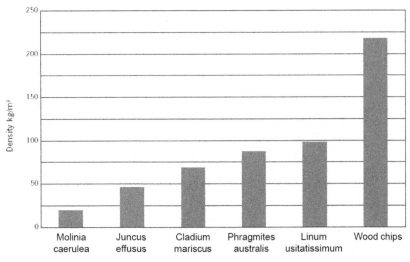

FIGURE 12.2

Absolute and apparent density.

Tensile strength results revealed similar properties for *M. caerulea* (19.53 MPa), *J. effusus* (22.24 MPa), *C. mariscus* (24.37 MPa), and *T. aestivum* (3.45−18 MPa) (Pullen and Scholz, 2011). *P. australis* is more resistant (70 MPa), with values comparable to *L. usitatissimum* (55.38 MPa) and *C. sativa* (46.63 MPa) (Fig. 12.3).

A comparison of thermal conductivity results with other bio-based materials already in use revealed that wetland plants can be used as insulation materials in loose-fill insulation systems and in earth-based composites: *J. effusus* (0.047 W/mK) showed properties similar to *T. aestivum* (0.041−0.0486 W/mK) (Ashour et al., 2010) and *C. sativa* (0.041 W/mK) (El-Sawalhi et al., 2016). *P. australis* (0.051 W/mK) and *M. caerulea* (0.055 W/mK) were comparable to *C. sativa* hurds (0.052 W/mK) (Balčiūnas et al., 2013). *C. mariscus* (0.071 W/mK), which presented the highest conductivity values, can be used as a thermal corrector in lime or in earth-based plasters.

Thermal conductivity of composite materials was also tested: the increase of content in fibers, by lightening the composites, used to decrease thermal conductivity. A high amount of fibers can make the composites more fragile, and unsuitable for prefabrication or transport. This is due to fiber morphology (too short or too smooth) and the use of highly liquid earth as a binder to keep them lighter but less resistant. Results revealed that wetland fibers have properties similar to composites such as *C. sativa* bricks and concrete and lime hemp, which are already in use and which have the same density (Fig. 12.4).

The effects of plant particle additions on compressive strength differed in the literature, showing that the influence of aggregate depends on the type of particle, soil composition, and testing method. The tests that have been done in this study revealed that the addition of fibers reduced the density and

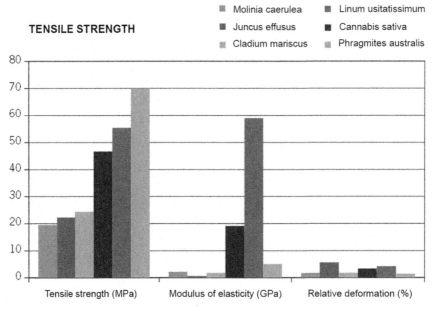

FIGURE 12.3

Tensile strength.

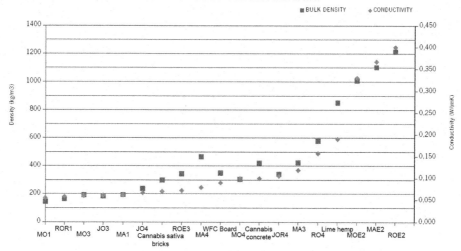

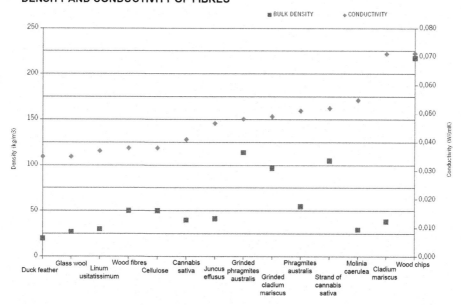

FIGURE 12.4

Density and conductivity of fibers and materials.

COMPRESSIVE STRENGTH

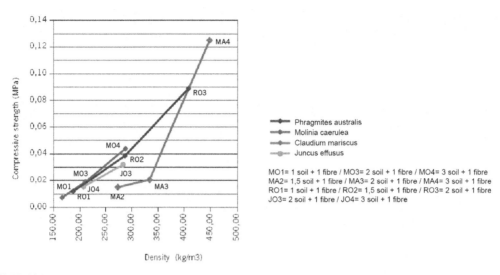

FIGURE 12.5

Compressive strength.

consequently the compressive strength with the results being similar among the fibers tested. Tests done on light cob revealed that for equal amounts of fibers, same type of soil and equipment, wetland fibers have similar results with other bio-based materials. For instance, with 2% fiber content, *C. mariscus* (0.67 MPa) and *P. australis* (0.64 MPa) obtained results similar to those of *C. sativa* hurds but higher than those of *T. aestivum* (0.55 MPa), while *M. caerulea* (1.57 MPa) has a compressive strength similar to composites with *L. usitatissimum* (1.72 MPa) (Fig. 12.5).

4. Discussion

As evidenced in the Results section, territorial analysis provided evidence of resources, agents, and dynamics, while experimental research explored the mechanical, physical, and hygrothermal properties of the analyzed plants. These two workflows were carried out in parallel so as to better direct value chain assessment.

One of the study's most significant results is the possibility of facilitating wetland maintenance, thus helping preserve their biodiversity. In some cases, the nature reserves pay an annual maintenance price that could be reduced if the plants had commercial value. On the other hand, stocks localization revealed the difficulty of quantifying a resource that grows spontaneously and is sometimes mixed with other species. Even though at the time of the study data were not yet complete (data on West Coast, Taut, Aure, and Vire valleys were missing), the comparison with stocks of other bio-based materials in the region such as *T. aestivum* (straw), *L. usitatissimum* (flax), and *C. sativa* (hemp) revealed that other supply sources need to be identified. The Park has a pivotal role in promoting partnerships, such as the one with the Boucles de la Seine Natural Regional Park, that would expand supply. Social enterprises and farmers will continue to oversee the maintenance of the site and play a role in the extraction of raw materials. One social enterprise expressed interest in participating in the production phase, for which they could provide manpower and infrastructure. Moreover, according to a study of the Norman Chamber of Agriculture (Legrain, 2017), farming is experiencing generational change: the younger generation's interest in wetland maintenance should be investigated, using a survey of a larger sample of farmers. Official reports revealed that 38% of surveyed enterprises in the region declared using bio-based materials for construction purposes (e.g., cellulose, *C. sativa*, *L. usitatissimum*, *T. aestivum* etc.). More than half of the businesses (58%) expressed interest in receiving more information on bio-based materials' technical characteristics. They pointed out a lack of product knowledge (32%), reflecting the need for further experimental research or professional training, in order to tackle variability in the resource performances and nonstandardization. A higher purchase price compared with other traditional materials was indicated as a barrier by 31% of enterprises. A comparison of the price of already-in-use bio-based materials (ranging from €1000/t for *L. usitatissimum*, €750/t for *C. sativa*, €450/t for *Miscanthus* and €150/t for *T. aestivum*) with the price of wetland plants (€40−50/t) shows the competitive potential of these resources. However, these prices could increase once the market to the construction sector was created. Moreover, harvesting difficulties (e.g., climatic conditions, accessibility, respect for biodiversity) which affect supply could generate extra costs.

Cultural capital analysis revealed the connection between the studied plants and the traditional habitat. The use of these materials in traditional architecture was lost because of changes in agricultural practices and thus the habitat, but it can be reintroduced via the renewed interest in cob construction (CobBauge project). The use of wetland plants can have a positive impact on local construction in the long term by reducing the need of imported materials and improving the thermal performance of traditional earth buildings.

Given this study's innovative nature, further experimental research is essential to test and characterize the properties of these resources, and to map product development. Laboratory tests showed the potential of these fibers as building materials, yielding results comparable to other bio-based materials and demonstrating that they can be used as insulation materials and earth-based composites. The tests carried out provided a preliminary characterization of the fibers. However, further research should be performed to test water vapor permeability, sorption-desorption, durability (biodegradation and microorganism development), and resistance to erosion and abrasion, water and fire, freezing and thawing. Due to time constraints these tests were not performed in this study.

5. Conclusion

The aim of the research presented in this chapter was to test PnrMCB wetland plants for their use as building materials and analyze how this value chain would improve territorial resilience.

PnrMCB wetlands occupy a remarkably large area considering the overall loss of French wetlands and their biodiversity during the 20th century. This has considerably increased interest in the conservation of these biodiversity-rich areas whose maintenance, assured by local communities through mowing and grazing, is necessary to prevent reforestation and avoid the risk of their disappearing.

Moreover, in recent years the appeal of wetlands has been enhanced by the recognition of their ecological value. Protection, restoration, and maintenance activities make it possible to find uses for the biomass produced by the wetland vegetation.

The experimental research in this study demonstrated that the selected wetland plants have the physical, mechanical, and hygrothermal characteristics necessary for use as insulation materials and in composite products, such as light cob.

Testing *M. caerulea, J. effusus,* and *C. mariscus,* in addition to *P. australis,* on which many studies have already been conducted, opens up a new research perspective for the use of these fibers in the construction sector.

The use of these plants as building materials, such as panels, earth-based composites, or loose-fill insulation systems, can improve the thermal performance of traditional cob houses, play an important role in heritage conservation, and provide a local market for the new materials.

The new market will make the maintenance of wetlands more rewarding for farmers and social enterprises, as they can provide the workforce and infrastructure without having to sacrifice European subsidies. This would be of interest to nature reserve managers, as it would facilitate the maintenance of wetlands and their biodiversity (lower maintenance cost, no need for biomass disposal, etc.) and consequently avoid the risk of reforestation.

Territorial analysis and experimental research results suggest that a building-construction value chain based on wetland plants could increase the resilience of the region from environmental, human, socioeconomic, and cultural perspectives. Enhanced territorial analysis and experimental research must overcome barriers including supply risks and variability of biomass availability in the extraction phase; heterogeneity of resources that could affect the properties of products in the production phase; lack of demand and competition from other already-in-use bio-based materials in the distribution phase; and lack of awareness of product characteristics by professionals in the implementation phase. In future research the establishment of a multidisciplinary working group with several stakeholders would help overcome organizational and procedural difficulties in project development.

Appendices
Survey questions

Surveyed farmers and social enterprises	
Topic	**Questions**
Resources	Which plant species are available?
	How much material is harvested?
	Do you know any of these plants: molinie (*Molinia caerulea*), marisque (*Cladium mariscus*), reed (*Phalaris arundinacea*), rush (*Juncus effusus*)?
	If so, do you know of any traditional use of these plants?
Maintenance	Which kind of maintenance do you perform (mechanical or manual?)
	How many times a year?
	Which kind of engine do you use?
	What are the main barriers to harvest?
	How much time do you need to mow an acre?
	How much do you get paid for mowing an acre?
	What traditional harvesting practices do you know?
Management	What is the current market for these plants?
	What's the price in case of sale? Within which radius (km)?
	How do you dry the plants?
	Where do you store them?
	How do you transport them?
	Do you think another output for these materials could be interesting for you?

Surveyed nature reserve manager	
Topic	**Questions**
Resources	Are *Molinia caerulea, Juncus effusus, Cladium mariscus, Phalaris arundinacea* available in the reserve?
	If so, which are the surfaces available for each plant?
	If so, is the stock monospecific or mixed?
	What are the main properties of these plants?
Maintenance	Who is in charge of the maintenance?
	How many times a year?
	Where do you stock or treat the harvested plants?
	Which are the stocks?
	How much material is harvested?
Management	Are there areas in the reserve that could possibly host one of this plant in the future?
	Could double use for a plant (e.g., phyto-purification and fiber production) be advantageous for the reserve management?

Surveyed: craftsmen	
Topic	**Questions**
Works	In which area do you work usually?
	What kind of work do you do?
	What are the most requested works?
	The number of construction sites use to increase, decrease, or stay stable?
Materials	On what criteria do you choose the materials you use?
	Do you use bio-sourced materials? If so, which ones?
	If so, what is the origin of these materials?
	Is the supply complicated?
	What is the cost of these materials?
	Are they more expensive compared to conventional materials?
	Have you ever used one of these plants: molinie (*Molinia caerulea*), marisque (*Cladium mariscus*), reed (*Phalaris arundinacea*), rush (*Juncus effusus*)?
	Do you know of any traditional use of these materials?
	Do you use raw earth for construction purposes?
Construction	Is the implementation of bio-sourced materials requested by the project manager?
	Do you find that the use of bio-sourced materials is more complicated than conventional materials? If so, why?
	What would facilitate the use of these materials?

Lab tests

Lab tests performed in the School of Building Engineering (ESITC) laboratory at Caen

Test: absolute density

Equipment: AccuPyc II machine 1340 (helium pycnometer)

Method:
The fibers were oven-dried and their mass (m) was measured with a high-precision balance. The volume (V) of a sample is determined by measuring the pressure difference of the injected helium in a calibrated volume (the reference cell) and absorbed into the small voids in the sample.
The fiber density is calculated by the equation: $\rho_{abs} = m/V$
For each fiber, three tests have been performed and for each test, three measurements are taken by the machine.

Note: this test is among the most accurate methods to calculate density, as it does not consider the fiber lumens, which allows the calculation of the absolute density of the sample.

Test: apparent density

Equipment: graduated cylinder

Method:
A graduated cylinder is filled with the sample; the cylinder is hit 20 times so that the material is tightly packed; the volume occupied by the sample is read; the sample weighed. The test is performed five times and the value deduced.
The density [ρ_{app}] is calculated by the equation: $\rho_{app} = m/V$

Note: this is an empirical method which allows comparative analysis between different resources. For this reason, the same graduated cylinder and the same protocol should be used for every test.

Test: absorption coefficient

Equipment: large plastic bins, small cotton bags, and a centrifuge

Method:
Empty bags are saturated with water and centrifuged at 500 rpm for 15 s.
2 g of the fibers (previously oven-dried) are weighed and put into each bag. Bags are then submerged in water. The bags are taken out of the water at the following time intervals: 1 min; 5 min; 15 min; 1 h; 4 h; 24 h, and 48 h. Six samples are picked up for each time point. Once removed, the full bags are briefly centrifuged at 500 rpm for 15 s and then weighed. The absorption coefficient (W) is calculated according to the equation: $W = m_{eau}/m_{fiber}$
It expresses the difference between wet and dry weight of the sample.
Once the data are plotted, these results allow us to ascertain each fiber water absorption curve over time, thus showing the variation in the water retention of the fibers. The ability of fibers to absorb water, which affects the water content of composite materials mixtures (e.g., raw earth and fibers), may produce variations on dry volume and modify the fiber/matrix interface.

Note: the absorption coefficient test has been carried out according to an experimental protocol developed by the group RILEM TC 236-BBM (Amziane et al., 2017).

Test: tensile strength

Equipment: SEM (scanning electron microscope) and INSTRON 3369 machine

Method:

SEM (scanning electron microscope) imaging was used to ascertain the internal structure of fibers and to calculate the area of their transverse section, which is required to estimate their tensile strength.

For the test, fiber samples of 100 mm length were prepared. Their extremities were protected by adhesive tape placed over a length of 30 mm. The length submitted to the test is therefore equal to 40 mm.

The sample is placed in the clamping jaws and its resistance tested at different speeds: 0.1 mm/min; 1 mm/min; 5 mm/min. The speed variation did not affect the results.

The test, repeated tens of times for every fiber, was carried out with a 50N sensor with the exception of *Molinia*, which was tested with a 10N sensor and plastic holders more apt to small fibers. The reed was not tested directly as the values were taken from the bibliography.

The tensile strength (R) is calculated by the equation: $\mathbf{R\ (MPa) = F_{max}/A}$, where F_{max} (N) is the maximum recorded force and A (m^2) is the area of the longitudinal section of the fiber.

The relative deformation (ε) is calculated by the equation: $\boldsymbol{\varepsilon\ (\%) = L - L0/L0}$

The modulus of elasticity (E) is calculated by the equation: $\mathbf{E\ (GPa) = R/\varepsilon}$

Note: tensile strength has been carried out according to similar studies which tested organic fibers with the INSTRON machine device. The modulus of elasticity depends directly to the slope of the linear part of the resistance/deformation curve, without considering the first stroke, which represents the preload applied by the machine to stretch the fiber and start the test.

Test: compressive strength

Equipment: uniaxial testing machine

Method:

A compressive force (50 kN) is applied to a given speed, perpendicularly to a cubic sample (150 × 150 × 150 mm) and the maximum stress supported is calculated at 10% of the relative deformation, since these materials are not subject to breakup.

The compressive strength [σ_{10}] was calculated with the equation $\boldsymbol{\sigma_{10} = 10^3\ F_{10}/A_0}$, where σ_{10} (kPa) is the compressive stress at 10% of the relative deformation, F_{10} (N) is the force corresponding to 10% of the relative deformation, and A_0 (mm^2) is the initial cross-sectional area of the sample.

The sample shall be positioned centrally between the compression plates of the testing machine, with a preload of 250 ± 10 Pa to ensure that the plates adhere to the sample. Compression is exerted by a plate moving at a constant speed of d/10 per minute with a tolerance of $\pm25\%$, where d is the initial thickness of the sample expressed in millimeters. The compression continues until the sample is broken or up to a relative deformation of 10%. Relative deformation is the quotient of the reduction in thickness of the sample by its initial thickness d_0, measured in the direction of the load.

Note: the compressive strength test has been carried out according to the European standard NF EN 826 (May 2013), which describes the equipment and procedures to be used to determine compression behavior of samples of thermal insulation products.

Test: thermal conductivity

Equipment: NETZSCH HFM 436 lambda machine (transient plane source device)

Method:

Thermal conductivity was tested with a pressure of 1.2 kPa and a ΔT of 20°C. Two types of samples were tested: Bulk fibers and composite materials with raw earth as a binder.

A 300 × 300 mm sample, with a thickness between 5 and 100 mm, is installed in the heat flow measuring device between two plates (upper and lower) stabilized at two different temperatures. Due to the gradient in temperature, the heat is transferred from the hot side to the cold side. Temperature and heat flow are measured once a minute by integrated sensors in both plates. The test continues until a balance is achieved.

For bulk fibers, a polystyrene frame (260 × 260 × 50 mm) has been used. It is necessary to weigh the frame before filling it, and to arrange the fibers inside without exerting any pressure.

Note: since bulk fibers tend to absorb moisture easily, it must be dried just before the test.

Stocks quantification

Source	Fiber type			
	Cladium mariscus	*Molinia caerulea*	*Phragmites australis*	*Juncus effusus*
Natura2000 Maps	- 12.53 ha from Cariçaie - 1.34 ha mixed with Ericaceae -heath	- 5.60 ha Moliniaie - 1.13 ha (mixed with *Cladium mariscus* and *Myrica gale*—sweetgale) - 4.64 ha (mixed with *Myrica gale*)	- 19.07 ha *Phragmition communis*; - 17.75 ha (mixed with *Solanum dulcamara*bbittersweet)	Not available
Nature Reserve of Mathon	- 700 m^2 - 9700 m^2 (mixed with phragmites).	- 3000 m^2	- 2000 m^2 - 3500 m^2 (mixed with *Carex paniculate* - Greater tussock-sedge)	
Espace Naturel Sensible in Saint Côme du Mont	—	—	3 ha	—
Nature Reserve of the Taute	—	5000 m^2	2000 m^2	—
The Hunting and Wildlife Reserve of the Bohons	73 ha	- 72 ha (monospecific) - 92 ha (mixed)	6 ha	Not available
National Nature Reserve of the Sangsurière and Adriennerie	6.75 ha	—	—	31 ha

References

Ademe, 2016. Inventory of bio-based products available on the market and identification of the targeted public ([Recensement des produits bio-sourcés disponibles sur le marché et identification des marchés publics ciblés]). Available at: https://www.ademe.fr/recensement-produits-biosources-disponibles-marche-identification-marches-publics-cibles.

Amziane, S., Collet, F., Lawrence, M., Magniont, C., Picandet, V., et al., 2017. Recommendation of the RILEM TC 236-BBM: characterisation testing of hemp shiv to determine the initial water content, water absorption, dry density, particle size distribution and thermal conductivity. Mater. Struct. 50. https://doi.org/10.1617/s11527-017-1029-3.

Aruan Efendy, M.G., Pickering, K.L., 2014. Comparison of harakeke with hemp fibre as a potential reinforcement in composites. Compos. Part Appl. Sci. Manuf. 67, 259–267. https://doi.org/10.1016/j.compositesa.2014.08.023.

Ashour, T., Wieland, H., Georg, H., Bockisch, F.-J., Wu, W., 2010. The influence of natural reinforcement fibres on insulation values of earth plaster for straw bale buildings. Mater. Des. 31, 4676–4685. https://doi.org/10.1016/j.matdes.2010.05.026.

Baggio, J.A., Brown, K., Hellebrandt, D., 2015. Boundary object or bridging concept? A citation network analysis of resilience. Ecol. Soc. 20 (2), 2. https://doi.org/10.5751/ES-07484-200202.

Balčiūnas, G., Vėjelis, S., Vaitkus, S., Kairytė, A., 2013. Physical properties and structure of composite made by using hemp hurds and different binding materials. Procedia Eng. 57, 159–166. https://doi.org/10.1016/j.proeng.2013.04.023.

Berkes, F., Folke, C., 1994. Linking social and ecological systems for resilience and sustainability. Beijer Discussion Paper Series 52.

Biggs, R., Schlüter, M., Bohensky, E., Burnsilver, S., et al., 2012. Towards principles for enhancing the resilience of ecosystem services. Annu. Rev. Environ. Resour. 37, 421–448.

Bouasker, M., Belayachi, N., Hoxha, D., Al-Mukhtar, M., 2014. Physical characterization of natural straw fibers as aggregates for construction materials applications. Materials 7, 3034–3048. https://doi.org/10.3390/ma7043034.

Bouguerra, A., Ledhem, A., de Barquin, F., Dheilly, R.M., Quéneudec, M., 1998. Effect of microstructure on the mechanical and thermal properties of lightweight concrete prepared from clay, cement, and wood aggregates. Cement Concr. Res. 28, 1179–1190. https://doi.org/10.1016/S0008-8846(98)00075-1.

COFREEN, 2010-2013. Guidebook of Reed Business, Interreg IV-A Programme. EU Investing. Available at: http://www.roostik.ee/trykivaljanded/COFREEN%20guidebook_korr_IK_2013.pdf.

Cui, B., Yang, Q., Yang, Z., Zhang, K., 2009. Evaluating the ecological performance of wetland restoration in the Yellow River Delta, China. Ecol. Eng. 35, 1090–1103.

El-Sawalhi, R., Lux, J., Salagnac, P., 2016. Estimation of the thermal conductivity of hemp-based insulation material from 3D tomographic images. Heat Mass Tran. 52, 1559–1569. https://doi.org/10.1007/s00231-015-1674-4.

Ghavami, K., Toledo Filho, R.D., Barbosa, N.P., 1999. Behaviour of composite soil reinforced with natural fibres. Cement Concr. Compos. 21, 39–48. https://doi.org/10.1016/S0958-9465(98)00033-X.

Hodge, I., Mcnally, S., 2000. Wetland restoration, collective action and the role of water management institutions. Ecol. Econ. 35, 107–118.

Köbbing, J.F., Thevs, N., Zerbe, S., 2013. The utilisation of reed (*Phragmites australis*): a review. Mires Peat 13, 1–14. Article 01.

Kroon, F.W., 2013. Saving reed lands by giving economic value to reed. Mires Peat 13 (10).

Kusler, J.A., Kentula, M.E. (Eds.), 1990. Wetland Creation and Restoration: The Status of the Science. Island Press, Washington DC, 616 pp.

Laborel-Préneron, A., Aubert, J.E., Magniont, C., Tribout, C., Bertron, A., 2016. Plant aggregates and fibers in earth construction materials: a review. Construct. Build. Mater. 111, 719–734. https://doi.org/10.1016/j.conbuildmat.2016.02.119.

Legrain, P., 2017. Agriculture and Agri-Food Industry in the Marais du Cotentin et du Bessin Natural Regional Park ([L'agriculture et l'agroalimentaire dans le Parc Naturel Régional des Marais du Cotentin et du Bessin]). Chambre d'Agriculture Normande.

Lokesh, K., Ladu, L., Summerton, L., 2018. Bridging the gaps for a 'circular' bioeconomy: selection criteria, bio-based value chain and stakeholder mapping. Sustainability 10, 1695. https://doi.org/10.3390/su10061695.

Meukam, P., Noumowe, A., Jannot, Y., Duval, R., 2003. Caractérisation thermophysique et mécanique de briques de terre stabilisées en vue de l'isolation thermique de bâtiment. Mater. Struct. 36, 453–460. https://doi.org/10.1007/BF02481525.

Millennium Ecosystem Assessment, 2005. Ecosystems and Human Well-Being: Wetlands and Water. World Resources Institute, Washington, DC, USA.

Nomadeis, A.M., 2014. Bio-based Building Materials Survey on the Perceptions, Practices and Expectations of Craft Companies (Upper and Lower Normandy) ([Matériaux de construction biosourcés Enquête sur les perceptions, pratiques et attentes des entreprises artisanales (Haute et Basse Normandie)]). Available at: http://www.nomadeis.com/EnqArtisans/RAPPORT-Haute-Basse-Normandie.pdf.

Observatoire Nationale de la Biodiversité (ONB), 2016. Remarkable Wetlands as Endangered Natural Areas What Land Use Within the Ramsar Sites of Metropolitan France? Retrospective 19752005 ([Les milieux humides remarquables, des espaces naturels menacés Quelle occupation du sol au sein des sites Ramsar de France métropolitaine ? Retrospective 19752005]). Retrieved from: http://indicateurs-biodiversite.naturefrance.fr/fr.

Pacheco-Torgal, F., Jalali, S., 2011. Cementitious building materials reinforced with vegetable fibres: a review. Construct. Build. Mater. 25, 575–581. https://doi.org/10.1016/j.conbuildmat.2010.07.024.

Parc des Marais du Cotentin et du Bessin, 2016. A Dynamic Territory 2010-2022. Mid-term Review ([Un territoire en mouvement 2010-2022. Bilan à mi-parcours]). Retrieved from: https://parc-cotentin-bessin.fr/sites/default/files/inline-files/2018_RAPPORT_MIPARCOURS_2010-2016.pdf. (Accessed 5 May 2020).

Parc des Marais du Cotentin et du Bessin, 2007. Atlas [Atlas cartographique]. Retrieved from: https://parc-cotentin-bessin.fr/sites/default/files/2021-04/PNRCMB_2007_ATLAS_CARTOGRAPHIQUE.pdf. (Accessed 5 May 2020).

Perrotti, D., Hyde, K., Otero Peña, D., 2020. Can water systems foster commoning practices? Analysing leverages for self-organization in urban water commons as social–ecological systems. Sustain, Sci 15, 781–795. https://doi.org/10.1007/s11625-020-00782-1.

Pullen, Q.M., Scholz, T.V., 2011. Index and engineering properties of Oregon cob. J. Green Build. 6, 88–106.

Sánchez-Zamora, P., Gallardo-Cobos, R., Ceña Delgado, F., 2016. The concept of resilience in the analysis of rural territorial dynamics: an approach to the concept through a territorial focus [La noción de resiliencia en el análisis de las dinámicas territoriales rurales: una aproximación al concepto mediante un enfoque territorial]. Cuadernos de Desarrollo Rural 13 (77), 93–116. https://doi.org/10.11144/Javeriana.cdr13-77.nrad.

Sánchez Zamora, P., Gallardo Cobos, R., 2019. Diversity, disparity and territorial resilience in the context of the economic crisis: an analysis of rural areas in southern Spain. Sustainability 11 (6), 1743. https://doi.org/10.3390/su11061743.

Sánchez-Zamora, P., Gallardo-Cobos, R., Ceña Delgado, F., 2014a. Rural areas face the economic crisis: analyzing the determinants of successful territorial dynamics. J. Rural Stud. 35, 11–25. http://10.1016/j.jrurstud.2014.03.007.

Sánchez-Zamora, P., Gallardo-Cobos, R., Ceña Delgado, F., 2014b. Rural areas of Andalusia facing economic downturn: analyzing territorial resilience factors, Economia Agraria y Recursos Naturales. Spanish Assoc. Agric. Econ. 14 (01), 1−30. https://doi.org/10.22004/ag.econ.180102.

Segetin, M., Jayaraman, K., Xu, X., Harakeke, 2007. Reinforcement of soil−cement building materials: manufacturability and properties. Build. Environ. 42, 3066−3079. https://doi.org/10.1016/j.buildenv.2006.07.033.

Tuan, A.P., 2018. Formulation and Characterization of Cob, an Earth-Fibres Composite Material ([Formulation et caractérisation d'un composite terre-fibres végétales : la bauge]). Génie civil. Normandie Université. Français. Available at: https://tel.archives-ouvertes.fr/tel-01938827/document.

Further reading

Agarwal, S., Rahman, S., Errington, E., 2009. Measuring the determinants of relative economic performance of rural areas. J. Rural Stud. 25 (3), 309e321. https://doi.org/10.1016/j.jrurstud.2009.02.003.

Alves Fidelis, M.E., Vitorino Castro Pereira, T., da Fonseca, O., Gomes, M., de Andrade Silva, F., Dias Toledo Filho, R., 2013. The effect of fiber morphology on the tensile strength of natural fibers. Jmr&T 2 (2), 149−157. https://doi.org/10.1016/j.jmrt.2013.02.003.

Asdrubali, F., Bianchi, F., Cotana, F., D'Alessandro, F., Pertosa, M., Pisello, A.L., Schiavoni, S., 2016. Experimental thermo-acoustic characterization of innovative common reed bio-based panels for building envelope. Build. Environ. 102 (217), e229. https://doi.org/10.1016/j.buildenv.2016.03.022.

Balázs István Tóth, 2014. Territorial Capital: Theory, Empirics and Critical Remarks. In: European Planning Studies. https://doi.org/10.1080/09654313.2014.928675.

Brunetta, G., Ceravolo, R., Barbieri, C.A., Borghini, A., de Carlo, F., Mela, A., Beltramo, S., Longhi, A., De Lucia, G., Ferraris, S., Pezzoli, A., Quagliolo, C., Salata, S., Voghera, A., 2019. Territorial resilience: toward a proactive meaning for spatial planning. Sustainability 11, 2286.

Bryman, A., Bell, E., 2007. Business Research Methods. Oxford University Press, Oxford.

Foster, K.A., 2007. A Case Study Approach to Understanding Regional Resilience, Working Paper 2007-08. Institute of Urban and Regional Development, University of California, Berkeley.

Fratesi, U., Perucca, G., 2018. Territorial capital and the resilience of European regions. Ann. Reg. Sci. 60, 241−264. https://doi.org/10.1007/s00168-017-0828-3.

Fröhlich, K., Hassink, R., 2018. Regional resilience: a stretched concept? Eur. Plann. Stud. 26 (9), 1763−1778. https://doi.org/10.1080/09654313.2018.1494137.

Gray, D.E., 2004. Doing Research in the Real World. SAGE Publications, London.

Häkkinen, T., Belloni, K., 2011. Barriers and drivers for sustainable building. Build. Res. Inf. 39 (3), 239−255. https://doi.org/10.1080/09613218.2011.561948.

Hulme, J., 2010. Sustainable Supply Chains that Support Local Development. The Prince's Foundation for the built environment.

INTERREG VA France (Channel) CobBauge. Available at: http://www.cobbauge.eu/en/.

Muizniece, I., Kazulis, V., Zihare, L., Lupkina, L., Ivanovs, K., Blumberga, D., 2018. Evaluation of reed biomass use for manufacturing products, Taking Into Account Environmental Protection Requirements. Agron. Res. 16 (S1), 1124−1132. https://doi.org/10.15159/AR.18.077.

OECD, 2001. Territorial Outlook 2001. Organization for Economic Cooperation and Development (OECD), Paris. http://www.oecd.org/newsroom/theoecdterritorialoutlook2001.htm.

Perrotti, D., Hyde, K., Otero Peña, D., 2020. Can water systems foster commoning practices? Analysing leverages for self-organization in urban water commons as social−ecological systems. Sustain. Sci. 15, 781−795. https://doi.org/10.1007/s11625-020-00782-1.

Rizzi, P., Graziano, P., Dallara, A., 2018. A capacity approach to territorial resilience: the case of European regions. Ann. Reg. Sci. 60, 285−328. https://doi.org/10.1007/s00168-017-0854-1.

Steiner, A., Attenton, J., 2015. Exploring the contribution of rural enterprises to local resilience. J. Rural Stud. 40, 30e45. https://doi.org/10.1016/j.jrurstud.2015.05.004.

Stockholm Resilience Centre, 2015. Applying Resilience Thinking: Seven Principles for Building Resilience in Social-Ecological Systems. Retrieved from: https://stockholmresilience.org/. (Accessed 10 February 2020).

Valentine, G., 2005. Tell me about …: using interviews as a research methodology. In: Flowerdew, R., Martin, D. (Eds.), Methods in Human Geography: A Guide for Students Doing a Research Project, second ed. Prentice Hall, Harlow, p. 110e126.

Walker, B., Holling, C.S., Carpenter, S.R., Kinzig, A., 2004. Resilience, adaptability and transformability in social-ecological systems. Ecol. Soc. 9 (2), 5.

Consequential life cycle assessment to promote the recycling of metallurgic slag as new construction material

Andrea Di Maria[1], Karel Van Acker[1,2]

[1]*Katholieke Universiteit Leuven (KU Leuven), Leuven, Belgium;* [2]*Center for Economics and Corporate Sustainability (CEDON), Katholieke Universiteit Leuven (KU) Leuven, Brussels, Belgium*

Chapter outline

1. Introduction

The implementation of circular economy strategies can be achieved by stimulating material recycling across different economic sectors. The energy and material exchanges among different industries can indeed mitigate resource consumption and lead to a closed material cycle. In this context, the symbiosis between steel and cement industries represents already a virtuous example. Steel and cement (the last as a component of concrete) are made from primary raw materials at high temperature in large industrial reactors, and are two of the most used materials in the world.

Traditional cement is used mainly as a binder in concrete production. The most common form of cement is the Ordinary Portland Cement (OPC), which is composed primarily of calcium silicate

Environmental Sustainability and Economy. https://doi.org/10.1016/B978-0-12-822188-4.00017-8

minerals, processed in a kiln at temperatures higher than 1400°C (Huntzinger and Eatmon, 2009). To reach those temperatures, high-energy consumption is required, either as electricity or as fuel, typically accounting for 30%–40% of total production costs (Szabó et al., 2006). According to van Oss and Padovani (2003), the energy consumption during OPC production produces an average of 0.48 t CO_2/t clinker, depending on the types and quantities of fuels used. A second essential and inevitable source of CO_2 originates from the calcination of the limestone raw material in the kiln. At temperatures above 900°C, the calcium carbonate ($CaCO_3$) contained in the limestone decomposes forming CaO and CO_2, which are released into the atmosphere. van Oss and Padovani (2003) proposed an average calcination emission factor of ≈ 0.51 t CO_2/t clinker, which is very similar to the one assumed by the Intergovernmental Panel on Climate Change (IPCC). In general, the cement industry produces a significant amount of greenhouse gas (GHG) emissions, and it contributes to 8%–10% of the global anthropogenic CO_2 emissions excluding those associated to land-use change (Hossain et al., 2018; Scrivener and Kirkpatrick, 2008).

Steel is an alloy made of almost pure iron, produced by reducing iron ores at high temperature using coal as a reducing agent. This reduction reaction produces CO_2 (Birat, 2012). The steel production industry has traditionally been a pioneer in the implementation of industrial symbiosis strategies, owing to the material and energy-intensive processes and a large amount of residues and energy released (Geiseler, 1996). A steel plant is well suited to be the core of an eco-industrial network, exchanging flows at a large scale (Dong et al., 2013).

The potential for synergy between the steel and cement industry is exploited when GGBFS, the Ground Granulated Blast-Furnace Slag, is produced as a coproduct in steel production and is reused in cement production as an additive (Tsakiridis et al., 2008). The use of GGBFS and some other steel slags in the cement industry is indeed one of the most well-established examples of beneficial synergy between steel and cement sectors (Johansson and Söderström, 2011; Van den Heede and De Belie, 2012). However, GGBFS represents only a portion of the total slag produced by the steel industry, and the potential valorization of other kind of steel slags in the cement industry is not fully explored at present.

Stainless Steel Slag (SSS), in particular, is produced during the stainless steel making process. Since chromium is an essential constituent of stainless steel, a fraction of it appears also in the slag, together with other heavy metals. The presence of those hazardous compounds, together with the huge volumes generated each year, poses environmental and health threats when managing the SSS (Huaiwei and Xin, 2011). Moreover, SSS occurs in a very fine texture (a few μm diameters), giving the slag the shape of a fine powder, which causes leaching risks, making the handling of the SSS problematic. The fine texture is due to a phenomenon called "dusting," which is caused by volume expansion of the slag during the cooling phase (Kim et al., 1992). To avoid the problem of dusting, boron oxide (B_2O_3) is commonly added during the cooling process of SSS in a quantity equal to 2% of the total mass of the slag. Boron addition prevents the formation of fine particles (Durinck et al., 2008). Stabilized SSS grains present a bigger texture (few mm) and a more stable chemical status, which allow their disposal in hazardous waste landfills or their reuse as low-quality aggregates. However, the valorization as aggregates represents a low-value application given the high-quality oxides (CaO, MgO, AlO_2) contained in the SSS, whose chemical potential can be activated and exploited.

Together with landfilling or reuse as low-quality aggregates after boron stabilization, a third possible end-of-life route for SSS is represented by its valorization as an alternative cement in construction materials. Recent research has shown that SSS can react with alkali activators and be used as a cement to create high-quality construction materials (Huaiwei and Xin, 2011; Motz and Geiseler, 2001; Panda et al., 2013; Salman et al, 2014, 2015; Sheen et al., 2016). Alkali activation can be simply described as a process where latent binding properties of slag are activated by mixing the slag with alkali activators (carbonates or silicates).

In addition to the technical feasibility, a sustainable environmental profile is another fundamental prerequisite for a successful large-scale deployment of new technologies (Di Maria et al., 2018). In this regard, a few studies have been published on the environmental performances of alkali activation technology applied to metallurgic slag. A short list of these studies and their findings is shown in Table 13.1. Some of the previous Life Cycle Assessment (LCA) studies focused only on some of the environmental aspects, as global warming potential (Duxson et al., 2007; Habert et al., 2011), abiotic resource depletion, and cumulative energy demand (Weil et al., 2009), while other studies provided a more complete overview of the environmental profile of alkali-activated material from metallurgic slag (Di Maria and Van Acker, 2018; Di Maria et al., 2018; Salman et al., 2016). All studies agreed that alkali-activated materials reduce the CO_2 emissions within a range of 40%−70% compared to OPC, while similar impacts are caused by abiotic resource depletion and cumulative energy demand, due to the production of the alkali activator solution, which gives the main contribution for the final environmental impact of alkali-activated materials. However, this result is only valid if no impact is allocated to the metallurgic slag acting as a precursor. If the impacts of the industrial process producing the slag are allocated by mass to the precursor, then the final results are entirely reverse, and the slag-based material resulted in having higher impacts than OPC concrete (Habert et al., 2011).

Previous environmental analysis is based on LCA, which is a technique to assess the potential environmental impacts and resources used throughout a product's life (Finnveden et al., 2009; Vázquez-Rowe et al., 2014). However, the efficacy of LCA when analyzing the consequences of a change into an economic sector (e.g., the introduction of a new recycled product into the market) has been recently criticized. While the traditional approach in LCA, so-called "attributional-LCA (ALCA)," presents many advantages in reporting and understanding the environmental impacts directly related to the system under study, it shows also limitations when it comes to tackling policy decision consequences, as it does not consider any indirect effect arising in the markets from changes in the supply or demand of a product (Brander et al., 2009; Vázquez-Rowe et al., 2013). However, all activities are interconnected through markets, and the change in the supply of one product may lead to consequences that propagate to other activities present in the same market. Therefore an alternative modeling technique, called consequential-LCA (CLCA), has been developed recently, aiming at including the consequences of a supply/demand change into the environmental analysis (Ekvall, 2002). While ALCA is a purely descriptive model, CLCA is a change-oriented model (Ekvall et al., 2005; Tillman, 2000). A change-oriented model means that the analyzed system includes all activities that are expected to change as a consequence of a change in supply/demand for the investigated product (Consequential-LCA, 2015). A more detailed analysis of the difference between ALCA and CLCA is presented in the next section.

The scope of this chapter is, therefore, to use a consequential approach to enlarge the current knowledge on LCA of alkali activated construction material from SSS. By considering also some market dynamics, the consequential LCA can assess better the overall environmental benefits of substituting OPC-based concrete with SSS-based construction blocks. The data used in the case study refer to the current situation in Flanders, the northern region of Belgium, and they have been collected

Table 13.1 Literature review of previous LCA studies on alkali activation of metallurgic slag.

Authors	Metallurgic slag (precursor)	LCA focus	Findings
Duxson et al. (2007)	Coal fly ash and metakaolin	Global warming potential	Geopolymers can, in general, deliver 80% reduction in CO_2 emissions compared to OPC-based concrete.
Weil et al. (2009)	Slag, fly ash, metakaolin	Global warming potential, abiotic resource depletion potential, cumulative energy demand	Compared to OPC, geopolymers showed comparable environmental impacts regarding ADP and CED, but 70% reduction in GWP.
Habert et al. (2011)	Fly ash, BFS, and metakaolin	Global warming potential	If no allocation of the residue is considered, geopolymers can save 45% of CO_2 emissions compared to OPC-based concrete. However, this result is completely reversed if the mass allocation is applied to the residue, and geopolymers resulted in having higher impacts than OPC-concrete.
Salman et al. (2016)	SSS	Overall technical, environmental, and economic analyses of alkali-activated construction material	Alkali-activated construction material presents potential environmental benefits, especially in CO_2 emissions reduction compared to traditional OPC. However, alkali activators production may represent an important limitation to the environmental and economic performances.
Di Maria and Van Acker (2018)	Industrial goethite	Overall analysis on environmental impact categories	A slag pretreatment process (e.g., fuming) drives the environmental profile of the slag valorization scheme, and it may offset the environmental benefits derived from the substitution of OPC with alkali-activated construction materials.
Di Maria et al. (2018)	SSS	Overall analysis on environmental impact categories	Alkali-activated materials from slag show significant CO_2 reduction compared to OPC-based concrete. In other environmental categories, alkali activators production represents the main environmental hotspot of the process.

during several collaborations between Flemish universities and local steel producers. The case study on SSS construction blocks will provide also a more general base to understand the differences between attributional and consequential approaches when analyzing the environmental performances of a new waste-derived product.

2. Consequential versus attributional approach

The attributional approach in LCA focuses exclusively on internal flows of a production system, and it aims at describing the environmental impacts directly related to the life cycle of the investigated product (Ekvall, 2000). Therefore, little consideration is paid to the effects that the system may have on other related economic systems (Vázquez-Rowe et al., 2014). However, when deciding industrial and policy level, the resulting change may lead to an increase or decrease in the use, and therefore in the demand, of the investigated product (market effects). Therefore, some decision can have significant consequences (or indirect effects) for the environmental impacts of activities outside the life cycle of the investigated product (Ekvall, 2000). An ALCA for recycled materials includes the possible consequences on the market by assuming that the recycled material replaces an equivalent virgin material, whose avoided impacts are given as credits (negative values) to the recycling process. However, the recycled material may also replace a different virgin material or other recycled material from an alternative waste source (Ekvall and Finnveden, 2001). The level of detail required to include all the possible consequences is exceptionally high. Moreover, most of the ALCA studies today simplify the analysis through a ceteris paribus condition (Ekvall, 2000). As described by Marvuglia et al. (2013) the ceteris paribus condition implies that the amount of product under assessment is not relevant to the study, as the results of the LCA can be directly scaled up or down, maintaining a linear relationship with the scaling factor. Because of the ceteris paribus condition, ALCA implicitly assumes that the increased use of the product in the investigated life cycle will automatically lead to (i) an increase in the demand and supply of the product, and (ii) a decrease in the demand and supply of a competing product (fully elastic market in demand and/or supply). Due to the missing information on possible consequences and the ceteris paribus assumption, the reliability of ALCA in answering policy-related questions has been recently questioned (Vázquez-Rowe et al., 2013).

Contrarily, the main assumption in the consequential approach is that only specific activities will be affected by a shift in the demand for a product. One of the first definitions of CLCA is given by Ekvall (2002), which defines CLCA as a methodology aiming "at describing the effect of changes in the life cycle." The term "changes" refers to modifications in some part of the inventory of the investigated life cycle, leading to consequences through a chain of cause-effect relationships (Curran et al., 2005). A more clear definition is given by Ekvall and Weidema (2004), where the consequential LCI methodology "aims at describing how the environmental relevant physical flows to and from the technological system will change in response to possible changes in the life cycle."

These relevant flows are related to (i) the suppliers providing the investigated product, and (ii) the suppliers producing the competing products in the same market. Such suppliers are defined as marginal suppliers (Ekvall and Weidema, 2004). A marginal supplier must be unconstrained and therefore able to adjust its production capacity to meet the new demand for its product (Weidema et al., 1999). The

CLCA methodology states clearly that only the marginal suppliers and the processes that are affected by the decision under analysis must be included in the system boundaries. Therefore, one of the main issues when applying CLCA is the identification of the processes and marginal suppliers to be included in the analyzed system.

As described above, the conventional practice in ALCA is to assume that all suppliers are unconstrained and that demand and supply are fully elastic. However, CLCA considers that suppliers may be constrained, and markets are imperfect. A change in demand or supply of a product leads to a consequent adjustment in the demand of all the related products in the same market. Therefore, market economic models could help to assess the magnitude of this adjustment. Among others, Bouman et al. (2000), Ekvall (2000), and Ekvall and Andrae (2006) started to investigate the feasibility of combining LCA and simple economic models.

The above-mentioned studies showed how the CLCA could complement ALCA and provide valuable knowledge on the possible environmental consequences in a broader context. On the other hand, the uncertainty of the CLCA results is considerable: data quality is usually poor, especially when estimating the demand/supply elasticities of a product; as such estimate is available only for a few goods. Although there is a lack of easily available marginal data and input for the economic model, all studies conclude that the integration between ALCA and economic models (and the consequential approach) can generate new knowledge, even when important input data are based on quantitative assumptions.

To overcome the limitation represented by data requirement in economic models, a simplified procedure for the identification of the marginal suppliers has been developed by Weidema et al. (1999, 2003, 2009). In its latest version, the procedure consists of four steps:

I. identify the scale and the time horizon of the change;
II. identify the geographical limits of the market;
III. identify the trends in the volume of the market;
IV. identify the growing suppliers, who are considered to be the most sensitive to a change in demand.

Instead of focusing on the consequences of price changes, the procedure focuses on the long-term real physical changes in supply (i.e., quantities). The main hypothesis is that if the demand for a product increases as a consequence of a policy decision, and there are no constraints on the supply of this product, then more product will be produced. In the short term, there can be many constraints on supply (suppliers need time to adapt their production to the new demand), and prices can go up with the demand. However, when no long-term constraints exist, the suppliers will adjust their production to the new demand, and therefore prices will fall back to the marginal production cost. On the contrary, if the supply of the demanded product is constrained both in the short and long term, an increase in demand will lead to an increase in price, but not to an increase in production.

Therefore, the main advantage of the procedure is that it does not require economic models and economic inputs (e.g., price elasticities). By using only physical data (e.g., quantities sold in the markets, production trends), the procedure allows to make a justifiable hypothesis on the possible marginal suppliers affected by a decision and to quantify the changes in demand and supply for the competing products.

3. Consequential LCA of SSS blocks production

3.1 Goal and scope

From the previous sections, it is evident that, with its limitations, CLCA can enlarge the scope of the environmental analysis to the effect that a new waste-based product may have on the market. It is therefore interesting to analyze how CLCA can take into account the possible effects on the market for construction materials when metallurgical slag substitutes OPC.

The goal is to assess the environmental consequences when a shift on the valorization route for SSS would occur, from a low-quality application (low-quality aggregates for road construction) to a high-quality application (new construction blocks, the SSS blocks). The study focuses on the potential substitution of traditional nonstructural construction blocks already present in the Flemish market with SSS blocks. Following the four-step procedure by Weidema, the study performs a simplified CLCA, with the identification of the marginal suppliers affected by the change in the SSS valorization route. Two markets are identified: (i) *the market for (non-structural) construction blocks* and (ii) *the market for low-quality aggregates,* as the markets are affected by the shift from low-quality to the high-quality valorization of SSS.

This study represents a first simplified attempt to demonstrate the importance that market dynamics may have when analyzing the environmental impacts of industrial slag valorization as new construction materials. Therefore, rather than building a new complete economic analysis of the markets for construction blocks and low-quality aggregates in Flanders, the study collects data from past market analysis and studies on the Belgian construction market, and identifies the potentially affected marginal suppliers.

3.2 Functional unit and system boundaries of the study

The CLCA expands the boundaries of the analyzed systems, to include also the markets that can be affected by the shift from low-quality to high-quality valorization. In particular, the market for nonstructural construction blocks and the market for aggregates are considered here.

The functional unit for the CLCA must be built differently compared to the attributional approach: while in the ALCA the functional unit represents a common function shared among all compared products, the functional unit in the CLCA must represent a marginal unit added to the current market stock, which will lead to market changes. Therefore, The functional unit represents a marginal unit of SSS blocks that are introduced in the market for construction blocks, namely the production of one SSS block, measuring $10 \times 20 \times 5$ cm^3 (0.001 m^3). Considering a density of 2220 kg/m^3, the weight of one block is ≈ 2.22 kg. Table 13.2 reports the material flows required to produce one SSS block.

The additional unit of SSS blocks affects the marginal suppliers already present in the market for construction blocks, causing a change Δx in the supply. At the same time, the amount of SSS needed to produce the SSS blocks are removed from the supply of the low-quality aggregates market, causing a change Δy in that supply. Fig. 13.1 clarifies the boundaries of the analysis and the considered potential consequences.

Table 13.2 Material flows required to produce one SSS block.

1 SSS block (10 × 20 × 5 cm^3; 2.22 kg)		
Input	**quantity**	**Unit /SSS block**
Slag	0.78	kg
NaOH	0.03	kg
Na silicate	0.06	kg
Aggregates	1.20	kg
Electricity	8E-04	kWh

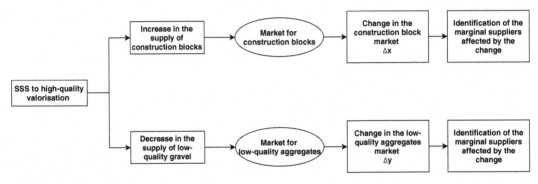

FIGURE 13.1

Boundaries of the analysis and potential consequences.

3.2.1 Identification of the marginal suppliers in the involved markets

For the sake of simplicity, this study identifies only a few products that can be affected by the shift from low-quality to high-quality recycling of SSS: OPC concrete and ceramic bricks in the market for construction blocks, and natural gravel in the market for low-quality aggregates. Therefore, the marginal suppliers affected by the change in the valorization of SSS will be identified among the suppliers of those products. It is assumed that the considered OPC concrete and ceramic bricks have similar applications compared to the SSS blocks, while natural gravel is a substitute of SSS as low-quality aggregate.

The affected marginal suppliers in the market should be identified taking into account the production's constraints and the supplier's capacity to adjust its production (Weidema, 2003). Within the four-step procedure, the study followed some modeling choices and calculations reported in Buyle et al. (2017) and explained hereafter.

The first step of the procedure of Weidema is the identification of the scale and the time horizon of the change. In this study, only marginal changes in supply and long-term effects are considered, assuming a fully elastic market in the long term. The second step is to define the geographical boundaries of the analysis. In this study, the geographical boundaries are set based on the volumes of traded products. Finally, the market trends and the most sensitive marginal suppliers are identified based on their potential for expanding (or decreasing) the supply of a product. Therefore, the increment in production volume in the last years is taken as a reference to identify the most sensitive suppliers: suppliers that incremented their production volume over the past few years are considered the most sensitive suppliers to a change (increase or decrease) in demand (Buyle et al., 2017; Schmidt and Thrane, 2009).

The markets for construction blocks and low-quality aggregates are rather complex and dynamic, and they involve many different suppliers of alternative materials. Together with the primary raw materials (e.g., natural aggregates), many products derived from secondary raw materials are already present in both markets. For instance, GGBFS and fly ash–based cements are well-established products in the construction blocks market, as well as low-quality aggregates derived from construction and demolition waste or other industrial slags. However, the supply of these byproducts is constrained, as a change in demand for the byproducts will not result in production increase. Therefore, byproducts are not included among the potential marginal suppliers (Buyle et al., 2017).

The next paragraphs report a brief description of the Flemish markets for construction blocks and low-quality aggregates and a discussion on the marginal supplier's identification for each market.

3.2.2 The marginal suppliers in the Flemish market for construction block

In the markets for construction blocks, the marginal supplier's identification is a rather uncertain procedure, because of the several potential alternatives present in the market. Several different materials can indeed perform as nonstructural construction blocks, such as OPC-concrete, ceramic bricks, plasterboard panels. In this study, only OPC concrete and ceramic bricks are considered as the products potentially affected by the introduction of one additional unit of SSS blocks. A description of the markets for OPC and ceramic bricks is reported in the next paragraphs.

3.2.2.1 The market for ordinary portland cement

OPC concrete is usually made directly in-situ, by mixing OPC, fine aggregates (sand), coarse aggregates (gravel), and water. Therefore, the study treats separately the marginal suppliers of the different components: OPC producers, fine and coarse aggregates producers. This paragraph describes the market for OPC. Section 3.2.3 describes the market for aggregates.

According to the information found in *The Global Cement Edition-12th Edition* (CEMNET, 2017) and in the report by FEBELCEM (2017), Belgium has seven cement production plants, six of them located in Wallonia and only one located in Flanders, in Ghent. According to data referring to 2015, Belgium consumes an average of 6.5 million tons of cement per year, with an import of 1.6 million tons and an export between 1.5 and 2 million tons (FEBELCEM, 2017). The market analysis considers as possible marginal suppliers only OPC producers, excluding cement produced from other technologies. OPC represents more than 90% of European cement production. Therefore the effect of this choice on the results is limited.

The cement market is traditionally a local market, with transport distances as a limiting factor. The importance of distances is however expected to decrease because of the reduction in transport costs. Therefore, the competitiveness of the European cement sector is expected to decline, especially in a region with good access to ports, such as Belgium. Due to the decreasing production trends of the European OPC producers, Buyle et al. (2017) identified only China and Turkey as potential OPC marginal suppliers. On top of that, although still positive, the Chinese exportation rate has significantly decreased compared to the last years (FEB, 2018), because of the growing internal demand. Therefore,

China is not considered as a stable supplier in the long term, and Turkey is considered as the only marginal supplier.

The present study assumes therefore that any modification of the cement demand in the Flemish market will affect the Turkish cement production and transport by ship from Turkey to Belgium.

3.2.2.2 The market for ceramic bricks

Ceramic bricks are a well-established and locally produced product in the market for construction blocks in Flanders. The local Belgian production in 2016 was 1801 m³, of which 26.4% is exported and only 5.5% imported (FBB, 2017). According to the *Rapport de l'industrie briquetière en Belgique* (Annual report 2016 for the brick industry in Belgium), the annual local production of ceramic bricks in Belgium is decreasing (−8% compared to 2015), while import rate is increasing, especially from the Netherlands (FBB, 2017). Consequently, the bricks producers from the Netherlands can be considered as the marginal suppliers in the Belgian market for bricks.

3.2.3 The marginal suppliers in the Flemish market for aggregates

Although aggregates are typically assumed as a local commodity, the local supply of aggregates in Flanders covers only partially the Flemish needs for primary surface minerals (De Smet et al., 2009). According to the report on the market for surface minerals in Flanders (De Smet, 2009), more than 70% of the marginal supply in Flanders is covered by import. The Flemish supply for aggregates can be reassumed as shown in Table 13.3.

Table 13.3 Aggregates import in Flanders (De Smet et al., 2009).

Region	Type of aggregates	kton/year	Year of data
Flanders (local)	Fine sand (construction and filling materials)	1793	2007
Wallonia (Belgium)	Gravel	1000	
	Sand	1000	
Netherlands	Fine sand	6525	2006
	Construction sand	4600	2006
	Gravel	300	2006
	Clay and loam	282	2008
United Kingdom	Coarse sand	1500	2007
	Sea gravel	1500	2007
	Gravel	320	2002−06
Germany	Coarse sand	2173	2006
	Gravel	933	2006
	Crushed stone	90	2002−06
	Clay and loam	472	2006−08
France	Gravel	53	2006
	Sand	42	2006
	Crushed stone	186	2006
Norway	Gravel	230	2007

The Netherlands and Germany are important suppliers of sand, while Norway supplies gravel. Trade of aggregates, therefore, is not limited by the national borders, but rather by the distance, with an average of 50 km for transport by truck. On the other hand, the high trade with Norway can be explained by the latest regulation on nature and landscape conservation, which is limiting the production of local Belgian gravel. A similar constraint occurs for sand and gravel from France, Germany, and the UK, and gravel from the Netherlands. Therefore, the suppliers of aggregates from Belgium (both Flanders and Wallonia), France, Germany, and UK are likely to be policy-constrained shortly, as well as the gravel suppliers from Belgium and the Netherlands. Suppliers from these countries are considered as nonsensitive suppliers.

In a growing market, the suppliers that are more sensitive to a change in demand are the ones who showed changes in production volumes over a certain period. Only two suppliers are likely to show increasing trends in the long term and are identified as the sensitive suppliers in the Flemish market aggregates, namely sand from the Netherlands and gravel from Norway. Suppliers from those countries are therefore considered as the only suppliers affected by a change in demand/supply in the Flemish market for aggregates. A more detailed analysis of the trends and statistical models used to support this conclusion is reported in Buyle et al. (2017).

3.2.4 Identifying the consequences: scenarios analysis

The identification of the marginal suppliers helps to assess the possible consequences of a shift from low-quality to high-quality recycling of SSS. To seek for simplicity, the country-specific electricity mix, the transport means, and distances are assumed as the only differences among the marginal suppliers. However, production techniques can also vary between different countries, and further data should be collected on the processes for each specific marginal supplier. Although this can be seen as a limitation of the study, gathering this information would have required an intensive data gathering, that would have significantly increased the complexity of the study. For future applications of consequential approach in the construction sector, however, further research should also be conducted on the specific production process used by the identified marginal supplier.

Moreover, as markets are interrelated, many other different consequences are possible, and suppliers of different materials can be involved. A more detailed economic analysis and the integration of partial and general equilibrium models could help the identification of all possible consequences. Based on the results for marginal suppliers in the Flemish aggregates and cement markets, the present study analyzes two possible scenarios.

3.2.4.1 Scenario 1

In this scenario, the high-quality valorization of SSS leads to a marginal reduction of the OPC concrete demand in the market for construction blocks. It is assumed that 0.001 m³ SSS blocks can substitute 0.001 m³ OPC concrete with a similar compressive strength (≈ 20 MPa). A reduction in the use of OPC leads to a consequent reduction in the import of OPC and aggregates, which are the elements that constitute the OPC concrete. Accordingly, the marginal suppliers affected by the decrease in demand are sand from the Netherlands, gravel from Norway, and OPC from Turkey. According to Dewar (2003), nonstructural OPC concrete with ≈ 20 MPa compressive strength has a density of 2600 kg/m³, and it is composed by 15 wt% of OPC and 80 wt% of aggregates, with 2/3 of gravel and 1/3 of sand. Therefore, to produce an OPC concrete block of 0.001 m³, there is a need of 0.39 kg of OPC, 1.39 kg of gravel, 0.69 kg of sand.

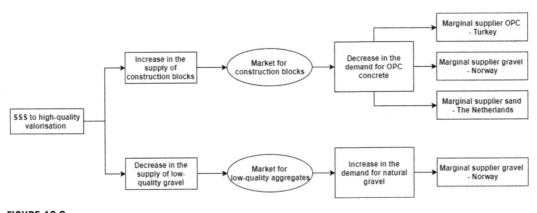

FIGURE 13.2

Consequences considered for scenario 1.

In the low-quality aggregates market, the lack of the supply of stabilized SSS causes a marginal increase of gravel import from Norway.

In the impact calculations, the productions from the different countries use the local electricity mix (the Netherlands for sand, Norway for gravel, and Turkey for OPC).

The sand from the Netherlands is assumed to be transported by truck, with an average distance of 100 km. The gravel from Norway is transported by freight ship, with an average distance of 1500 km. Finally, the OPC from Turkey is also transported by a freight ship, with an average distance of 3000 km.

The changes ΔX_1 and ΔY_1 in the two markets for scenario 1 are calculated as follow:

$\Delta X_1 = $ (SSS blocks production + transport of SSS blocks to the local market)

$\quad -$ (production and transport of OPC from Turkey

$\quad +$ production and transport of sand for OPC concrete from the Netherlands

$\quad +$ production and transport of gravel for OPC concrete from Norway)

$\Delta Y_1 = $ (Production and transport of gravel from Norway for the aggregates market)

$\quad -$ (SSS stabilization with boric oxide + stabilized SSS

$\quad -$ based aggregates transport to low $-$ quality aggregates market)

Fig. 13.2 and Table 13.4 reassume the main assumptions for scenario 1.

3.2.4.2 Scenario 2

Scenario 2 assumes that the SSS blocks will substitute an equivalent amount of ceramic bricks from the Netherlands. According to Salman et al. (2016) ceramic bricks having a density of 2400 kg/m³ present a similar compressive strength compared to the SSS blocks. The avoided bricks production uses the local Dutch electric mix. The ceramic bricks are assumed to be transported by truck for an average

Table 13.4 Physical flows in scenario 1.

Caused impacts	Market affected	Quantity
SSS blocks production	construction blocks	1 block: 2.22 kg
Transport of SSS blocks	construction blocks	50 km (truck)
Gravel production from Norway	low-quality aggregates	0.78 kg
Transport of gravel from Norway to Belgium	low-quality aggregates	1500 km (ship)
Avoided impacts	**Market affected**	**Quantity**
OPC production from Turkey	construction blocks	0.39 kg
Transport of OPC from Turkey to Belgium	construction blocks	3000 km (ship)
Sand production from the Netherlands	construction blocks	0.69 kg
Transport of sand from the Netherlands to Belgium	construction blocks	100 km (truck)
Gravel production from Norway	construction blocks	1.39 kg
Transport of gravel from Norway to Belgium	construction blocks	1500 km (ship)
SSS stabilized with boric oxide	low-quality aggregates	0.78 kg
Boric oxide	/	0.0156 kg
Transport of stabilized SSS-based aggregates	low-quality aggregates	50 km (truck)

distance of 100 km. The changes ΔX_2 and ΔY_2 in the two markets for scenario 1 are calculated as follow:

$$\Delta X_2 = (\text{recycling process for SSS} + \text{transport of SSS blocks to the local market})$$
$$- (\text{production and transport of ceramic bricks from the Netherlands})$$

$$\Delta Y_2 = (\text{Production and transport of gravel from Norway for the aggregates market})$$
$$- (\text{SSS stabiliszation with boric oxide} + \text{stabilized SSS transport to low})$$
$$- \text{quality aggregates market})$$

Fig. 13.3 and Table 13.5 reassume the main assumptions for scenario 2.

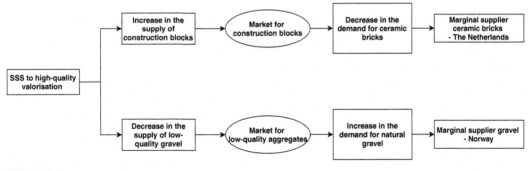

FIGURE 13.3

Consequences considered for scenario 2.

Table 13.5 Physical flows in scenario 2.

Caused impacts	Market affected	Quantity
SSS blocks production	construction blocks	2.22 kg
Transport of SSS blocks	construction blocks	50 km (truck)
Gravel production from Norway	low-quality aggregates	0.78 kg
Transport of gravel from Norway to Belgium	low-quality aggregates	1500 km (ferry)
Avoided impacts	Market affected	Quantity
Ceramic bricks production from the Netherlands	construction blocks	2.4 kg
Transport of ceramic bricks from the Netherlands to Belgium	construction blocks	100 km (truck)
SSS stabilized with boric oxide	low-quality aggregates	0.78 kg
Transport of stabilized SSS-based aggregates	low-quality aggregates	50 km (truck)

To facilitate the comparison among the different processes, all processes have been grouped into four macroprocesses:

- *SSS blocks*: includes the processes of production and transport of SSS blocks.
- *Market for construction blocks:* includes the avoided production and transport of gravel from Norway, sand from the Netherlands, and OPC from Turkey (Scenario 1); the avoided production and transport of bricks from the Netherlands (scenario 2).
- *Avoided SSS to low-quality*: includes the avoided production of boric oxide to stabilize the SSS and avoided transport of stabilized SSS-based aggregates to low-quality aggregates market.
- *Market for low-quality aggregates:* includes the production and transport of gravel from Norway to the low-quality aggregates market.

Fig. 13.4 clarifies the subprocesses included in each macroprocess.

3.3 Environmental impact assessment

The environmental impact assessment phase in LCA is not affected by the choice of attributional or consequential approaches, as the main difference between ALCA ad CLCA refers to the data collection and system definition phase.

The impact assessment calculation is used to translate the collected data into specific environmental impact categories. Impact assessment models usually refer to two main approaches: the "problem-oriented" or midpoint, and the "damage-oriented" or endpoint. The midpoint analysis assesses the contribution of each material/energy flow to several environmental categories, which analyze different aspects of environmental effects (from global warming to acidification or land use). Midpoint results are usually presented for up to 20 different impact categories, depending on the selected midpoint calculation method considered. On the other hand, endpoint analysis aggregates the midpoint results in only three main areas of impacts for the natural environment, namely human health, ecosystem, and natural resources. As reported by many authors (see Kägi et al., 2015), midpoint analysis provides scientifically reliable results, but it can hardly answer questions such as "is product A

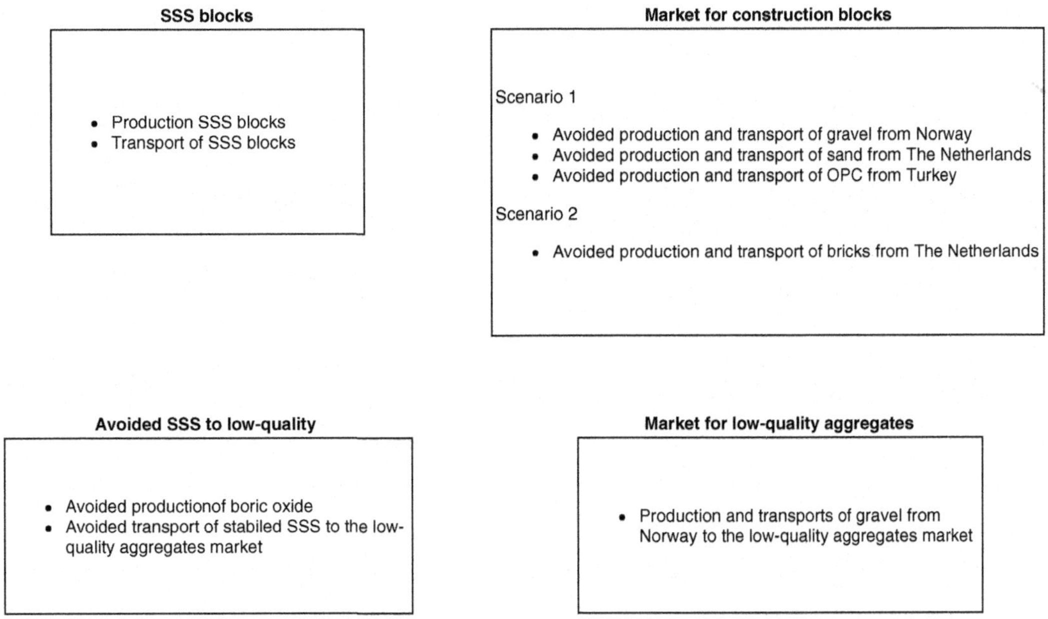

SSS blocks

- Production SSS blocks
- Transport of SSS blocks

Market for construction blocks

Scenario 1

- Avoided production and transport of gravel from Norway
- Avoided production and transport of sand from The Netherlands
- Avoided production and transport of OPC from Turkey

Scenario 2

- Avoided production and transport of bricks from The Netherlands

Avoided SSS to low-quality

- Avoided productionof boric oxide
- Avoided transport of stabiled SSS to the low-quality aggregates market

Market for low-quality aggregates

- Production and transports of gravel from Norway to the low-quality aggregates market

FIGURE 13.4

Macroprocesses.

environmentally better than product B?" as some midpoint categories may show different tendencies than others. The results from the endpoint analysis are easier to be compared, but they carry high uncertainty due to the aggregation process. Therefore, the combination of midpoint and endpoint results can assist in better interpretation and transparency. More detail explanation on midpoint and endpoint environmental impacts calculation can be found in JRC (2011).

The results of the impact assessment showed in the next section have been calculated using Recipe v, implemented in the LCA software Gabi v. 9.1.0.53. The environmental impacts for the background processes have been calculated using the ECOINVENT v.3.5 database.

3.4 Results

Fig. 13.5 shows the comparison of the two scenarios, with the results for each category of Recipe midpoint with a hierarchical perspective, normalized to 100%. For each category, a negative value of the final result means that the avoided impacts (OPC concrete production and transports in scenario 1, ceramic bricks production in scenario 2) are higher than the caused impacts (SSS block production and transport, and gravel production and transport for both scenarios).

Scenario 2 presents a higher impact in the categories of agricultural land occupation, freshwater and human ecotoxicity, metal and water depletion. At the same time, it presents higher environmental benefits (as avoided impact) in the category climate change, fossil depletion, marine eutrophication, particulate matter formation, photochemical oxidant formation, terrestrial ecotoxicity, and urban land

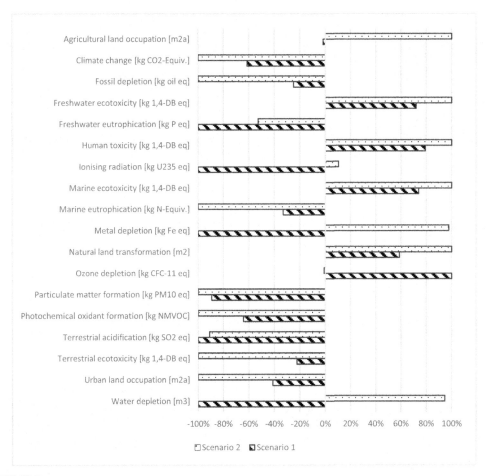

FIGURE 13.5

Midpoint results: scenarios comparison.

occupation. Scenario 1 presents an impact higher than scenario 2 only in the category of ozone depletion.

Table 13.6 shows the contribution analysis for the Recipe midpoint results for scenario 1 and scenario 2. The only difference between the two scenarios is represented by the marginal suppliers in the market for construction blocks: avoided production of OPC concrete components (OPC, gravel, sand) for scenario 1 and avoided bricks production for scenario 2. Therefore, the first four columns of Table 13.6 (SSS blocks, market for low-quality aggregates, avoided SSS to low quality, and subtotal) are equal for both scenarios. The highest impact for all categories is caused by the production of the SSS blocks. The impact from the production of SSS blocks is always higher than the impact for the production and transport of the gravel from Norway, which is the marginal substitute of SSS in the market for low-quality aggregates. In scenario 1, the highest avoided

Table 13.6 Process contribution to midpoint results for scenario 1 and scenario 2.

Impact category	Unit	SSS block	Market LQ aggregates	Avoided SSS to low quality	Subtotal	Avoided OPC blocks	Total scenario 1	Avoided bricks	Total scenario 2
Agricultural land occupation	m2a	5.25E-03	2.79E-04	−1.93E-03	3.60E-03	−3.28E-03	3.19E-04	−2.22E-02	−1.86E-02
Climate change	kg CO_2-Equiv.	9.80E-02	1.68E-02	−4.14E-02	7.34E-02	−4.06E-01	−3.33E-01	−6.16E-01	−5.43E-01
Fossil depletion	kg oil eq	2.71E-02	5.29E-03	−1.62E-02	1.62E-02	−4.94E-02	−3.32E-02	−1.50E-01	−1.34E-01
Freshwater ecotoxicity	kg 1.4-DB eq	1.36E-02	7.61E-04	−4.05E-03	1.03E-02	−4.85E-03	5.46E-03	−2.74E-03	7.57E-03
Freshwater eutrophication	kg P eq	4.09E-05	2.42E-06	−1.77E-05	2.56E-05	−3.29E-05	−7.28E-06	−2.95E-05	−3.88E-06
Human toxicity	kg 1.4-DB eq	1.08E-01	5.97E-03	−4.25E-02	7.15E-02	−5.66E-02	1.49E-02	−5.25E-02	1.90E-02
Ionizing radiation	kg U235 eq	1.82E-02	1.78E-03	−5.53E-03	1.45E-02	−1.65E-02	−2.05E-03	−1.43E-02	1.50E-04
Marine ecotoxicity	kg 1.4-DB eq	1.12E-02	6.51E-04	−3.38E-03	8.47E-03	−4.19E-03	4.28E-03	−2.70E-03	5.77E-03
Marine eutrophication	kg N-Equiv.	2.67E-05	8.23E-06	−1.30E-05	2.19E-05	−5.35E-05	−3.16E-05	−1.18E-04	−9.61E-05
Metal depletion	kg Fe eq	1.06E-02	1.37E-03	−6.50E-03	5.47E-03	−6.92E-03	−1.45E-03	−4.20E-03	1.27E-03
Natural land transformation	m^2	2.46E-05	1.67E-05	−1.63E-05	2.50E-05	−6.31E-05	−3.81E-05	−8.96E-05	−6.46E-05
Ozone depletion	kg CFC-11 eq	4.21E-08	2.51E-09	−6.15E-09	3.85E-08	−1.84E-08	2.01E-08	−3.86E-08	−1.40E-10
Particulate matter formation	kg PM10 eq	2.32E-04	9.29E-05	−2.54E-04	7.09E-05	−5.47E-04	−4.76E-04	−6.06E-04	−5.35E-04
Photochemical oxidant formation	kg NMVOC	3.82E-04	2.23E-04	−3.00E-04	3.05E-04	−1.30E-03	−9.95E-04	−1.86E-03	−1.56E-03
Terrestrial acidification	kg SO_2 eq	6.00E-04	2.85E-04	−5.12E-04	3.73E-04	−1.45E-03	−1.08E-03	−1.36E-03	−9.87E-04
Terrestrial ecotoxicity	kg 1.4-DB eq	2.08E-05	1.89E-06	−1.03E-05	1.24E-05	−2.19E-05	−9.51E-06	−5.48E-05	−4.24E-05
Urban land occupation	m^2a	2.00E-03	4.38E-04	−1.17E-03	1.27E-03	−2.55E-03	−1.28E-03	−4.38E-03	−3.11E-03
Water depletion	m^3	4.08E-01	1.14E-01	−1.25E-01	3.97E-01	−7.41E-01	−3.44E-01	−7.05E-02	3.27E-01

impact is always represented by the market for construction blocks, due to the avoided production of OPC, gravel, and sand. Also in scenario 2, the market from construction blocks (the avoided production and transport of bricks) is the highest avoided impact, except in the categories freshwater ecotoxicity, marine ecotoxicity, metal depletion, and water depletion. In these categories, the highest avoided impact is represented by the avoided SSS to low quality (mostly the avoided production of boric oxide).

To allow a more straightforward comparison between the two scenarios, Fig. 13.6 shows the results in Recipe endpoint. For all endpoint damages, both scenarios present a negative value, meaning that in both cases the avoided impacts are higher than the caused impacts. Some differences can, however, be highlighted between the two scenarios: in all three damages, the results of scenario 2 are higher than the results for scenario 1, meaning that, since the final result is negative, scenario 2 has better environmental performances compare to scenario 1.

To better understand the results in Fig. 13.6, the process contribution analysis is reported in Table 13.7 (see Fig. 13.4 for a more detailed explanation of the macroprocesses).

As before, the results for the macroprocesses *SSS-blocks, avoided SSS to low quality,* and *market for low-quality aggregates* are identical for both scenario, as there are no differences in the inputs for these macroprocesses in the two scenarios. The only differences are reported for the macroprocess *market for construction blocks,* namely avoided OPC concrete versus avoided bricks production. For all endpoint damages, the negative impacts of the market for construction blocks in scenario 2 are higher than the results for scenario 1. To better highlight the differences between the two scenarios, Fig. 13.7 breaks down the processes contributing to the macroprocess *market for construction blocks* in the two scenarios.

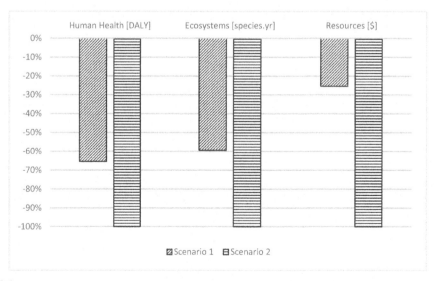

FIGURE 13.6

Endpoint results: scenarios comparison.

Table 13.7 Process contribution to midpoint results for scenario 1 and scenario 2.

	SSS blocks	Avoided SSS to low quality	Market for low-quality aggregates	Subtotal	Avoided OPC blocks	Total scenario 1	Avoided bricks	Total scenario 2
Human Health [DALY]	2.73E-07	-1.54E-07	5.19E-08	1.71E-07	-7.50E-07	-5.79E-07	-1.06E-06	-8.85E-07
Ecosystems [species.yr]	9.47E-10	-4.14E-10	1.79E-10	7.12E-10	-3.44E-09	-2.73E-09	-5.31E-09	-4.59E-09
Resources [$]	5.23E-03	-3.13E-03	9.70E-04	3.08E-03	-8.65E-03	-5.57E-03	-2.49E-02	-2.18E-02

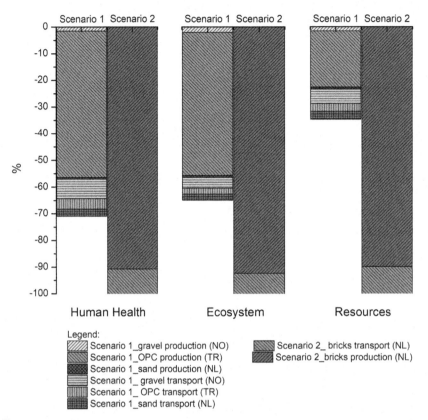

FIGURE 13.7

Breakdown of the subprocesses contribution to the macroprocess "market for construction blocks," at the endpoint level. *NL*, the Netherlands; *NO*, Norway; *TR*, Turkey.

Compared to scenario 2, the market for construction blocks in scenario 1 presents a lower environmental benefit in all endpoint damages (−71% human health, −65% ecosystem, −35% resources), with the avoided OPC production presenting the highest environmental benefits in all three endpoint damages (−55% human health, −54% ecosystem and −21% resources). The avoided gravel and sand productions have a low contribution (−3% human health, −3% ecosystem, −3% resources). The avoided transports of OPC from Turkey, gravel from Norway, and sand from the Netherlands also play a significant role in the overall environmental benefit for scenario 1 (in total: −14% human health, −8% ecosystem, −11% resources), with the avoided transport of gravel from Norway having the highest avoided impact in all endpoint damages. In scenario 2, the highest environmental benefits in all endpoint damages are given by the avoided ceramic bricks production (91% human health, 92% ecosystem, 90% resources). From the analysis described above, it appears clear that at an endpoint level the avoided impacts of bricks production produced a higher environmental benefit compared to the avoided production of the three concrete components OPC, gravel, and sand.

3.4.1 Results interpretation

The midpoint and endpoint results in Figs. 13.5 and 13.6 confirm the findings of the attributional LCA presented in previous studies: the valorization of SSS as new binder substituting OPC or ceramic bricks can lead to overall environmental benefits.

In the midpoint analysis in Fig. 13.5, some of the categories present a negative value, meaning that the changes in the market for construction blocks (in both OPC concrete and ceramic bricks cases) are the ones with the highest effect on final environmental impacts. In the midpoint categories presenting a positive impact in Fig. 13.5, the avoided impacts cannot offset the caused impact of SSS block production and low-quality gravel substitution.

The endpoint analysis in Fig. 13.6 reports a more explicit comparison between the two scenarios. For all endpoint damages, both scenarios present a negative value. This leads to the conclusion that, at an endpoint level, when the SSS valorization route switches from low-quality to high-quality valorization, the consequent increase of gravel import from Norway does offset the benefits of the avoided production of OPC concrete or bricks. The substitution of OPC concrete or clay bricks with SSS blocks in the Flemish market can lead to environmental benefits, compared to the current situation. Moreover, the substitution of clay bricks may lead to higher environmental benefits compared to the substitution of OPC concrete. The endpoint analysis quantified these benefits as +35% avoided impact on human health, +41% avoided impact for ecosystem, and +75% avoided impact for resources. The contribution analysis at midpoint and endpoint levels, reported in Tables 13.6 and 13.7, showed that this difference is mostly due to the different avoided impacts of the clay bricks production compared to the concrete components production OPC, while a less significant role is played by the different avoided transports.

4. Conclusions

The alternative use of SSS blocks made from SSS can avoid the production of traditional materials, as OPC concrete or ceramic bricks. Since today SSS are stabilized and used as low-quality aggregates, the use of SSS in SSS blocks production will lead to a replacement of SSS by natural aggregates. Therefore, the shift from the present situation to the potential recycling as SSS blocks can have consequences involving two markets: the market for construction blocks and the market for low-quality aggregates.

Based on previous studies on the construction market in Belgium, some marginal suppliers are identified as the ones affected by the change: suppliers producing different components for OPC concrete production (gravel producers from Norway, sand producers from the Netherlands, OPC producers from Turkey) and ceramic brick producers from the Netherlands. The two scenarios analyzed are based on the hypothesis that the SSS blocks substitute (i) the OPC concrete, or (ii) the ceramic bricks in the market for construction blocks. The stabilized SSS-based aggregates are replaced by gravel from Norway in both scenarios.

The results for both scenarios show that the increased use of gravel from Norway does not offset the environmental benefits obtained thanks to the avoided production of traditional OPC concrete components or ceramic bricks. Therefore, the changes in the construction blocks market are the ones driving the results of the whole system. Comparing the results of the two scenarios, the substitution of OPC concrete components leads to lower benefits in some of the environmental categories if compared to the substitution of ceramic bricks (e.g., climate change). However, for some other categories (e.g., metal depletion), the substitution of OPC leads to higher environmental benefits.

It is important to notice that this study represents only a first exercise in including market dynamics in the case of SSS valorization. The affected products and the marginal suppliers have been identified qualitatively according to previous studies and market analysis, and the scenarios analyzed represent only some of the potential consequences led by the introduction of the SSS blocks in the market. For the future development of the research and the methodology, more rigorous identification of the market dynamics (affected technologies and marginal suppliers) is recommended, and a broader analysis of the potentially affected marginal suppliers in the Flemish construction market should be carried out. Nonetheless, the CLCA approach allowed the identification of the most probable geographic location of the suppliers, which will be affected by the change in the market, leading to a more precise calculation of the transport distances and on the electricity mix in the country where the substitute is produced.

This represents a potential step forward in policy support exercise compared to ALCA, where the selection of the substituted products is based on mere hypothesis and market dynamics are not included. When policy interventions aim at supporting a specific technology, for instance, through a subsidy or a tax on competing products, it is fundamental to have a reliable prediction on the potential impacts of such policy on future markets. In the presented case of SSS valorization, for instance, the relevant information derived by CLCA could support policymakers in pursuing an action at European level, rather than only at national level, since all identified marginal suppliers of the Belgian market for construction material are suppliers from foreign countries. Another relevant indication is the higher environmental benefits when substituting ceramic bricks rather than OPC-concrete. It is also important to consider that ALCA and CLCA are not be seen as alternatives, but rather as complementary approaches in analyzing policy effects. When ALCA identifies the SSS valorization of construction blocks as a cleaner technology compared to traditional products, CLCA can first confirm the findings including market dynamics, and then suggest policymakers on which niches of the market a policy should target to favor the diffusion of SSS-based blocks.

References

Birat, J.P., 2012. Sustainability footprint of steelmaking byproducts. Ironmak. Steelmak. 39, 270–275.

Bouman, M., Heijungs, R., van der Voet, E., van den Bergh, J.C.J.M., Huppes, G., 2000. Material flows and economic models: an analytical comparison of SFA, LCA and partial equilibrium models. Ecol. Econ. 32, 195–216. https://doi.org/10.1016/S0921-8009(99)00091-9.

Brander, M., Tipper, R., Hutchinson, C., Davis, G., 2009. Consequential and Attributional Approaches to LCA: A Guide to Policy Makers with Specific Reference to Greenhouse Gas LCA of Biofuels. Econom. Press Technical.

Buyle, M., Pizzol, M., Audenaert, A., 2017. Identifying marginal suppliers of construction materials: consistent modeling and sensitivity analysis on a Belgian case. Int. J. Life Cycle Assess. 1–17. https://doi.org/10.1007/s11367-017-1389-5.

CEMNET, 2017. The Global Cement Report, twelfth ed.

Consequential-LCA, 2015. Why and When?.

Curran, M.A., Mann, M., Norris, G., 2005. The international workshop on electricity data for life cycle inventories. J. Clean. Prod. 13, 853–862. https://doi.org/10.1016/j.jclepro.2002.03.001.

De Smet, L., Bogaert, S., Vandenbroucke, D., Van Hyfte, A., De Coster, K., 2009. Onderzoek Duurzame Bevoorrading: Gebruik Lokale Oppervlakte- Delfstoffen of Import Van Minerale Grondstoffen ([Research Sustainable Supply: Use of Local Surface Minerals or Import of Mineral Raw Materials]) (in Dutch).

Dewar, J., 2003. In: Choo, B.S. (Ed.), 1 — Concrete Mix Design A2 — Newman, John. Butterworth-Heinemann, Oxford, pp. 3—40.

Di Maria, A., Van Acker, K., 2018. Turning Industrial Residues into Resources: An Environmental Impact Assessment of Goethite Valorization. Engineering. https://doi.org/10.1016/J.ENG.2018.05.008.

Di Maria, A., Salman, M., Dubois, M., Acker, K. Van, 2018. Life cycle assessment to evaluate the environmental performance of new construction material from stainless steel slag. Int. J. Life Cycle Assess. 1—19. https://doi.org/10.1007/s11367-018-1440-1.

Dong, L., Zhang, H., Fujita, T., Ohnishi, S., Li, H., Fujii, M., Dong, H., 2013. Environmental and economic gains of industrial symbiosis for Chinese iron/steel industry: Kawasaki's experience and practice in Liuzhou and Jinan. J. Clean. Prod. 59, 226—238. https://doi.org/10.1016/j.jclepro.2013.06.048.

Durinck, D., Engström, F., Arnout, S., Heulens, J., Jones, P.T., Björkman, B., Blanpain, B., Wollants, P., 2008. Hot stage processing of metallurgical slags. Resour. Conserv. Recycl. 52, 1121—1131. https://doi.org/10.1016/j.resconrec.2008.07.001.

Duxson, P., Provis, J.L., Lukey, G.C., van Deventer, J.S.J., 2007. The role of inorganic polymer technology in the development of 'green concrete'. Cement. Concr. Res. 37, 1590—1597. https://doi.org/10.1016/j.cemconres.2007.08.018.

Ekvall, T., 2002. Cleaner production tools: LCA and beyond. J. Clean. Prod. 10, 403—406. https://doi.org/10.1016/S0959-6526(02)00026-4.

Ekvall, T., 2000. A market-based approach to allocation at open-loop recycling. Resour. Conserv. Recycl. 29, 91—109. https://doi.org/10.1016/S0921-3449(99)00057-9.

Ekvall, T., Andrae, A., 2006. Attributional and consequential environmental assessment of the shift to lead-free solders (10 pp). Int. J. Life Cycle Assess. 11, 344—353. https://doi.org/10.1065/lca2005.05.208.

Ekvall, T., Finnveden, G., 2001. Allocation in ISO 14041—a critical review. J. Clean. Prod. 9, 197—208. https://doi.org/10.1016/S0959-6526(00)00052-4.

Ekvall, T., Tillman, A.-M., Molander, S., 2005. Normative ethics and methodology for life cycle assessment. Life Cycle Assess. 13, 1225—1234. https://doi.org/10.1016/j.jclepro.2005.05.010.

Ekvall, T., Weidema, B.P., 2004. System boundaries and input data in consequential life cycle inventory analysis. Int. J. Life Cycle Assess. 9, 161—171. https://doi.org/10.1007/BF02994190.

FBB, 2017. Le secteur en quelques chiffres [WWW Document]. URL: http://www.brique.be/secteur-briquetier/le-secteur-en-quelques-chiffres/.

FEB, 2018. Focus Conjoncture — Une croissance de 2% possible en 2018 [WWW Document]. URL: http://www.feb.be/publications/focus-conjoncture–une-croissance-de-2-possible-en-2018/.

FEBELCEM, 2017. Febelcem: Présentation du secteur du ciment, de la production à la consommation. Evolution du marché du ciment [WWW Document]. URL: http://www.febelcem.be//fr/informations-economiques/presentation-du-secteur-du-ciment-de-la-production-a-la-consommation-evolution-du-marche-du-ciment/#c978.

Finnveden, G., Hauschild, M.Z., Ekvall, T., Guinée, J., Heijungs, R., Hellweg, S., Koehler, A., Pennington, D., Suh, S., 2009. Recent developments in life cycle assessment. J. Environ. Manag. 91, 1—21. https://doi.org/10.1016/j.jenvman.2009.06.018.

Geiseler, J., 1996. Use of steelworks slag in Europe. Waste Manag. 16, 59—63. https://doi.org/10.1016/S0956-053X(96)00070-0.

Habert, G., d'Espinose de Lacaillerie, J.B., Roussel, N., 2011. An environmental evaluation of geopolymer based concrete production: reviewing current research trends. J. Clean. Prod. 19, 1229—1238. https://doi.org/10.1016/j.jclepro.2011.03.012.

Hossain, M.U., Poon, C.S., Dong, Y.H., Xuan, D., 2018. Evaluation of environmental impact distribution methods for supplementary cementitious materials. Renew. Sustain. Energy Rev. 82, 597–608. https://doi.org/10.1016/j.rser.2017.09.048.

Huaiwei, Z., Xin, H., 2011. An overview for the utilization of wastes from stainless steel industries. Resour. Conserv. Recycl. 55, 745–754. https://doi.org/10.1016/j.resconrec.2011.03.005.

Huntzinger, D.N., Eatmon, T.D., 2009. A life-cycle assessment of Portland cement manufacturing: comparing the traditional process with alternative technologies. Present Anticip. Demands Nat. Resour. Sci. Technol. Polit. Econ. Ethical Approaches Sustain. Manag. 17, 668–675. https://doi.org/10.1016/j.jclepro.2008.04.007.

Johansson, M.T., Söderström, M., 2011. Options for the Swedish steel industry – energy efficiency measures and fuel conversion. Energy 36, 191–198. https://doi.org/10.1016/j.energy.2010.10.053.

JRC, 2011. ILCD Handbook: Recommendations for Life Cycle Assessment in the European Context. Publication Office of the European Union doi: 10.278/33030.

Kägi, T., Dinkel, F., Frischknecht, R., Humbert, S., Lindberg, J., De Mester, S., Ponsioen, T., Sala, S., Urs, Schenker, W., 2015. In: Conference Session Report: SETAC Europe 25th Annual Meeting Session "Midpoint, Endpoint or Single Score for Decision-Making?"—SETAC Europe 25th Annual Meeting. https://doi.org/10.1007/s11367-015-0998-0.

Kim, Y.J., Nettleship, I., Kriven, W.M., 1992. Phase transformations in dicalcium silicate: II, TEM studies of crystallography, microstructure, and mechanisms. J. Am. Ceram. Soc. 75, 2407–2419. https://doi.org/10.1111/j.1151-2916.1992.tb05593.x.

Marvuglia, A., Benetto, E., Rege, S., Jury, C., 2013. Modelling approaches for consequential life-cycle assessment (C-LCA) of bioenergy: critical review and proposed framework for biogas production. Renew. Sustain. Energy Rev. 25, 768–781. https://doi.org/10.1016/j.rser.2013.04.031.

Motz, H., Geiseler, J., 2001. Products of steel slags an opportunity to save natural resources. Waste Manag. 21, 285–293. https://doi.org/10.1016/S0956-053X(00)00102-1.

Panda, C.R., Mishra, K.K., Panda, K.C., Nayak, B.D., Nayak, B.B., 2013. Environmental and technical assessment of ferrochrome slag as concrete aggregate material. Construct. Build. Mater. 49, 262–271. https://doi.org/10.1016/j.conbuildmat.2013.08.002.

Salman, M., Cizer, Ö., Pontikes, Y., Snellings, R., Vandewalle, L., Blanpain, B., Balen, K.V., 2015. Cementitious binders from activated stainless steel refining slag and the effect of alkali solutions. J. Hazard Mater. 286, 211–219. https://doi.org/10.1016/j.jhazmat.2014.12.046.

Salman, M., Cizer, Ö., Pontikes, Y., Vandewalle, L., Blanpain, B., Van Balen, K., 2014. Effect of curing temperatures on the alkali activation of crystalline continuous casting stainless steel slag. Construct. Build. Mater. 71, 308–316. https://doi.org/10.1016/j.conbuildmat.2014.08.067.

Salman, M., Dubois, M., Maria, A., Di Van Acker, K., Van Balen, K., 2016. Construction materials from stainless steel slags: technical aspects, environmental benefits, and economic opportunities. J. Ind. Ecol. 20, 854–866. https://doi.org/10.1111/jiec.12314.

Schmidt, J., Thrane, M., 2009. Life Cycle Assessment of Aluminium Production in New Alcoa Smelter in Greenland. Aalborg Univerisity.

Scrivener, K.L., Kirkpatrick, R.J., 2008. Innovation in use and research on cementitious material. Cem. Concr. Res. 38, 128–136. https://doi.org/10.1016/j.cemconres.2007.09.025. Special Issue — the 12th International Congress on the Chemistry of Cement. Montreal, Canada, July 8–13, 2007.

Sheen, Y., Huang, L.-J., Sun, T.-H., Le, D.-H., 2016. Engineering properties of self-compacting concrete containing stainless steel slags. Proc. Sustain. Dev. Civil Urban Transp. Eng. 142, 79–86. https://doi.org/10.1016/j.proeng.2016.02.016.

Szabó, L., Hidalgo, I., Ciscar, J.C., Soria, A., 2006. CO_2 emission trading within the European Union and Annex B countries: the cement industry case. Energy Pol. 34, 72–87. https://doi.org/10.1016/j.enpol.2004.06.003.

Tillman, A.-M., 2000. Significance of decision-making for LCA methodology. Environ. Impact Assess. Rev. 20, 113–123. https://doi.org/10.1016/S0195-9255(99)00035-9.

Tsakiridis, P.E., Papadimitriou, G.D., Tsivilis, S., Koroneos, C., 2008. Utilization of steel slag for Portland cement clinker production. J. Hazard Mater. 152, 805–811. https://doi.org/10.1016/j.jhazmat.2007.07.093.

Van den Heede, P., De Belie, N., 2012. Environmental impact and life cycle assessment (LCA) of traditional and 'green' concretes: Literature review and theoretical calculations. Cement Concr. Compos. 34, 431–442. https://doi.org/10.1016/j.cemconcomp.2012.01.004.

van Oss, H.G., Padovani, A.C., 2003. Cement manufacture and the environment Part II: environmental challenges and opportunities. J. Ind. Ecol. 7, 93–126. https://doi.org/10.1162/108819803766729212.

Vázquez-Rowe, I., Marvuglia, A., Rege, S., Benetto, E., 2014. Applying consequential LCA to support energy policy: land use change effects of bioenergy production. Sci. Total Environ. 472, 78–89. https://doi.org/10.1016/j.scitotenv.2013.10.097.

Vázquez-Rowe, I., Rege, S., Marvuglia, A., Thénie, J., Haurie, A., Benetto, E., 2013. Application of three independent consequential LCA approaches to the agricultural sector in Luxembourg. Int. J. Life Cycle Assess. 18, 1593–1604. https://doi.org/10.1007/s11367-013-0604-2.

Weidema, B., Frees, N., Nielsen, A.-M., 1999. Marginal production technologies for life cycle inventories. Int. J. Life Cycle Assess. 4, 48–56. https://doi.org/10.1007/BF02979395.

Weidema, B.P., 2003. Market Information in Life Cycle Assessment.

Weidema, B.P., Ekvall, T., Heijungs, R., 2009. Guidelines for Application of Deepened and Broadened LCA.

Weil, M., Dombrowski, K., Buchwald, A., 2009. 10 – Life-cycle analysis of geopolymers. In: Woodhead Publishing Series in Civil and Structural Engineering. Woodhead Publishing, pp. 194–210.

Market and sustainability

Material and energy services, human needs, and well-being

Kai Whiting[1], Luis Gabriel Carmona[2,3,5], Angeles Carrasco[4]

[1]*Faculty of Architecture, Architectural Engineering and Urban Planning, Université Catholique de Louvain, Louvain-la-Neuve, Belgium;* [2]*MARETEC—LARSyS, Instituto Superior Técnico, Universidade de Lisboa, Lisboa, Portugal;* [3]*Faculty of Environmental Sciences, Universidad Piloto de Colombia, Bogotá, Colombia;* [4]*Mining and Industrial Engineering School of Almadén, Universidad de Castilla—La Mancha, Almadén, Spain;* [5]*Institute ForWARD (For Worldwide Alternative Research and Development), Bogotá, Colombia*

Chapter outline

1. Introduction

Energy and material resources constitute the physical foundation of the socioeconomic system. They also represent the physical basis for the creation of gross domestic product (GDP), and as such, if GDP is taken as a proxy for life satisfaction and national development, their consumption (or accumulation) is indirectly linked to personal and societal well-being, respectively (Haberl et al., 2019; Schandl et al., 2018). On the negative side, unsustainable levels of material production and consumption have caused humankind to overshoot various planetary boundaries (Rockström et al., 2009; Steffen et al., 2015). Di Giulio and Fuchs (2014), in particular, argue that "sustainable development" provides the conceptual framing required to effectively manage resource provision (and by extension, human well-being) in such a way that socioeconomic activity does not occur to the detriment of planetary health. Using the original Brundtland (1987) definition as a basis, they state:

Environmental Sustainability and Economy. https://doi.org/10.1016/B978-0-12-822188-4.00008-7

The goal of sustainable development is providing human beings in the present and in the future with the resources necessary to meet their objective needs and therefore to be able to live a good life according to their individual choices.

Brand-Correa and Steinberger (2017) have made a case that there are two intermediate steps between resource consumption (or accumulation) and well-being (specifically that linked to *eudaimonia,* also known as the "good life"). The first intermediate step is "energy and material services." For this chapter, we use Whiting et al.'s (2020) material service concept to discuss well-being from a service perspective. The latter is helpful because it can be used to identify and quantify the personal or societal function that a given energy flow, material flow, or material stock provides to an individual or group for the fulfillment of a human need, or the achievement of a desired state or end goal (Haberl et al., 2017).

The second intermediate step Brand-Correa and Steinberger (2017) identified is that of the "need satisfiers," a concept first described by Doyal and Gough (1991) in their "Theory of Human Needs" (not be confused with Max-Neef's, 1991 theory of the same name). Within Doyal and Gough's framework, "need satisfiers" refer to the spatial-temporal specific means and ways that intermediate needs, including energy and material services, are made manifest in society, so as to adequately ensure the meeting of basic human needs. The concept of human needs has been researched by various scholars across the sciences and the humanities because of its significance in determining the human condition and societal development.

Consequently, the specific aims of this chapter are (1) insert material services into Doyal and Gough's Theory of Human Needs; and (2) explore the implications and challenges that might occur when presenting material services as a quantifiable and useful intermediate step between resource consumption (or accumulation) and certain aspects of well-being. In Section 2, we give a brief overview of some of the key definitions, observations, and contemporary human need frameworks. In Section 3, we define and characterize "material services." In Section 4, firstly we explicitly connect material services to Gough's Theory of Human Needs. Secondly, and building upon Brand-Correa and Steinberger (2017), we distinguish between material services and the need satisfiers that support the material aspects of human well-being. We then explore the potential application of material services for the establishment of consumption thresholds. Finally, in Section 5, we discuss material services and well-being within the scope of sustainable development and reflect upon some of the challenges that may occur upon the adoption of the material services concept by policymakers and other sustainability professionals and academics.

2. Defining human needs

While it is beyond the scope of this present chapter to provide a detailed description of all the human need or well-being theories or examine the similarities and differences between them (which is something that has already been done by various authors, e.g., Alkire, 2002; Brand-Correa, 2018; Gough, 2014), this literature review section provides some conceptualization of what constitutes a human need, before delving more deeply into the link between resource provision and the meeting of human needs, as expressed by Doyal and Gough's (1991) Theory of Human Needs.

One way to understand human needs is to view them as the fundamental drivers of people's behavior, expressed as thoughts or actions undertaken to achieve a specified purpose, goal, or set of goals (Guillen-Royo, 2014). However, this definition fails to distinguish "wants" or "preferences" from "needs," which is essential if we are to grapple with Di Giulio and Fuchs (2014) definition of sustainable development, which would hold that the meeting of everyone's wants is "unsustainable" while the meeting of everyone's needs "must be sustained." Of course, by adopting such a position, one is affirming a priori that any such distinction between a "need" and a "want" holds meaning for human well-being and societal development, beyond the mere imposition of someone else's values, principles, or opinion (see Stern, 1989). For Reinert (2018), the distinction between wants and needs can be clarified by the way of an example:

> Nearly six million infants and children perish before the age of five. It would seem strange and disrespectful to say that these individuals perish because of a failure to meet their wants or preferences … preference failure does not lead to death.

Aside from the physiological needs that any human being must meet in order to survive (e.g., food, clean water, healthcare, and shelter), there are various other factors that researchers have put forward with regards to how one might identify a "human need" (see an in-depth discussion by McGregor et al. (2009), for example). On the most basic level, we can, based on our own experience, understand that human needs have two components: the physiological and the psychological. We can also appreciate that both must be properly considered, and satisfied, for the proper functioning and, consequently, the flourishing of a human being (Kim and Kollak, 2006).

Reinert (2018) asserts that basic needs are *developmentally related to the human condition and verifiably so*. For him, basic needs include food, potable water, sanitation, healthcare, shelter, and a level of education which facilitates some degree of modern social participation. Braybrooke (1987) considers something to be a "basic need" if an individual could not undertake any one of the following roles without it: citizen, parent, householder, or worker, which suggests, that on some level, what constitutes a psychological need can be self-determined. He proposes that basic needs include food, water, exercise and rest, companionship, education, social acceptance and recognition, sexual activity, recreation, and freedom from harassment. Deci and Ryan's (2000) self-determination approach identifies needs by considering the prerequisites to psychological growth, integrity, and well-being, which they state are *the need for competence, relatedness, and autonomy*. Similarly, Nussbaum's "functional capabilities" with its categories of "senses, imagination, thought," "emotions," and "bodily integrity" tie human needs to notions of an individual's capacity to live a "good life." For Nussbaum (and by extension Sen, cf; Sen, 1999, 1985, 1994), this necessarily involves the personal freedom to be able to positively choose one's lifestyle and preferences. Max-Neef's (1991) own "Theory of Human Needs" meanwhile is founded upon a matrix of nine axiological categories (those derived from a set of values) of subjectively determined human needs including affection, creation, freedom, identity, and participation crossed by the four existential categories (those which define the nature of existence) of "being," "having," "doing,"and "interacting." Costanza et al. (2007) combined the structure of Max-Neef's human need framework with Nussbaum and Glover's (1995) development of the capabilities approach for their Quality of Life Assessment (QLA), which aims to measure the meeting of human needs at a policy level via a broad set of indicators (e.g., "subsistence," "reproduction and care," and "spirituality"). Costanza et al. (2007) state that their QLA can be the means through which subjective notions of well-being are objectively analyzed in the context of sustainable development and national policy more generally.

Doyal and Gough (1991), through their independently formulated "Theory of Human Needs," established a hierarchy of needs that goes from universal goals onto basic needs and then to intermediate needs, which, by including material prerequisites, tie resource provision to well-being. For them, humankind shares the universal goals of avoidance of serious harm, social participation, and critical participation. In turn, these goals are fulfilled through the meeting of three basic universal needs: physical health, critical participation, and autonomy. The latter refers to *the ability to make informed choices about what should be done and how to go about doing it* (Gough, 2017b, p. 4). Such needs then should be prioritized over resource consumption for the meeting of whims and preferences (Gough, 2017a).

The theoretical underpinning of Doyal and Gough's (1991) well-being framework asserts that while survival is the most primal of needs, all individuals require a reasonable level of physical and mental health in order to participate in society, that is to say, operate according to their human nature. Higher levels of independence and self-sufficiency, which Doyal and Gough refer to as "critical autonomy" in turn reflect the human capacity to consider cultural constructs, norms, and values—both one's own and those of others. According to Gough (2017b), the universality of physical health and autonomy hinges upon his belief that if these needs are not satisfied then a measurable serious harm will befall those affected. Such harm does not constitute the emotions evoked when a person is said to feel anxious or unhappy. Instead, such harm can be identified by a severe infringement upon an individual's (or group's) ability to function. It implies obstacles to effective social participation, which given humankind's social nature is detrimental to our most basic interests. It then follows that, in Gough's (2017b, p. 4) words, the meeting of *basic needs is the universal precondition for effective participation in any form of social life.*

Although basic human needs are the same for everyone, the way in which they are met differs in line with cultural norms and values, which in turn influence the way in which Doyal and Gough's intermediate needs of "adequate nutritional food and water," "adequate protective housing," "non-hazardous work environment," "non-hazardous physical environment," "appropriate health care," "security in childhood," "significant primary relationships," "physical security," "economic security," "safe birth control and childbearing," and "appropriate basic and cross-cultural education" are made manifest in society. Furthermore, and given the complexity of intermediate needs, their provision requires the meeting of social preconditions or, in other words, a whole host of civic and political structures. These include those associated with energy and material production; the transmission of culture; the constitution of freedoms, rights, obligations, and wider societal participation. The relationship between universal goals through to social preconditions is depicted in Fig. 14.1, which shows the Theory of Human Needs framework, as it was originally conceived.

3. Material services

The material aspects of human need fulfillment are met through a specific combination of resource flows and stocks that result from the interaction between the ecosystem and the socioeconomic system. The link between resource extraction and use from source to services has been captured by the ecosystem service community in various ways, including the ecosystem service cascade developed by Haines-Young and Potschin (e.g., Haines-Young and Potschin, 2018; Potschin and Haines-Young, 2011). It has also been investigated and accounted for by industrial ecologists and ecological economists under the scope of "energy services" or "material services." For example, Cullen and Allwood (2010), Knoeri et al. (2016), and Kalt et al. (2019) trace energy flows through to energy services. Baccini and Brunner (2012) and Whiting et al. (2020) link material consumption and accumulation to material services.[1]

---

[1] While a discussion of the differences/similarities between ecosystem services and material services is merited, the extension that required to tackle this issue lies beyond the scope of this chapter.

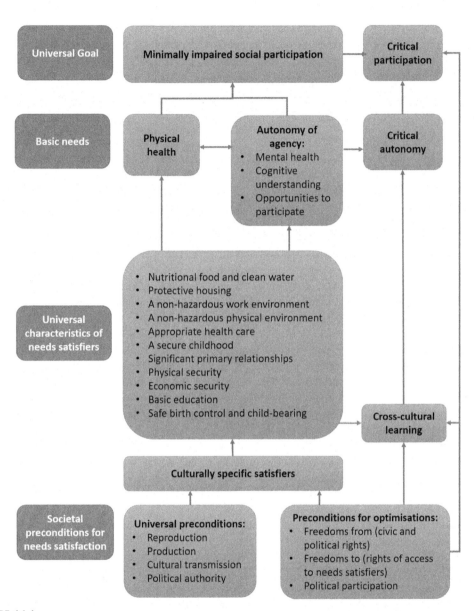

FIGURE 14.1

Theory of Human Needs framework.

Source: Doyal, L., Gough, I., 1991. A Theory of Human Need. Palgrave Macmillan, New York, USA.

The material service concept borrows heavily from energy services, which accounts for society's reliance on energy to support and sustain social metabolism, through the conversion of energy sources into energy services (Nakićenović et al., 1993; Grünbühel et al., 2003; Fell, 2017). It is defined by Whiting et al. (2020) as:

> Those functions that materials contribute to personal or societal activity with the purpose of obtaining or facilitating desired end goals or states, regardless of whether or not a material flow or stock is supplied by the market.

The distinction between energy services and material services is in function of the scope, i.e., whether service provision accounting and analysis is restricted to energy flows or expanded to also include material flows (e.g., lubricants, fertilizer, salt and detergents) and material stock (e.g., engines or buildings). The inclusion of stocks permitted under a material service analysis addresses the frequently overlooked role that material accumulation plays in supporting societal activities and economic growth (see Krausmann et al., 2017; Pauliuk and Müller, 2014; Weisz et al., 2015). That said, it is worth noting that any distinction between energy services and materials services is merely theoretical—a methodological step taken to facilitate the tracing of resources from extraction to service delivery. In reality, of course, most services have both energy and material components because in the processes that form social metabolism, all materials have embedded energy and all energy is converted/transported through materials.

A flow (or stock) provides a service when upon interacting with an end-user it fulfills a defined purpose, measurable in physical units (such as lumen-hours or joules). As material service provision *necessarily* involves such interaction, a radiator left on accidentally in an empty room does not constitute a service. Not all material services contribute to GDP because wealth creation is not indicative of service provision. Some forms of permanent shelter, for example, may never have entered the market, nor have a monetary value, but that does not make them any less of a shelter to those that choose to live there. This idea is captured in the final part of the material services definition. It highlights the fact that not all flows or stocks come from the market and that it is possible to separate economic activity from the need to provide individuals and society with services. This opens the concept to traditional or alternative forms of community and trade, including those existing historically or prehistorically, which did require material services but did not have what we would recognize as a market mechanism for their provision.

At the same time, some flows, which contribute to GDP, do not provide a service but instead solely contribute to social status, wealth accumulation, redundant/obsolete stock or waste (Shue, 1993). The production of fake "lifejackets" that are so heavy that they actively contribute to a person drowning is a poignant example (Miliband, 2017; Tzafalias, 2016). These kinds of flows do not convert into a material service because the jackets do not fulfill their purpose as a buoyancy aid, nor do they keep a person warm. Other material flows that do not support any of the material service categories identified by Whiting et al. (2020) include packaging that does not contribute to product protection, but is instead manufactured for marketing and branding, for the sake of monetary gain, prestige, or other arbitrary purposes (see Rundh, 2009; Van Rompay et al., 2012). This observation highlights the tensions that can arise when some individuals and corporations prioritize resource use for the maximization of wealth creation over a sustainable provision and optimization of material services.

One should also clearly differentiate between material services and the immaterial aspects that support societal end goals, such as heightened social participation. Job creation, caregiving, and education are three such examples which depend on a number of material services for their delivery but are not material services in their own right. While our beliefs and socio-political structures, of which norms and values form part, are in many ways responsible for the distribution and use of resources (Baccini and Brunner, 2012; Lent, 2017; Lewis and Maslin, 2018), they are not in themselves flows, stocks, or material services, but rather immaterial aspects that govern how and why resources are consumed.

3.1 Material service characteristics

The concept of *material services* can be applied to humankind universally across spatial-temporal barriers and cultural differences. While not the same, material services are aligned to the characteristics of "basic needs" as defined by Gough (2017a, pp. 45−47) in the sense that they are:

1) **Objective:** they represent the physical material aspects that contribute to the physiological and psychological requirements of human beings, measurable in physical units (e.g., lumen-hour, kcal, and m^2). While the specific satisfier (i.e., the exact way a material service is made manifest in society via the technology used), the frequency of its use, and the exact number of units consumed will differ relative to local geography, social customs, and personal expectations and preferences, the need for material service provision is universal. For example, the number of lumen-hours in Norway during winter will be much greater than the number of lumen-hours consumed by those living at the equator, due to differences in natural light intensity and duration. Likewise, the activities facilitated by artificial lighting may be vastly different, but both populations will require artificial illumination to provide visual comfort to one degree or another. Material services can be measured in terms of quality and quantity depending on the unit(s) used and the aim of the study. A set of possible units is provided in Whiting et al. (2020). One must be careful to select appropriate units in order to measure and compare properly the variables of interest. For example, "sustenance" can be approached with quantity metrics such as kcal and grams of protein or kg of food. However, such indicators do not discriminate between two types of the same unit, e.g., the number of kcal provided by a doughnut and the exact same number of kcal provided by pulses. Furthermore, the number of kcals does not support value judgments, nor can it be used to explain how or why these kcals of sustenance transform into well-being. Thus, where appropriate, quantity can be complemented with the use of quality indicators such as *Shannon Entropy* or *Modified Functional Attribute,* which measure nutritional diversity (Remans et al., 2014). In order to select an appropriate indicator(s), collect sufficient data, frame the results, and undertake a balanced and nuanced evaluation of the material service being assessed, a detailed contextual analysis is advised. Such an analysis would need to take into consideration those factors associated with the study's spatial-temporal scope and the corresponding local preferences and values. When looking at "sustenance," for example, one might evaluate data on the most consumed food types, most frequented food chains, average food prices, and typical food basket for the average household. One might also consider factors linked to exercise and employment (whether it's sedentary or well-paid, etc.) and the type of ingredients favored by the food sector, as all these will have a bearing on the number of kcal consumed and culinary preferences. One might also consider undertaking surveys in order to better understand how people perceive food, in terms of wants and needs. This contextual analysis will then provide insights on the nature of the service and help a researcher understand why, for example, American and Italian food per capita intake is very

similar (3682 kcal and 3579 kcal, respectively) but the number of obesity-related deaths in the United States (71.9 deaths per 100,000) is almost double that of Italy (36.25 deaths per 100,000) (Roser and Ritchie, 2013; Ritchie and Roser, 2017). This phenomenon cannot be explained if one restricts the material service approach to merely comparing kcal.

2) **Plural:** material services cannot be aggregated or summarized under one single unit. The only way to do this would be to introduce the subjective criteria that already exist in the weighting and normalization stages in a life cycle analysis (cf. Bruijn et al., 2002).

3) **Nonsubstitutable:** in the sense that greater unit of m^2 for shelter, as a material service, will not make up for a lack of kcal of sustenance.

4) **Satiable:** material service provision will plateau as additional lumens, food, m^2 of shelter reach diminishing returns and/or may actively hinder human well-being. In the case of visual comfort, too much illumination will blind a person. Too much food will lead to morbidity. Once a person reaches their destination, they don't require more transport.

5) **Cross-generational** in the sense that all people past, present, and of the future require the same material services, although the manifestation of those services may differ (e.g., tallow or spermaceti candle in the Georgian period vs. the light-emitting diode (LED) lamps of the 21st century).

6) Material services have an **ethical component** if human flourishing is considered to be a desired end of state and if society agrees that well-being is facilitated by adequate service provision and hindered when one or more services are not available to a person or a community. Resource scarcity and the overshooting of planetary boundaries add to this ethical dimension. Di Giulio and Fuchs (2014) argue that environmental degradation and scarcity necessitates the creation of defined minimum and maximum levels of consumption. The minimum represents the lowest amount necessary for a person to meet a "fulfilled life." The maximum level is one that does not infringe on other people's ability to achieve the same or better standard of living (either now or into the future). The range between these two values is something they refer to as "sustainable consumption corridors." The latter are defined by cultural, geographical, and historical context and evolve as technology, customs, social norms, and values change.

It is important to note that a single material service unit does not capture all relevant aspects of service provision. Passenger kilometers (pkm), for example, quantify the distance that a user travels but do not capture the quality of the experience. This metric, as with many others that could be used to measure "mobility" as a material service, is a proxy and thus can be misleading if not properly contextualized. For example, a high pkm may not necessarily correspond to high service quality because while it may mean that the transport network is large, allowing a person to travel further, it could equally signify that the transport system is so overcrowded that a person cannot enter a carriage when it stops at their station. This is why we must carefully interpret service units prior to recommending a course of action. One should also measure service in a whole range of units in order to ensure that the correct aspect is being evaluated. A high pkm is not contributing to well-being if it means that road users are habitually stuck in heavy traffic and thus do not have time to visit friends and family, so as to enjoy Gough's satisfier of "significant primary relationships." Likewise, a bigger service network (a feature captured in a high pkm) is not always better. In fact, one could argue that a transport service of the highest quality enables a person to travel fewer kilometers and still achieve their end goal.

In this respect, the material service approach can be used to highlight where higher passenger kilometers start to detrimentally affect the quality of the service provision. Again, this needs to be done with care because, in calculating the average, one can easily neglect those people who do not have sufficient access to a material service. An able-bodied person, for example, may not require as many square meters of shelter to navigate their home compared to someone in a wheelchair. Furthermore, in focusing solely on average, decision-makers may fail to identify where users experience diminishing returns (i.e., where more kcal, m^2 or pkm do not actively further well-being).

The application of material services for the purpose of understanding how material flows and stocks contribute to well-being supports the idea of Di Giulio and Fuchs' (2014) sustainable consumption corridors and the setting of an upper and lower bound for a given set of metrics (e.g., kcals, lumens-sec, m^3 water at 40 C). Although the material service concept does not dictate what the threshold should be, we know that if a person wishes to read, write, or simply navigate a room, for example, they will require a certain number of lumen-hours, lux, or candela per square meter, in the form of either natural light or illumination as a material service. This is true whether that person was born into the European Union or into the Roman Empire, as reading a papyrus would have required the same (or a similar) amount of light as reading a letter written today. In this respect, illumination needs are analogous, although the technology has changed. This is significant because it demonstrates that lumen-hours and other physical units are directly comparable. The same holds true, albeit with some caveats, for other services such as "sustenance," which is directly comparable when one analyzes the service thresholds for people of the same gender, age group, and level of activity.

4. Integrating materials services to the Theory of Human Needs framework

Gough (2017a, 2017b) on widening the scope of his theory, so as to frame it in the language of Brundtland's (1987) definition of sustainable development, began integrating resource consumption and environmental dimensions into aspects of well-being, particularly for the meeting of intermediate needs. However, he was predominantly focused on climate breakdown, and he stopped short of creating a fully integrated framework. In response to this, in Fig. 14.2, we update the Theory of Human Needs to more explicitly capture the environmental aspects that support the meeting of human needs. Following on from Lamb and Steinberger (2017) and Gough's (2017a) discussions on the value of incorporating the natural system into well-being frameworks, we add the category "environmental preconditions" to Doyal and Gough's (1991) theoretical model. This category constitutes the biophysical components and processes that support life on earth and which, by extension, provide the physical foundation of the socioeconomic system.

The addition of environmental preconditions is justified by the fact that our planet and its natural processes, including ecosystem services, provide the physical limit for societal well-being, which does not exist in a socioeconomic vacuum. Likewise, a person's physical and mental capacities are ultimately subject to, and dependent on, the physical limits imposed by the environment. Consequently, societal and personal goals will only result in sustainable well-being if they are pursued within planetary boundaries. On the contrary, physical survival/health and personal autonomy, as the most basic human needs, will not be achievable on a global scale.

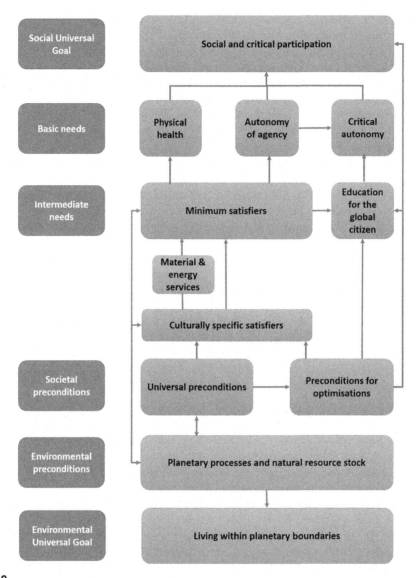

FIGURE 14.2

Introducing natural resources into Gough's Human Needs Theory as shown by the *green* (gray in print version) boxes.

One way to analyze the relationship between aspects of well-being and sustainability that connects well to Gough's Theory of Human Needs is through the environmental efficiency of well-being introduced by Dietz et al. (2009). This concept permits the evaluation of those activities deemed detrimental to the environment because it juxtaposes them to improvements in quality of life. It has been applied by Lamb et al. (2014), for example, to measure national human development trajectories, using life expectancy as a proxy, relative to consumption-based carbon emission intensity. However, it could also measure environmental impact relative to an increased access to, or quality of, material services. For example, a think tank might consider that organic farming practices and food production will automatically result in a higher quality "sustenance" and a higher level of "environmental protection and restoration," as two material services. To evaluate their claim, one could use the material service concept to identify how a policy shift away from environmentally damaging pesticides and toward organically produced fruits and vegetables would influence consumer preferences in a way that enhances (or reduces) access to "sustenance." One might demonstrate that this policy proposal will not necessarily invoke an improvement in "sustenance" as higher prices may cause a larger proportion of the population to eat a less varied diet and would thus reduce their access to nutritional food. In which case, a reduced environmental impact might detrimentally affect "sustenance" service provision, a tradeoff which would then need to be debated. In Fig. 14.2, material services are located between "minimum satisfiers" and "cultural specific satisfiers." This is because although material services are universal, the quantities and the way in which they are made manifest in society are dependent on cultural norms and values. We changed the terminology of "cross-cultural learning" to "education for the global citizen" to capture the critical lens offered by the global educative framework used by the United Nations, among others, to effectively communicate the values and practices required for societal transformation and the facilitation of sustainable development (see Misiaszek, 2018; Whiting et al., 2018b).

4.1 Connecting material services to needs satisfiers

Some researchers have extended the scope beyond services and gone onto investigate the relationship between resource use and aspects of well-being (Smith et al., 2013). For example, Day et al. (2016) evaluated the provision of energy services through the capabilities approach (as defined in Sen, 1994; Nussbaum, 2011), by differentiating energy service, secondary capabilities, and basic capabilities. Brand-Correa and Steinberger (2017) undertook an assessment of energy services within the scope of Max-Neef's (1991) Theory of Human Needs, differentiating between energy service and intermediate needs' satisfiers. Rao and Baer (2012) and Rao and Min (2018) integrated some material service categories to the well-being concepts developed by Doyal and Gough (1991) and Nussbaum (2000). In Table 14.1, we review various studies that link energy or material services to the Capability Approach (CA) and/or Theory of Human Needs (THN).

As shown in Table 14.1, Rao and Min (2018) is the most readily comparable study relative to this present chapter, as it is the only one that used contemporary well-being frameworks to identify the role of material services in fulfilling human needs. Rao and Min proposed a minimum value of material dependence for the meeting of basic human needs, a threshold which they call "Decent Living Standards" (DLS). The purpose of the latter is "to identify what universal material satisfiers are required by people everywhere." They propose the "lowest common denominator" of basic material requirements that are instrumental (but not sufficient) to achieve physical, and to an extent social, dimensions of human wellbeing, whether conceived as basic needs or basic capabilities, and independent of peoples values or relative stature in society." For example, when it comes to "living conditions," Rao and Min suggest that proper fulfillment of this material service requires a minimum floor space of 30 m^2 for a family of three, adequate lighting, which they argue should be electric, basic comfort provided by modern heating/cooling equipment, an adequate freshwater supply of 50 L per capita per day, and safe waste disposal via in-house improved toilets.

Table 14.1 Connections between energy, material services, and specific well-being frameworks.

Well-being framework	Resource stage	Study	Application	Observations
Capability Approach (CA)	Energy service	Day et al. (2016)	Energy poverty	Assesses individual and household access to energy services and the potential for energy poverty, within the capability approach.
		Walker et al. (2016)	Participatory processes	Puts forward the case that participatory processes are the most legitimate way of defining energy needs in any given society and are the most relevant for identifying how capabilities (well-being) are supported by energy services.
		Wood and Roelich (2019)	Well-being, energy, and climate change	The authors argue that climate change mitigation and energy use (particularly fossil fuels) should be analyzed through a well-being lens that captures the complex relationships between resource provision, capabilities, and environmental deterioration.
	Material consumption or accumulation (flows and/or stocks)	Holland (2014, 2008)	Resource use, well-being, and socioenvironmental justice	Argues that society must identify and understand the ecological conditions and the appropriation of resources and services in a way that allows humans to flourish.
		Peeters et al. (2015a, 2015b)	Functioning constraints and capability thresholds	Using capability thresholds and functioning constraints, these papers advocate for the ethos of restraint, as a precondition for sustainable development. There is an emphasis on the fairness of personal resource budgets.
		Hirvilammi et al. (2013)	Material footprint and capabilities	Using the Material Input Per Unit of Service (MIPS) method, the authors explain the connections between capabilities and material footprints, with the aim of understanding human dependency on natural resources.
Theory of Human Needs (THNs)	Energy service	Brand-Correa and Steinberger (2017)	Energy services and human needs decoupling	Proposed a set of mixed methods (quantitative and qualitative) to capture insights on the cultural particularities for delivering energy services and needs satisfiers.

Table 14.1 Connections between energy, material services, and specific well-being frameworks.—cont'd

Well-being framework	Resource stage	Study	Application	Observations
		Brand-Correa et al. (2018)	Participatory processes	Argues that community-based approaches and methodologies are particularly relevant for the identification and understanding of needs satisfiers and their role in energy service provision.
		Rao and Baer (2012)	Decent living standards	Differentiates those activities that lead to human need satisfaction and those that are predominately linked to furthering social status and wealth. Quantifies the energy consumption and carbon emissions of those activities that support decent living standards.
	Material consumption or accumulation (flows and/or stocks)	Briceno and Stagl (2006)	Participatory processes	Argues for a more socially oriented perspective of product design to enable individuals to experience increased levels of social participation, creativity, autonomy, and friendship.
		Gough (2017b, 2017a)	Well-being, the problem of growth and climate change	Argues that emissions can be reduced, and well-being increased by switching from high to low-carbon services and goods and in stabilizing absolute levels of consumer demand.
		Koch et al. (2017)	Degrowth	Proposes a deprioritization of "happiness" (which they consider to be a subjective form of well-being) and the prioritizing of human needs satisfaction within environmental limits.
Mixed approach (THN + CA)	Energy service	Darby and Fawcett (2018)	Energy sufficiency	A mix of a quantitative and qualitative energy need assessment that connects services with well-being.
	Material service (including flows and stocks)	Rao and Min (2018)	Decent living standards	Proposes universal minimum thresholds at the individual and societal level for the meeting of those material aspects that support well-being.

The problem with Rao and Min's (2018) "universal satisfiers" is that what they are predominantly describing is a scenario which those in the West would consider to be the barebones required for material-derived well-being. While such scenarios may be "universal" for the majority of communities alive today (with a few notable exceptions such as indigenous populations or the Amish, who do not use electricity), they are not universal in the true sense in that they are not applicable to any population prior to the invention of electricity or mass sanitation systems.

For this reason, we maintain that it is better to establish a range for material service thresholds similar to the sustainable consumption corridor concept established by Di Giulio and Fuchs (2014). This range will depend on geographical location, available technology, and existing cultural norms and values for a defined nation, region, or population. To establish it would require an in-depth contextual analysis. Such a process removes the inconsistencies that occur spatial-temporally or across cultural differences. It means that the concept can be applied to any human community from the Neanderthals to the breath of the last of the *Homo sapiens*. This is effectively what Whiting et al. (2020) did when comparing illumination, as a material service, for Pompeii and Herculaneum in 79 BCE with that of Georgian London (c.1820). Based on the respective average inhabitant's access to lumen-hours, their case study showed that the average Roman living around Mount Vesuvius had greater access to lighting than the average Georgian living in London, or in fact any average inhabitant of Great Britain prior to the Industrial Revolution. While these results do not state whether the average Roman felt happier than the average Georgian, it does say something about their quality of life and the priorities of Ancient Roman society relative to Britain's in the 1820s.

Fig. 14.3 illustrates how material services represent the interconnection between stocks and flows and the fulfillment of intermediate needs. If one takes illumination as a material service example for the 21st century, the flow corresponds to electricity, while the LED bulb, cables, switches, and the electricity production infrastructure correspond to the stocks. Both stocks and flows are required to provide lumen-hours in a conventional house that obtains kWh from the grid.

Of course, visual comfort is only one aspect of day-to-day life, and, as Rao and Min (2018) rightly observe, to be able to experience "decent living standards" one would also need a certain degree of space and thermal comfort, be free from hunger, and so on. Their reflection supports our assessment as

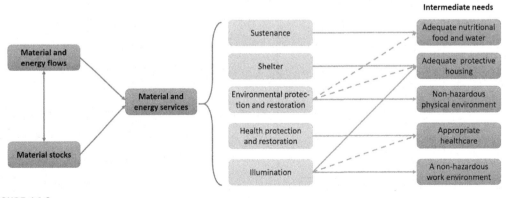

FIGURE 14.3

Some example relationships between material services and intermediate needs. Note: *Arrows* indicate a direct link between material services and intermediate needs. *Dashed arrow* = indirect link.

to the usefulness of identifying, measuring, and assessing material services, even when actual or perceived well-being is not directly linked to efficient material service delivery. National and local governments, for instance, may prioritize employment, in order to achieve specific aspects of well-being, over more efficient means of service provision (Schaffartzik, 2019a). An example of this includes the expansion of the car industry, even though private vehicles are not the most material efficient form of transport, nor the most sustainable when one considers their carbon footprint (Carmona et al., 2021a; Schaffartzik 2019b). The same logic applies to private jets or other forms of extravagant consumption, which offer a material service at a considerable cost to the environment. That said, the efficiency of such transport can be assessed by multiple service units, and the results could be used to identify political or corporate priorities regarding whether this form of travel is desirable.

In this respect, one can also use the material service concept to identify *who* is using the resources and thus determine appropriate upper and lower thresholds for resource consumption. This could be done via the social metabolic classes proposed by Otto et al. (2019) and Otto and Schuster (2019), in conjunction with some of Rao and Min's (2018) Decent Living Standard Dimensions and/or Di Giulio and Fuchs' (2014) sustainable consumption corridors. The material service concept could also be used to extend Fanning and O'Neill (2019) research on the relationship between life expectancy, carbon emissions, and perceived happiness. Arguably, this would help policymakers, government officials, corporate executive boards, and NGOs to identify potential opportunities for degrowth, whereby service provision is prioritized over new production and collective ownership favored over individualistic accumulation.

5. Discussion

Despite the significance of both energy and materials in supporting material aspects of well-being, energy remains almost exclusively at the forefront of major international policy initiatives such as the Sustainable Development Goals (SDGs). Innovative explorations into sustainable resource use has likewise often overlooked the role that materials play in the economy. Kate Raworth's (2017) *Doughnut Economics*, for instance, fails to mention materials when rightly critiquing classic economic theory for failing to understand the role of energy as a driver of socioeconomic growth and development. Gough's (2017a) evolution of his *Theory of Human Needs* focuses almost exclusively on the role of energy consumption and carbon emissions as contributing factors to climate breakdown, as well as the overshooting of planetary boundaries. In addition, energy access as opposed to service access appears in alternative political frameworks such as the Foundational Economy Collective (2020) *Manifesto,* which proposes a COVID-19 response that considers societal development and collective well-being from a resource-based rather than income-based perspective. However, there is also a growing recognition that Universal Basic Services (as an alternative to Universal Basic Income) might be a more equitable initiative with which to ensure that resource consumption prioritizes the meeting of human needs through the lens of sustainable development (Coote and Percy, 2020; Cotte, 2021).

The material services concept elaborated upon in Carmona et al. (2021a, 2021b, 2017) and Whiting et al. (2018a, 2020) is one way of quantifying the intention behind, and the sustainability of, material flows and stocks. It expands on the concept of energy services, which identifies that end users do not demand energy per se, but rather require it to achieve certain end states or conditions that have the potential to further their well-being (Day et al., 2016; Kalt et al., 2019; Wood and Roelich, 2019).

To connect material consumption and accumulation with human needs, one could use the material service concept to trace the value production chain from the specified need back to material sourcing (e.g., mineral extraction, agriculture). This is useful when critically analyzing the viability and sustainability of national policy. For example, a government could want to bring in a "clean air policy" to improve physical health by reducing the number of children per 1000 diagnosed with asthma. In which case, the material services approach could be used to evaluate how the ppm concentration of air pollutants reduces relative to the forecast number of electric cars and nonmotorized forms of transport. This can be achieved because the concept connects the material service of "environmental protection and restoration" to a vehicle's resource flows and stocks. One would also have to take into account the significant flows and stocks utilized by an electric car that are not needed in the manufacture and operation of a conventional vehicle. Lithium, a rare earth metal required to produce electric car batteries, is one such example and is as critical as it is rare (Valero and Valero, 2014). Its extraction has been linked to both environmental and socioeconomic concerns (Hindery, 2013; Kushnir and Sandén, 2012). Here too, material services can play a role by determining whether the amount of lithium required by the "clean air policy" is higher than that which can be extracted without severely impacting on the "environmental protection and restoration" service around the rare earth mine. These two results, from the urban area and the mine, will then indicate whether the "clean air policy" should be adjusted to promoting carpooling or expanding/improving electrified public transport provision instead of electric car ownership. Of course, there is likely to be tensions that material services cannot address. For one thing, material service units do not instruct policymakers, corporate leaders, or the general public to take a specific action or to prioritize the service provision of one group or individual over another. In this respect, a passenger-km or kcal is simply data that can influence a decision but cannot state whether the reasoning or the value judgments behind it are sound.

5.1 Challenges to material service adoption

The application of material services to well-being is not without its challenges, especially given the number of assumptions required to estimate the level of service that the average end user experiences. One also has to recognize that quantification on its own is not necessarily indicative of the reality on the ground. Many passenger kilometers, for example, do not necessarily translate into a high-quality service. While they may signify that the transport network is large, which allows a person to cover a considerable distance, they might equally mean that the transport system is so overcrowded that a person cannot access a bus when waiting at a stop. Likewise, one could mistakenly infer that a larger number of lumen-hours per capita means that society has invested considerably in lighting technologies, when in fact a small elite group experiences a very large number of lumen-hours while the majority of the population have no (or very limited) access to artificial lighting, and thus limited visual comfort during the hours of darkness. In this case, the average would become skewed and would not represent the "average user" at all. No individual service unit can capture all relevant aspects that constitute a material service, which is why multiple units are beneficial and unit justification is essential when defining a case study's scope and method. It is also important to acknowledge the inherent uncertainty of estimated values, even if one accurately interprets the data collected and offers reasonable assumptions in light of the contextual analysis.

As with any metric, there is no guarantee that one really has measured what they set out to do in the first place. If a practitioner does not select appropriate units, makes poor assumptions, or does not sufficiently understand the context, then it is difficult to infer that we do in fact understand something new about someone's way and quality of life. One way to overcome this issue could be in the development and distribution of questionnaires, with the idea of establishing the relationship a given community has to a set of material services. This would enable a dialogue that explores the different ways that people perceive the role of energy or materials at the individual and societal level. It would also identify what they tend to focus on, or measure, when it comes to those aspects of well-being that are facilitated when one has access to certain resources. For example, it might be that kWh of energy are predominantly valued in terms of hours of leisure time (cultural material services) or the ability to facilitate "sustenance" or "thermal comfort." Such dialogues would also prevent miscalculations as to what is regarded as a "service" rather than a "disservice." This is important when one considers that the material service "communication and information storage," for example, can be measured in bytes, which will come in bundles of both service (e.g., the bandwidth used to show a TV program) and disservice (e.g., unwanted advertisements).

Having access to different kinds of end users facilitates conversations regarding sustainable resource consumption because one is able to talk about energy or materials in terms of the benefits that individuals would gain (or lose) should their access change. It also provides an opportunity to discuss the objectivity of material service units and how the establishment of upper and lower thresholds, when integrated into resource policy, can result in an improved quality of life.

Jolibert et al. (2014) and Brand-Correa et al. (2018) provide evidence that questionnaires or focus groups provide direct and in-depth consultation into the way individuals and communities perceive resources and the services into which those resources are transformed in order to meet their human needs. However, such initiatives are not without their challenges, as the questions asked and the way the questionnaire is structured might inadvertently provide information regarding a community's perception of a material service rather than the nature of the material service itself. There is also the issue of overcoming, or at least minimizing, noise that results from the way that a question is framed, or how a concept is explained prior to community, or even expert, engagement with the topic at hand. A narrow interpretation of material services, based solely on the researcher's definition, may be a considerable problem given that a service perspective is not the lens through which most people typically discuss, or policymakers design, sustainable resource use initiatives. Within the sustainable development discourse (including the Sustainable Development Goals), sustainable energy is typically described in terms of energy efficiency rather than service provision, even though the latter is more directly linked to human well-being and planetary health than efficiency is per se. As with any paradigm shift, there will be some people who are reluctant to consider concepts through yet another lens. That said, the fact that the service perspective provides an alternative vision that supports progress toward sustainable development means that it holds significant potential for policy at the global, national, and local level.

6. Conclusions

This chapter reviewed Doyal and Gough's (1991) Human Needs framework and subsequent updates, such as Gough (2017b, 2017a) that expanded the theory's scope to incorporate environmental concerns. Through the integration of this theory with the material service concept, as developed by Carmona et al. (2017) and Whiting et al. (2018a, 2020), this chapter explicitly linked natural resources with certain aspects of well-being, through the provision of what Doyal and Gough refer to as "intermediate needs" and "need satisfiers."

Brand-Correa and Steinberger's (2017) argument that there are in fact two intermediate steps (services and satisfiers) between resource consumption (or accumulation) and well-being is upheld. The acknowledgment of services and need satisfiers, as the material precursors that form the physical aspects that contribute to well-being, has implications for sustainable development. We have discussed the merits regarding the establishment of resource consumption and material service thresholds to ensure that both present and future needs are met (lower bound) and that excessive consumption undertaken for the meeting of wants at the expense of planetary health (upper bound) are curtailed. One way to do this is via the sustainable corridor concept proposed by Di Giulio and Fuchs (2014). The exact range between the bounds will depend on various characteristics including location, technological preferences, and existing cultural norms and values for a defined nation, region, or population. In this respect, the global tackling of the most urgent challenges of the 21st century will not be a one-size-fits-all service model of well-being but rather one anchored into a local sense of identity imbued by a universal set of values. Coming up with universal values when it comes to sustainable resource use is not as impossible as it might sound, given the existence of the Universal Bill of Human Rights and the successful phasing out of CFCs as agreed upon by the ratification of the Montreal Protocol in 1989. Future research will need to evaluate how, in practice, one might use material service metrics to open up a debate into the morality and ethics of material consumption and accumulation, especially when it comes to matters related to socioenvironmental justice, including what exactly should constitute a "need" or a "want" in the context of sustainable development.

Acknowledgments

We thank Julia Steinberger for her helpful comments, suggestions, and advice. We also appreciate the time taken by Hanna Murray-Carlsson and Leonidas Konstantakos to comment on this manuscript. L.G.C. acknowledges the support of Colciencias.

Funding

K.W. acknowledges the financial support of UCLouvain's "FSR Incoming Post-doc 2020 as above". L.G.C. acknowledges the financial support of Fundação para a Ciência e a Tecnologia (FCT) and MIT Portugal Program through grant PD/BD/128038/2016.

References

Alkire, S., 2002. Dimensions of human development. World Dev. 30 (2), 181−205.
Baccini, P., Brunner, P.H., 2012. Metabolism of the Anthroposphere: Analysis, Evaluation, Design. MIT Press, Cambridge, US.
Brand-Correa, L.I., 2018. Following the Golden Thread: Exploring the Energy Dependency of Economies and Human Well-Being. University of Leeds.
Brand-Correa, L.I., Steinberger, J.K., 2017. A framework for decoupling human need satisfaction from energy use. Ecol. Econ. 141, 43−52.
Brand-Correa, L.I., Martin-Ortega, J., Steinberger, J.K., 2018. Human scale energy services: untangling a "golden thread". Energy Res. Soc. Sci. 38, 178−187.
Braybrooke, D., 1987. Meeting Needs. Princeton University Press.
Briceno, T., Stagl, S., 2006. The role of social processes for sustainable consumption. J. Clean. Prod. 14 (17), 1541−1551.
Bruijn, H., Duin, R., Huijbregts, M.A.J., 2002. Handbook on Life Cycle Assessment.

Brundtland, G.H., 1987. Report of the World Commission on Environment and Development: 'Our Common Future'. United Nations, New York, NY, USA.

Carmona, L.G., Whiting, K., Haberl, H., Sousa, T., 2021a. The use of steel in the United Kingdom's transport sector: a material stock-flow-service nexus case study. J. Ind. Ecol. 25 (1), 125–143. https://doi.org/10.1111/jiec.13055.

Carmona, L.G., Whiting, K., Carrasco, A., Simpson, E., 2021b. What can the stock-flow-service Nexus offer to corporate environmental sustainability? In: Cotte Poveda, A., Pardo Martinez, C.I. (Eds.), Environmental Sustainability and Development in Organizations: Challenges and New Strategies. CRC Press, Boca Raton, US, pp. 50–69.

Carmona, L.G., Whiting, K., Carrasco, A., Sousa, T., Domingos, T., 2017. Material services with both eyes wide open. Sustainability 9 (9), 1508.

Coote, A., Percy, A., 2020. The Case for Universal Basic Services. Polity press, Cambridge, UK.

Costanza, R., Fisher, B., Ali, S., Beer, C., Bond, L., Boumans, R., Danigelis, N.L., Dickinson, J., Elliott, C., Farley, J., 2007. Quality of life: an approach integrating opportunities, human needs, and subjective well-being. Ecol. Econ. 61 (2–3), 267–276.

Cotte, A., 2021. Universal basic services and sustainable consumption. Sustainability: Sci. Pract. Policy 17 (1), 32–46. https://doi.org/10.1080/15487733.2020.1843854.

Cullen, J.M., Allwood, J.M., 2010. The efficient use of energy: tracing the global flow of energy from fuel to service. Energy Pol. 38 (1), 75–81.

Darby, S., Fawcett, T., 2018. Energy Sufficiency: An Introduction. EECE, European Council for an Energy Efficient Economy, Stockholm.

Day, R., Walker, G., Simcock, N., 2016. Conceptualising energy use and energy poverty using a capabilities framework. Energy Pol. 93, 255–264.

Deci, E.L., Ryan, R.M., 2000. The "what" and "why" of goal pursuits: human needs and the self-determination of behavior. Psychol. Inq. 11 (4), 227–268.

Di Giulio, A., Fuchs, D., 2014. Sustainable consumption corridors: concept, objections, and responses. GAIA Ecol. Perspect. Sci. Soc. 23 (3), 184–192.

Dietz, T., Rosa, E.A., York, R., 2009. Environmentally efficient well-being: rethinking sustainability as the relationship between human well-being and environmental impacts. Hum. Ecol. Rev. 114–123.

Doyal, L., Gough, I., 1991. A Theory of Human Need. Palgrave Macmillan, New York, US.

Fanning, A.L., O'Neill, D.W., 2019. The wellbeing–consumption paradox: happiness, health, income, and carbon emissions in growing versus non-growing economies. J. Clean. Prod. 212, 810–821.

Fell, M.J., 2017. Energy services: a conceptual review. Energy Res. Soc. Sci. 27, 129–140.

Foundational Economy Collective, 2020. 2020 Manifesto for the Foundational Economy. https://foundationaleconomycom.files.wordpress.com/2020/04/2020-manifesto-for-the-foundational-economy.pdf.

Gough, I., 2014. Lists and thresholds: comparing the Doyal-Gough theory of human need with Nussbaum's capabilities approach. Capabil. Gender Equal. 357–381.

Gough, I., 2017a. Heat, Greed and Human Need: Climate Change, Capitalism and Sustainable Wellbeing. Edward Elgar Publishing.

Gough, I., 2017b. Recomposing consumption: defining necessities for sustainable and equitable well-being. Phil. Trans. R. Soc. A 375 (2095), 20160379.

Grünbühel, C.M., Haberl, H., Schandl, H., Winiwarter, V., 2003. Socioeconomic metabolism and colonization of natural processes in Sangsaeng village: material and energy flows, land use, and cultural change in northeast Thailand. Hum. Ecol. 31 (1), 53–86.

Guillen-Royo, M., 2014. 'Human Needs'. Encyclopedia of Quality of Life and Well-Being Research, pp. 3027–3030.

Haberl, H., Wiedenhofer, D., Erb, K.H., Görg, C., Krausmann, F., 2017. The material stock–flow–service nexus: a new approach for tackling the decoupling conundrum. Sustainability 9 (7), 1047.

Haberl, H., Wiedenhofer, D., Pauliuk, S., Krausmann, F., Müller, D.B., Fischer-Kowalski, M., 2019. Contributions of sociometabolic research to sustainability science. Nat. Sustain. 1.

Haines-Young, R., Potschin, M., 2018. Common International Classification of Ecosystem Services (CICES) V5.1: Guidance on the Application of the Revised Structure. Fabis Consulting Ltd, Nottingham, UK. www.cices.eu.

Hindery, D., 2013. From Enron to Evo: Pipeline Politics, Global Environmentalism, and Indigenous Rights in Bolivia. University of Arizona Press, Tucson, Arizona, EUA.

Hirvilammi, T., Laakso, S., Lettenmeier, M., Lähteenoja, S., 2013. Studying well-being and its environmental impacts: a case study of minimum income receivers in Finland. J. Human Dev. Capabil. 14 (1), 134−154.

Holland, B., 2008. 'Justice and the environment in Nussbaum's "capabilities approach" why sustainable ecological capacity is a meta-capability'. Polit. Res. Q. 61 (2), 319−332.

Holland, B., 2014. Allocating the Earth: A Distributional Framework for Protecting Capabilities in Environmental Law and Policy, first ed. Oxford University Press, Oxford, UK.

Jolibert, C., Paavola, J., Rauschmayer, F., 2014. Addressing needs in the search for sustainable development: a proposal for needs-based scenario building. Environ. Val. 23 (1), 29−50.

Kalt, G., Wiedenhofer, D., Görg, C., Haberl, H., 2019. Conceptualizing energy services: a review of energy and well-being along the energy service cascade. Energy Res. Soc. Sci. 53, 47−58.

Kim, H., Kollak, I., 2006. Nursing Theories: Conceptual and Philosophical Foundations. Springer Publishing Company.

Knoeri, C., Steinberger, J.K., Roelich, K., 2016. End-user centred infrastructure operation: towards integrated end-use service delivery. J. Clean. Prod. 132, 229−239.

Koch, M., Buch-Hansen, H., Fritz, M., 2017. Shifting priorities in degrowth research: an argument for the centrality of human needs. Ecol. Econ. 138, 74−81.

Krausmann, F., Wiedenhofer, D., Lauk, C., Haas, W., Tanikawa, H., Fishman, T., Miatto, A., Schandl, H., Haberl, H., 2017. Global socioeconomic material stocks rise 23-fold over the 20th century and require half of annual resource use. Proc. Natl. Acad. Sci. U. S. A. 114 (8), 1880−1885.

Kushnir, D., Sandén, B.A., 2012. The time dimension and lithium resource constraints for electric vehicles. Resour. Pol. 37 (1), 93−103.

Lamb, W.F., Steinberger, J.K., 2017. 'Human well-being and climate change mitigation'. Wiley Interdisciplin. Rev.: Climate Change 8 (6).

Lamb, W.F., Steinberger, J.K., Bows-Larkin, A., Peters, G.P., Timmons Roberts, J., Ruth Wood, F., 2014. Transitions in pathways of human development and carbon emissions. Environ. Res. Lett. 9 (1), 014011.

Lent, J.R., 2017. The Patterning Instinct: A Cultural History of Humanity's Search for Meaning. Prometheus Books, New York, US.

Lewis, S., Maslin, M., 2018. The Human Planet: How We Created the Anthropocene. Penguin, London, UK.

Max-Neef, Manfred, A., 1991. Human Scale Development: Conception, Application and Further Reflections. The Apex Press, New York and London.

McGregor, J.A., Camfield, L., Woodcock, A., 2009. Needs, wants and goals: wellbeing, quality of life and public policy. Appl. Res. Q. Life 4 (2), 135−154.

Miliband, D., 2017. The Refugee Crisis Is a Test of Our Character. In: Presented at the TED2017: The Future You, Vancouver, CA. https://www.ted.com/talks/david_miliband_the_refugee_crisis_is_a_test_of_our_character.

Misiaszek, G.W., 2018. Educating the Global Environmental Citizen: Understanding Ecopedagogy in Local and Global Contexts. Routledge.

Nakićenović, N., Grübler, A., Inaba, A., Messner, S., Nilsson, S., Nishimura, Y., Rogner, H.-H., Schäfer, A., Schrattenholzer, L., Strubegger, M., 1993. Long-term strategies for mitigating global warming. Energy 18 (5), 401.

Nussbaum, M.C., 2000. Women and Human Development: The Capabilities Approach. Cambridge University Press.

Nussbaum, M.C., 2011. Creating Capabilities: The Human Development Approach. Harvard University Press.

Nussbaum, M.C., Glover, J., 1995. Women, Culture, and Development: A Study of Human Capabilities. Oxford University Press, Oxford, UK.

Otto, I.M., Schuster, A., 2019. Socio-Metabolic Class Theory. In: Presented at the 13th Conference of the International Society for Industrial Ecology (ISIE) — Socio-Economic Metabolism Section. 13th-15th May 2019, Berlin.

Otto, I.M., Kim, K.M., Dubrovsky, N., Lucht, W., 2019. Shift the focus from the super-poor to the super-rich. Nat. Clim. Change 9 (2), 82.

Pauliuk, S., Müller, D.B., 2014. The role of in-use stocks in the social metabolism and in climate change mitigation. Global Environ. Change 24, 132—142.

Peeters, W., Jo, D., Sterckx, S., 2015a. The capabilities approach and environmental sustainability: the case for functioning constraints. Environ. Val. 24 (3), 367—389.

Peeters, W., Jo, D., Sterckx, S., 2015b. Towards an integration of the ecological space paradigm and the capabilities approach. J. Agric. Environ. Ethics 28 (3), 479—496.

Potschin, M.B., Haines-Young, R.H., 2011. Ecosystem services: exploring a geographical perspective. Prog. Phys. Geogr. 35 (5), 575—594.

Rao, N.D., Jihoon, M., 2018. Decent living standards: material prerequisites for human wellbeing. Soc. Indicat. Res. 138 (1), 225—244.

Rao, N.D., Paul, B., 2012. "Decent living" emissions: a conceptual framework. Sustainability 4 (4), 656—681.

Raworth, K., 2017. Doughnut Economics: Seven Ways to Think Like a 21st-Century Economist. Chelsea Green Publishing, Vermont, US.

Reinert, K.A., 2018. No Small Hope: Towards the Universal Provision of Basic Goods. Oxford University Press.

Remans, R., Wood, S.A., Saha, N., Lee Anderman, T., Ruth, S., DeFries, 2014. Measuring nutritional diversity of national food supplies. Glob. Food Secur. 3 (3—4), 174—182.

Ritchie, H., Roser, M., 2017. 'Obesity'. Our World in Data. https://ourworldindata.org/obesity.

Rockström, J., Steffen, W., Kevin, N., Persson, Å., Stuart Chapin III, F., Lambin, E., Lenton, T.M., Scheffer, M., Folke, C., Joachim Schellnhuber, H., 2009. Planetary boundaries: exploring the safe operating space for humanity. Ecol. Soc. 14 (2).

Roser, M., Ritchie, H., 2013. 'Food Supply'. Our World in Data. https://ourworldindata.org/food-supply.

Rundh, B., 2009. Packaging design: creating competitive advantage with product packaging. Br. Food J. 111 (9), 988—1002.

Schaffartzik, A., 2019a. International Metabolic Inequalities Mediated by the Distribution of Stocks. In: Presented at the 13th Conference of the International Society for Industrial Ecology (ISIE) - Socio-Economic Metabolism Section. 13th-15th May 2019, Berlin.

Schaffartzik, A., 2019b. Personal Correspondence.

Schandl, H., Fischer-Kowalski, M., West, J., Giljum, S., Dittrich, M., Eisenmenger, N., Geschke, A., Lieber, M., Wieland, H., Schaffartzik, A., 2018. Global material flows and resource productivity: forty years of evidence. J. Ind. Ecol. 22 (4), 827—838. https://doi.org/10.1111/jiec.12626.

Sen, A., 1985. Well-being, agency and freedom: the dewey lectures 1984. J. Philos. 82 (4), 169—221.

Sen, A., 1994. Well-being, capability and public policy. G. Degli Econ. Ann. Econ. 333—347.

Sen, A., 1999. Commodities and Capabilities. In: OUP Catalogue.

Shue, H., 1993. Subsistence emissions and luxury emissions. Law Pol. 15 (1), 39—60.

Smith, L.M., Case, J.L., Harwell, L.C., Smith, H.M., Summers, J.K., 2013. Development of relative importance values as contribution weights for evaluating human wellbeing: an ecosystem services example. Hum. Ecol. 41 (4), 631—641.

Steffen, W., Richardson, K., Rockström, J., Cornell, S.E., Fetzer, I., Bennett, E.M., Biggs, R., Carpenter, S.R., de Vries, W., Cynthia, A., de Wit, 2015. Planetary boundaries: guiding human development on a changing planet. Science 347 (6223), 1259855.

Stern, N., 1989. The economics of development: a survey. Econ. J. 99 (397), 597–685.

Tzafalias, M., 2016. Fake Lifejackets Play a Role in Drowning of Refugees. In: Bulletin of the World Health Organization, 2016. https://doi.org/10.2471/BLT.16.020616.

Valero, A., Valero, A., 2014. Thanatia: The Destiny of the Earth's Mineral Resources: A Thermodynamic Cradle-to-Cradle Assessment. World Scientific Publishing, Singapore.

Van Rompay, Thomas, J.L., Peter, W., De Vries, Bontekoe, F., Tanja-Dijkstra, K., 2012. Embodied product perception: effects of verticality cues in advertising and packaging design on consumer impressions and price expectations. Psychol. Market. 29 (12), 919–928.

Walker, G., Simcock, N., Day, R., 2016. Necessary energy uses and a minimum standard of living in the United Kingdom: energy justice or escalating expectations? Energy Res. Soc. Sci. 18, 129–138.

Weisz, H., Suh, S., Graedel, T.E., 2015. Industrial ecology: the role of manufactured capital in sustainability. Proc. Natl. Acad. Sci. U. S. A. 112 (20), 6260–6264.

Whiting, K., Konstantakos, L., Carrasco, A., Carmona, L.G., 2018a. Sustainable development, wellbeing and material consumption: a stoic perspective. Sustainability 10 (2), 474.

Whiting, K., Konstantakos, L., Misiaszek, G., Simpson, E., Carmona, L., 2018b. Education for the sustainable global citizen: what can we learn from stoic philosophy and Freirean environmental pedagogies? Educ. Sci. 8 (4), 204.

Whiting, K., Luis Gabriel Carmona, L.I., Brand-Correa, Simpson, E., 2020. Illumination as a material service: a comparison between ancient Rome and early 19th century London. Ecol. Econ. 169C, 106502.

Wood, N., Roelich, K., 2019. Tensions, capabilities, and justice in climate change mitigation of fossil fuels. Energy Res. Soc. Sci. 52, 114–122.

Assessment of mechanisms and instruments of climate finance: a global perspective

15

Vaishali Kapoor[1], Medha Malviya[2]

[1]*Deen Dayal Upadhyaya College, University of Delhi, New Delhi, Delhi, India;* [2]*Independent Researcher in Environmental Economics and Economics of Climate Change*

Chapter outline

Environmental Sustainability and Economy. https://doi.org/10.1016/B978-0-12-822188-4.00004-X

1. Introduction

Climate change refers to a change in the earth's atmosphere through anthropogenic activities. It is garnering the attention of a wide range of individuals and institutions, including business houses, NGOs, governments, financing institutions, multilateral institutions, and whom not, when the brunt is even felt at the individual level, especially by the poor individuals in the developing countries. The nature of this problem is global, and thus the response to it cannot be local. Therefore, to reduce its adverse impacts, all nations have stood in unity and are helping each other to mitigate and adapt to the impacts of climate change. As it hits the very survival and livelihood of the people in the developing nations, it is imperative to resolve it at the earliest by giving them greater access to Climate Finance since it is a matter of justice.

Upon understanding climate change's impact, the United Nations Framework Convention on Climate Change (UNFCCC) was formed in 1992, and the buzzword that emerged after its deliberations was "Climate Finance." Climate finance initially referred to the flow of public funds from developed nations to developing nations. However, later the need for private funds was recognized, and thus the concept of climate finance was extended to include mobilized private finance. The word "mobilized" indicates the role of public intervention in increasing private flows toward adaptation and mitigation activities.

This chapter is divided into five sections. The first section provides an overview of why climate finance is needed and how it came into being. The second section discusses the mechanism, i.e., the creation and management of funds and role of various organizations formed under the Convention. The third section deals with various instruments and channels of funds. The fourth section revolves around the debate whether developing countries must be given greater access to climate finance or not. The last section is devoted to analyzing the challenges ahead and the scope of research in climate finance.

2. From climate change to climate finance: an overview

The impact of climate change has been felt by and large by everyone in the world economy. The measurement of such costs has also attracted policymakers, academicians, researchers, and others. After measuring its impacts, the solution lies in mitigating them and creating projects that help reduce its impacts. Thus, there arose the need for financing the projects with climate change attributes. For explaining climate finance, a bit of background is required, and for the same this section is divided into three subsections. The first one raises concerns about climate change, and the second one helps us quantify its impacts in economic terms. The third section amalgamates the two, and describes the emergence and definition of climate finance.

2.1 Concerns about climate change

It is known that post-industrial revolution temperature levels have been increasing beyond sustainable levels (IPCC, 2014; Stern, 2008), and the warming of the climate system has been at an unprecedented scale over the past three decades such that the period of 1983–2012 has been the warmest of last 1400 years (IPCC, 2014). The climate sensitivity has been projected to be in the range of 1.5–5.3 degrees and at 5 degrees higher than preindustrial era (around the 1850s) "most of the world's snow will disappear … [and] this would eventually lead to sea-level rises of 10 meters or more" (Stern, 2008). Such high levels of temperature rises will hit the very survival of living beings. The Paris Agreement signed by 196 countries—which is about 96% of emissions—is a milestone in fighting climate change, as, without this pledge, temperatures could have been 3–4 degrees higher than preindustrial levels (Farid et al., 2016), which are now pledged to keep low to 1.5–2 degrees.

The impact of climate change is wide-ranging and threatens all nations globally but has a greater degree of negative impacts on developing nations (Farid et al., 2016). The adverse impacts of climate change are witnessed in developing countries and entail huge cost on the economies. Let us first understand the issue, of the adverse impacts of climate change. These include biological risks like extinction of species, ocean acidification's negative impact on marine organisms, risks to human health and physical risks like melting snow and ice, which is altering hydrological systems and causing sea-level rise, floods, wildfires, water scarcity owing to reduced fresh surface water and underground water levels (Labbat and White, 2007; IPCC, 2014). At the ongoing rate of climatic change, with no adaptive measure being taken, the next threat to human survival emerges from risks to food security (Abeygunawardena et al., 2009; IPCC, 2014), as agricultural production is projected to fall with rising temperatures and gains from excess rainfall would not be able to compensate for the loss (Labbat and White, 2007).

2.2 Economics of climate change

Estimation for GDP loss due to climate change ranges between 1% (Stern, 2008) and 2% (Tol, 2002). If required steps are not taken, it might trigger abrupt and irreversible changes that will continue for centuries even if the global mean temperature is stabilized (Stern, 2008; IPCC, 2014). Regarding agricultural production loss, this fall is even more peculiar to developing nations as they are already in hotter zones (Mendelsohn, 2008). Then there are threats of natural disasters like cyclones and storm surges (Stern, 2008; IPCC, 2014), which had cost in the range of 0.1%–0.9% of GDP per annum during 1990–2018. It also fell heavily on the low- income developing countries, and the number of deaths due to disasters has also been higher each year in developing nations (IMF, 2019; Abeygunawardena et al., 2009; Eboli et al., 2010). Such disasters hamper economic growth and exacerbate poverty (Abeygunawardena et al., 2009; Farid et al., 2016). If costs of uncertainty are also taken into consideration, then the estimates of even "worst case" scenarios will fall short (Burke, 2015).

The second aspect is related to the role of developed countries. The industrial revolution brought about an excessive rise in CO_2 levels, and countries that had benefited are developed nations today, and affected countries are developing countries because of worsened climatic condition spillovers (Stern, 2008, Abeygunawardena et al., 2009). With the proximity to the equator; sea levels rising threats to small island countries and have limited readiness to leverage adaptation, therefore on social justice grounds and following their common but differentiated responsibilities and respective capabilities, developed countries are now expected to help developing countries financially in mitigation and adaptation activities (Farid et al., 2016).

Eboli et al. (2010) estimates that developed nations like Russia, Europe, and Japan will tend to moderately gaining in terms of real potential GDP in the range of 1.2%–2.4% by the end of the century, while on the other hand, countries of East Asia will be down by as much as 12.6% and 10.3% for the Middle East and North Africa. This difference is attributed to labour productivity—impacted by human health and hot and humid conditions—which is the more significant cause of the projected global damage hinging greatly on East Asia, the Middle East, and North Africa. The benefits for developed nations are mainly due to a pattern of tourist flows skewed toward developed nations at higher latitudes (Farid et al., 2016; Eboli et al., 2010).

Climate change poses potential risks to macro-financial stability (Farid et al., 2016). The insurance sector would have to consider the risks and uncertainties of climate change while modelling premiums and computing risks. Microinsurance faces additional challenges arising from catering to poorer households (Abeygunawardena et al., 2009). The investment decisions will be altered, as cost-benefit equations would change and also, the carbon footprint of a company may alter company's assets' valuations (Farid et al., 2016).

Excessive carbon dioxide has been added to the atmosphere over the years since private production costs and services do not include the social costs of climatic changes. It is known as "externality," and to internalizing carbon should be priced like other goods and added to the cost of goods and services like any other input's price, known as "Carbon pricing" (Stern, 2008), which acts as a disincentive to producing carbon and incentive to shift to low carbon technologies and infrastructure (Steckel et al., 2017).

2.3 Origin and definition of climate finance

Though carbon pricing could be a feasible solution, it would be an added burden on the poor/low-income households in developing nations (Farid et al., 2016). Thus, under the United Nations Framework Convention on Climate Change's (UNFCCC) obligations that developed countries pledged to help developing countries with an annual commitment of $100 billion by 2020 (UNFCCC, 1992). It was not before COP 17 at Durban that the word "Climate finance" was used formally in the UNFCCC climate finance decisions booklet, 2020. The Copenhagen accord was indicative of only the public flow of funds from North to South and when "developed countries pledged to provide new and additional resources approaching $30 billion for the period 2010−2012 with balanced allocation between mitigation and adaptation" to developing countries when they referred to "Climate Finance" (UNFCCC, 1992). It is not only the public funds that could suffice; the knowledge and change had to be brought to even the private sector. Thus, it was realized that private investment would have to make a shift from "brown" to "green" projects to contain rise in temperatures to 2 degrees above preindustrial levels (Falconer and Stadelmann, 2014; Kato et al., 2014). The authors recognized it well before COP 21 in 2015 (Paris Agreement). Now, the role of the government will be to mobilize private climate finance while helping private sectors in the maintenance of return-risk in green projects comparable to other private projects.

Climate finance is defined as "finance that aims to reduce emissions, and enhance sinks of greenhouse gases and aims to reduce the vulnerability of, and maintain and increase the resilience of, human and ecological systems to negative climate change impacts" by the UNFCCC Standing Committee on Finance. The Climate Policy Initiative (CPI) defines *climate finance* as "financial resources required to cover the cost of transition to a low carbon global economy and to adapt to, or build resilience against, current and future climate change impacts." One greatest challenge is making the definition of "Climate Finance" comprehensive, is that there is no globally accepted single definition; it is alternatively defined as total resources devoted toward climate change or flow of funds from developed countries to developing countries to tackle climate change (IPCC, 2014). It is imperative to decide which activities are included and labelled as "Climate Finance," which flows will be treated *additional* and counted toward $100 billion. There are also issues regarding tracking private investment, instruments, channels, and actors (Falconer and Stadelmann, 2014; Clapp et al., 2012). There is a lack of clarity whether guarantees or insurances will be included in $100 billion. All these are the challenges faced by the COP at the meetings and while submitting biennial reports on climate finance. The Standing Finance Committee (SFC) report clarifies that COP 25 is unclear about the operational definition. It states that the committee "invites Parties to submit via the submission portal, 6 by 30 April 2020, their views on the operational definitions of climate finance for consideration by the Standing Committee on Finance in order to enhance its technical work on this matter in the context of preparing its 2020 Biennial Assessment and Overview of Climate Finance Flows" (UNFCCC Climate Finance Decisions Booklet, 2020).

In addition to the challenges already discussed, there are issues about how international community relates "additional" flows to $100 billion if this includes both private and public funds or not, how is "mobilized finance" counted and tracked, and how transformational change is defined in this context (Brown et al., 2010; Clapp, 2012; Mersmann et al., 2014). The answers to all of these questions pave the way for an operational definition.

3. Organizations and their roles

Climate change refers to change in the earth's atmosphere through direct or indirect anthropogenic activities. As this is one of the biggest challenges faced by the human race in recent times, there was a need for an international organization to help channel the efforts of all countries together to fight climate change. In this light, the United Nations Framework for Climate Change (UNFCCC) was established in 1992 to help countries set up national targets for limiting or reducing global warming. Financial resources specifically aimed at tackling climate change are known as "Climate Finance"; although there is no single definition, they can be broadly classified as sources of funds required to support projects toward mitigation or adaptation to climate-related risks. The UNFCCC defines *climate finance* as "local, national or transnational financing—drawn from public, private and alternative sources of financing—that seeks to support mitigation and adaptation actions that will address climate change." The long-term effects of climate change will be seen in the future, but efforts must begin right now. Climate finance will require a stable flow of funds to face the challenges of climate mitigation and adaptation projects. Different sources and ways of financing need to be identified to increase the funds over the decades.

The organizations which play an essential role in coordinating the climate finance flows from both public and private finance sources are: (a) United Nations Framework for Climate Change Convention (UNFCCC); (b) the United Nations Environment Program (UNEP); (c) the Organization of Economic Cooperation and Development (OECD), and (d) the G20.

3.1 United Nations Framework for Climate Change Convention (UNFCCC)

The UNFCCC is also known as "the Convention," helps the member countries in their efforts to combat climate change (UNFCCC, n.d.). It is headquartered in Bonn, Germany, with a membership of 197 countries known as Parties to the Convention or Conference of the Parties (COP). The main goal of the convention is to achieve the target of limiting the rise of global average temperature to 1.5°C above preindustrial levels.

3.1.1 Standing Committee on Finance

The Standing Committee on Finance (SCF) helps the COP in carrying out the activities related to the Financial Mechanism of the Convention to manage climate financing to the developing countries through coordination and exchange of information (UNFCCC, 2019). Further, it uses expert ideas and views for periodic reviews of the Financial Mechanism, making recommendations for the smooth operation of the entities. Other functions include preparing a biennial assessment (BA) of the financial flows from the developed countries to the developing countries (UNFCCC, 2020a).

3.2 International treaties and agreements under UNFCCC

3.2.1 The Kyoto Protocol (1997)

The Kyoto Protocol was adopted on 11 December, 1997 but was enforced on 16 February, 2005. A total of 192 Parties entered into the first legally binding agreement, which was aimed to tackle the adverse impacts of climate change. By adopting individual national policies and measures, each country had agreed to limit or reduce greenhouse gas (GHG) emissions (UNFCCC, 2020d).

Under the principle of "common but differentiated responsibility and respective capabilities," the developed countries had pledged to reduce their annual GHG emissions up to an average of 5% emission reduction compared to 1990 levels (UNFCCC, 2020d) over the five years of the first commitment period (2008−12) (European Commision, n.d.).

The protocol focuses on the trading of emission permits for countries to reach the GHG level targets. Even though it promotes the meeting of the primary targets by the countries through national measures, it allows for market-based mechanisms, which are the following:

3.2.1.1 International emissions trading

Under Article 17 of Kyoto Protocol, the Annex II Parties (Table 15.1) have been allowed emission units or "assigned amount units" (AAUs), wherein the excess or the unused units can be traded. As carbon dioxide or CO_2 is the principal greenhouse gas, it is traded as a commodity on the "carbon market." Similarly, removal units (RMUs) are traded based on land use, land-use change, and forestry (LULUCF) such as reforestation (UNFCCC, 2020b).

3.2.1.2 Joint implementation (JI)

As per Article 6 of Kyoto Protocol, a developed country (Annex I Party) can undertake a project in developing countries for emission reduction and earn emission reduction units (ERUs) which are equivalent to one tonne of CO_2 each (UNFCCC, 2020c). It contributes to the Kyoto target and is a flexible and cost-efficient mechanism of fulfilling individual Parties' commitments. JI can be implemented in two ways:

- A host nation meeting the eligibility requirement uses its approaches to establish project baselines, verify emission reductions, and allocate ERUs.
- The JI supervisory committee (an official international body) sets international procedures for baselines, verification, and other procedures (Mullins, 2002).

3.2.1.3 Clean Development Mechanism (CDM)

The Clean Development Mechanism (CDM) is one of three "flexibility mechanisms" identified in the Kyoto Protocol (the other two being the Joint Implementation (JI) and International Emissions Trading), which are used to achieve the global reduction targets along with promoting sustainable development in developing countries.

The industrialized nations can trade certified emission reduction (CER) credits, each equivalent to one tonne of CO_2, to meet a part of their emission reduction targets through projects in developing nations (UNFCCC, n.d.). These projects must demonstrate the "additionality" criteria proving that it will result in a more significant long-term reduction in emissions (Earth Journalism Network, 2016).

Table 15.1 Parties to Kyoto Protocol.

Annex I	Annex II	Nonannex parties
Industrialized countries who were members of the OECD (Organization for Economic Cooperation and Development) in 1992, plus countries with economies in transition (the EIT), including the Russian Federation, the Baltic States, and several central and eastern European States.	Members of the OECD in Annex I, excluding EIT countries. These countries were obligated to support the developing countries through provision of financial resources in reducing carbon emissions to meet their climate mitigation and adaptation targets. Other measures include promoting development through transfer of environment friendly technologies through the Global Environment Facility (GEF), Green Climate Fund (GCF), the Adaptation Fund (Special Climate Change Fund & Least Developed Countries Fund).	All the developing countries are included in this category are not subject to any binding commitments due to their limited capacity to deal with climate change impacts. Other groups include the Alliance of Small Island States (AOSIS), Small Island Developing States (SIDS) and the Least Developing Countries (LDCs).

Modified from Source: UNFCCC; Green Facts, n.d. Parties & Observers of the UNFCCC. https://www.greenfacts.org/en/climate-change-ar4/toolboxes/6.htm

3.2.2 Copenhagen Accord

The Copenhagen Accord (2009) acknowledged limiting the rise in global temperature to below 2°C, increasing the frequency of reporting on the national emission inventories, and limiting emissions every 2 years by the developing countries. The developed countries committed $30 billion for the next 3 years to reduce deforestation, deploy clean technologies, and adapt to climate change in developing countries and a commitment of $100 billion toward the GCF (Union of Concerned Scientists, 2009).

3.2.3 The Paris Agreement

The Paris Agreement (2015) builds upon the Convention to keep the global temperature rise below 2°C above preindustrial levels. The developed nations committed to submit indicative information on future support every 2 years, including projected levels of public finance (UNECE, 2016). They had agreed to mobilize $100 billion per year, to help developing nations' needs and priorities for combating climate change. Beyond 2025, they have intended to continue this existing collective mobilization goal.[1]

---

[1]United Nations Framework on Climate Change Convention (UNFCCC) (2020). Climate Finance in the negotiations.https://unfccc.int/topics/climate-finance/the-big-picture/climate-finance-in-the-negotiations.

3.3 United Nations Environment Program (UNEP)

The main focus of UNEP is to support the private sector financial institutions like banks, investors, and insurers in understanding and identifying the commercial opportunities from climate action (UNEP, 2020). The developing countries have easy access to these funds (directly and through accredited entities) from the Green Climate Fund (GCF) through the *UN Readiness Program*, the Global Environment Facility (GEF), and the Adaptation Fund (AF) as well as through other bilateral or multilateral public sources.

UNEP launched the Finance Initiative (UNEP FI), a global partnership between the UN environment and the financial sector for mobilizing private sector finance for sustainable development based on three frameworks:

- *Principles for Responsible Banking* (*PRB*) launched on 22 September, 2019, aligning banking businesses to contribute to individual and society's goals mentioned in sustainable development goals and the Paris Climate Agreement.
- *Principles for Sustainable Insurance* (*PSI*), established in 2012 by UNEP FI, include environmental, social, and governance (ESG) issues relevant to insurance business.
- *Principles for Responsible Investment* (*PRI*), established in 2006 by UNEP FI and the UN Global Compact, includes ESG issues in investment decision-making.

3.4 Organization of Economic Cooperation and Development (OECD)

The Organization for Economic Cooperation and Development (OECD) is a group of 37 developed countries that addresses challenges like improving economic performance and job creation to foster strong education and fighting international tax evasion.[2]

In 2017, an OECD analysis showed that climate finance reached $71.2 billion, up from $58.6 billion in 2016. This result was broadly consistent with estimated public climate finance levels from developed nations at $66.8 billion in 2020, excluding export credits (Fig. 15.1, OECD, 2019). However, the report concluded that even more efforts were required to scale up public finance and improve the effectiveness of private finance channels.

3.5 G20

The G20, established in 1999, comprises 20 countries such as Argentina, Australia, Brazil, Canada, China, France, Germany, India, Indonesia, Italy, Japan, Mexico, Russia, Saudi Arabia, South Africa, South Korea, Turkey, United Kingdom, United States, and the European Union. All these countries are responsible for over 80% of current greenhouse gas emissions while representing 85% of global GDP and 51% of the population. The world's 99% of historical CO_2 emissions (excluding land-use) from 1850 to 2013 come from them, but the respective responsibility to reduce emissions vary based on their levels of cumulative historical and current emissions and obligations under UNFCCC.

Increased subsidies fossil fuels consumption have posed a significant level of threat in climate change impacts. The G20 has now been pushing for phasing out of these subsidies, focusing on sustainable investments and decreasing climate risks. They must promote innovative public finance

---

[2]Organization of Economic Cooperation and Development (OECD). About the OECD. https://www.oecd.org/about/.

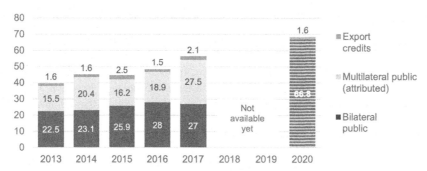

FIGURE 15.1

Public finance 2013–17 estimates and 2020 projection ($billion).

sources to support climate action in developing countries and generate additional financial resources. Even though the estimated price of adaptation could be as high as $500 billion at a 2°C scenario, the developed countries have committed to mobilizing $100 billion, by 2020, wherein only 16% of climate finance will adapt. The G20 countries need more concrete ambitions to limit global temperatures below 2°C and stick to their national pledges for their fair share of contributions to emissions. Focus on building climate resilience, promoting and respecting gender equality, and human rights in all climate actions should be the main criteria (CARE, 2017).

3.6 Various financial mechanisms under the Convention

3.6.1 Global Environment Facility (GEF)

GEF was established in 1992 to provide support to developing countries, including EIT's for meeting their international obligations. It administers several trust funds—the Least Developed Countries Fund (LDCF), the Special Climate Change Fund (SCCF), Capacity-building Initiative for Transparency (CBIT), Nagoya Protocol Implementation Fund (NPIF), and the Adaptation Fund. These trust funds are replenished every four years from 40 donor developed countries to adapt and transfer technology through voluntary pledges. The funds are administered through the GEF Trust Fund, administered by the World Bank along with other implementing agencies, the United Nations Development Program (UNDP) and the United Nations Environment Program (UNEP).

3.6.2 Green Climate Fund (GCF)

GCF is the largest climate fund to support developing countries for climate mitigation and adaptation projects. It was set up in 2010 by UNFCCC and focuses on promoting paradigm shift to low-emission and climate-resilient development. Countries and groups which are highly vulnerable to the effects of climate change, like the Least Developed Countries (LDCs), Small Island Developing States (SIDS), and African States (GCF, n.d.) are supported through this jointly mobilized financial resource mechanism from the developed nations. A direct access modality is also established to help these vulnerable groups easily access climate change funding and integrate it with their national climate action plans.

The National Designated Authority (NDA) acts as the interface between their government and GCF and must approve all GCF project activities within the country. The aim is to have a balanced portfolio (50:50) by supporting paradigm shifts in both mitigation and adaptation investments over time. GCF launched its initial resource mobilization in 2014 and gathered pledges worth $10.3 billion and additional funds of $9.8 billion as GCF's first replenishment (2020−23). Through its Private Sector Facility (PSF) into its framework, it engages the private sector to offer a wide range of financial products including grants, concessional loans, subordinated debt, equity, and guarantees.

3.6.3 Adaptation Fund

The Adaptation Fund, set up in 2001, helps finance projects and programmes to help vulnerable communities in developing countries adapt to climate change. The funding is primarily sourced from the government and private donors and of a 2% share of proceeds of Certified Emission Reductions (CERs) (Adaptation Fund, 2019). The Adaptation Fund Board (AFB) manages the fund, which consists of 16 members and 16 alternates and meets two times a year. The World Bank is the interim trustee of the Fund and helps sell the CER certificates, and overlooks the management of the Adaptation Fund trust fund. Other fund initiatives include enhancing of the Readiness Program, including its South-South mentoring channel, mandatory compliance for implementing agencies of the fund's environmental, social safeguards and gender policy, promoting microfinance schemes, supporting local industries and farmers, and voluntarily tracking mobilization of climate finance (UNFCCC, 2018b).

It has been estimated that as of 30 June 2019, there have been cumulative receipts of $887.1 million into the Adaptation Fund Trust Fund, comprising of $ 201.4 million from the monetization of CER's, $657.9 million from additional contributions and $27.8 million from investment income earned on the Trust Fund balance (UNFCCC, 2019).

3.6.4 National Climate Funds

UNFCCC mentions the National Climate Fund (NCF) as a "country-driven mechanism that supports the collection, blending, coordination of, and accounting for climate finance at the national level." It directs from public, private, and multilateral and bilateral sources toward climate change projects that align and promote the national objectives. Therefore, every structure of the national climate fund depends on the stated national objectives, areas of the climate change problem, scope, funding sources, governing bodies, a trustee, and implementing agents (Meirovich et al., 2013).

3.6.5 Climate Investment Fund (CIF)

The CIF, established in 2008, is an $8 billion fund aimed at helping the developing and middle-income countries to combat climate change by accelerating investments in clean energy, climate resilience, and sustainable forests. These resources are held in a trust and managed by the World Bank and disbursed as grants, highly concessional loans, and risk mitigation instruments through multilateral development banks (MDBs) (CIF, 2018).

CIF has the following funding programs:

- Clean Technology Fund (CTF)

The fund comprises $5.4 billion to provide resources to the developing countries for access to low carbon technologies, out of which over $4 billion is used toward renewable energy. The role of the private sector in mobilizing funding is acknowledged, through the Dedicated Private Sector Programs (DPSP), launched in 2013, which have provided "risk-appropriate capital to finance high-impact, large-scale private sector projects in clean technology, such as geothermal power, mini-grids, energy efficiency, and solar PV."

- Strategic Climate Fund (SCF)

The SCF helps more vulnerable countries adapt their development programs to climate change impacts, including prevention of deforestation. Other initiatives include encouraging discussions between donors and recipient countries about climate-related investment and funding support from a range of "bilateral donors, private sector and civil society contributors."

- Pilot program for climate resilience

The $1.2 billion programs allow vulnerable developing countries access to concessional and grant funding by integrating climate resilience into strategic development planning across sectors and stakeholder groups.

- Scaling-up renewable energy program

The $720 million program helps the world's poorest nations access funds to implement renewable energy like solar, geothermal, and biomass. It also helps align the country's priorities, and demonstrates the economic, social, and environmental viability of renewable energy like the renewable energy mini-grid systems for isolated and off-grid communities in 14 countries with over $200 million investment.

- Forest investment program

The aim of the program is to help the developing countries access financial resources at low interest rates and other grants for prevention of deforestation and forest degradation.

- Private sector

As the importance of the private sector is being recognized in mobilizing climate finance, the Private Sector Set-Asides (PSSAs) allocate concessional financing to help engage the private sector in sustainable forestry (FIP), climate resilience (PPCR), and energy access through renewable energy in low income countries (SREP).

3.7 Other supporting organizations

3.7.1 United Nations Development Program (UNDP)

UNDP helps poverty eradication in developing countries through development programs and policies, including building institutional capabilities and resilience for achieving sustainable development. The Global Environmental Finance (UNDP-GEF) Unit covers areas in: (a) sustainable management of ecosystem goods and services; (b) scaling-up of climate change adaptation and mitigation; (c) sustainable, affordable, and accessible energy services; (d) sustainable management of chemicals and waste; and (e) improved water and ocean governance. The developing countries can achieve their development goals along with catalyzing environmental finance for sustainable development.

3.7.2 World Bank (WBG)

The main aim of WBG is to eradicate extreme poverty and foster growth of income. Climate change is a significant risk to the development targets of developing countries; the WBG's Climate Change Action Plan (2016−20) helps countries move toward more sustainable energy systems, strengthen their resilience, capacity building, and achieve sustainable development goals. The focus has been increasing on clean energy systems and infrastructure to face climate shocks in the future. The Nationally Determined Contributions, or NDCs, are the nationally determined pledges for climate actions defined in the Paris Agreement in December 2015. To support these NDCs, the WBG's Climate Change Action Plan will help these developing nations deliver on their targets. It has already helped by adding 30 GW of renewable energy, put in place early warning systems for 100 million people, and developed climate-smart agriculture investment plans for at least 40 countries; increased direct adaptation climate finance to reach $50 billion over FY 21−25 (World Bank, 2020). WBG also helps in the provision of improved weather data and forecasting, access to drought-resistant crops, climate-resilient roads, disaster insurance, and cyclone-resistant houses.

4. Instruments

Lack of national and international climate finance constraints is the constraining factor in the effective scale of investment in adaptation and mitigation (IPCC, 2014). Thus, it is crucial to understand how funds can be mobilized. To get funds and deploy them in appropriate projects, governments need financial instruments that they might use with private players, institutions, and other countries' governments. Thus, financial instruments are divided into broadly two groups, viz. instruments to mobilize funds and instruments to deploy them (WWF, 2018). The green bonds, carbon performance bonds, and credit default swaps are examples of the former, while equity capital, debt capital, and risk management belong to the latter category. Apart from these, instruments are required to enhance credit, provide insurance, and support the revenue of the private players. We shall discuss all of these in this section, along with their suitability, advantages, and disadvantages.

4.1 Instruments for raising funds

The instruments have to be used to raise funds for climate finance projects. Following instruments have been used:

4.1.1 Green bonds

The word "Green" is indicative of *Green Finance,* of which climate finance is a subset. Green finance is a broader term which covers climate change projects and other projects to achieve sustainability and environmental goals. The climate change projects aim solely at GHG emissions and climate mitigation aspects. The green bonds are like standard bonds in terms of fixed (sometimes, market linked) income stream for the holder or buyer, risks borne, risk return profile, credit ratings, and also, the seller promises to pay back after, usually longer, maturity. Green bonds are different from regular bonds only in respect of usage of funds; funds from green bonds are to be invested in projects that aim to address climate change or other environmental concerns. The scaling-up for "Green bonds" is high while its reliability is low (Torvanger et al., 2016). The green bond market started in 2007 with the issuance by the World Bank has grown since 2014, with issuance amounting to $37 billion in 2014 to $257.5 billion in 2019 (CBI, 2019). The issuers include large corporations like Vasakronan, SNCF, ICBC, Apple; government or government agency-led nations like France, Australia, United States, Mexico, etc.; and municipalities and city-level agencies like Massachusetts, City of Johannesburg, etc. (WWF, 2018; CBI, 2017). The GCLF report 2019 reported that "corporations and municipal governments spent an average of $3 billion annually on projects outside renewable energy, using the proceeds from green bond issuances." There are as many as 22 stock exchanges worldwide facilitating green bond issuance until June 2020 (CBI, 2019).

They are deemed to be fit for financing climate change projects, as these projects require high upfront capital that can be paid back after a longer horizon (WWF, 2018), and a study on 121 European Bonds for the period 2013−17 found that green bonds are more convenient than their nongreen counterparts (Gianfrate and Peri, 2019). CBI (2017) report indicated that green bonds are heavily oversubscribed, bond prices are tight, and on average, perform better than even standard bonds by the same issuer (Partridge and Medda, 2020). The studies have also found that there is a premium in the issuance of green bonds issued by municipal agencies (Baker et al., 2018) and a green premium termed as "greenium" of 5 basis points in 2018 is also found in the secondary markets of green bonds (Partridge and Medda, 2020).

The only thing is tracking climate finance through issuance could be misleading, as usually the proceeds flow to refinancing or in other words, to already existing projects that may create the problem of double-counting (Buchner et al., 2019). There are challenges in making it successful. These bond markets should be made accessible to smaller projects to benefit from green premiums, reducing their cost of capital; the government should create and endorse green bond labels aligned with long-term decarbonization pathways (Shishlov et al., 2016). Also, governments need to look at the regulatory framework, legitimacy, and transparency of bonds, and take necessary actions pertaining to governance so as to avoid greenwashing (Park, 2018). All the advantages mentioned above make the green bond a potent instrument for raising funds for climate finance (Gerard and Wilson, 2009; Gianfrate and Peri, 2019).

4.1.2 Climate policy performance bonds

The climate policy performance bonds, as the name suggests, are linked to the performance in terms of carbon emissions reduction as per the commitment or else increase in production of renewable sources of energy. The government issuing it promises to pay less than the market rate of interest if commitments are met, and if the government is unable to meet its commitment, then interest paid is greater than the market rate. It is effective as governments face pressure to achieve commitments; as not achieving them is costlier now, and the additional burden is equivalent to the difference between the two interest rates announced. Thus, CPPBs help translating mere promises into action (Michaelowa et al., 2016).

The issues with green bonds have been overcome with the introduction of climate policy performance bonds. These require less transparency, less accountability, and there is lesser risk of green-washing, as these have been linked to observable outcomes (CO_2 levels and production of renewable sources of energy). While all other institutions benefit from mitigation and decarbonized economy, there is an additional advantage for insurance companies as they can use CPPBs to hedge funds (Michaelowa et al., 2016).

4.1.3 Debt for climate swaps

Upon witnessing Japan's environmental policies that led Japanese private companies to contribute to the destruction of Southeast Asia's forests, in 1984, WWF developed a way to use the debt of least developed countries to finance conservation activities known as debt for nature swaps (Asiedu-Akrofi, 1991). The investor—bilateral or multilateral donors, private investors, or NGOs—may write off the debt of the highly indebted country in return for ecological bonds or local currency and raise funds in environmental projects in the recipient country (WWF, 2018). The investors do it at a profit margin.

Similarly, debts for climate finance swaps have been devised with the only difference that funds are used for adaptation and mitigation projects. It is especially advantageous for the countries with huge debts, small islands, and least developed countries since these entail no extra financial cost to the recipient countries (Meirovich et al., 2013; Warland and Michaelowa, 2015). Though debt for climate swaps may seem a viable option even for poverty reduction in the least developed countries, there are specific issues with debt for climate swaps. Firstly, it is highly likely that the lower the governance quality in a nation, the greater will be debt swaps and vice-versa. Secondly, it may crowd out local funds for activities with climate change attributes. Thirdly, this might create the problem of moral hazard that they might incur debts recklessly, as they might find swap a cheaper and easier way to get rid of future debts (Michaelowa et al., 2016) and, the scaling-up effect for debt for climate swaps is predicted to not so high (Torvanger et al., 2016). The commonwealth proposes that donors should write off 100% of small state's debt stock held at various multilateral institutions contingent on the agreement by a debtor to make annual payments to a trust fund equal to existing debt in local currency, while at the same time multilateral institutions ensure that donor participates in proportion to climate finance pledges (Mitchell, 2015).

4.1.4 Debt for Climate Policy performance swaps

The idea of "Debt for Climate Policy performance swaps" was pioneered by Michaelowa et al. (2016) in their report, where they combined the two instruments, viz., "Debt for Climate swaps" and "climate policy performance bonds." Each has its advantages. The former assures strict governance, and the latter relieves the least developed countries of their debt burden. Thus, combining the two instruments yields yet greater positive outcomes in the direction of climate finance. The donor here could be the country that holds government bonds of the recipient country. It could work in two ways: the donor country reduces interest rate on the debt held or reduces the total debt burden contingent on the fact that the least developed country achieves its commitment of GHG mitigation target and otherwise not. A third party independent of both of countries which entered into the swap could audit the performance (Michaelowa et al., 2016).

4.1.5 Structured funds

The structured funds are tranches of different projects pooled into one, tailored and customized to suit the appetite of risk and return of specific investors. Regarding climate finance, pooling helps diversification and usually, public donors pick tranche with greatest risk (Lindenberg, 2014). The structured funds are better suited than green bonds for the smaller individual projects and smaller institutional investors. According to Lindenberg (2014), structured funds have the following advantages:

An expert from a development bank that set up such funds goes even further, claiming that structured funds can solve for most problems that hinder more private engagement. The knowledge gap is met with the public-private partnership structure of such a fund, and country risk is reduced by diversification, currency risk by hedging, transparency risk by the regulation under a Luxembourg regime, and the viability risk is reduced by demonstrated profitability.

Besides all advantages mentioned, there are certain problems that this tailored instrument cannot be replicated elsewhere, so it entails substantial transaction costs. The issuer is generally commercial banks and public donors take high risk funds without any control over the decisions, creating risk of principal-agent. Since structured funds may crowdin funds from local partner banks, leverage ratio is estimated to be high (Lindenberg, 2014).

4.2 Instruments to deploy investments

Following are the set of instruments that can be used to deploy raised funds into projects:

4.2.1 Equity

Government invests as owners in the climate mitigation and adaptation projects, which serves the twin purpose of providing an initial capital base and signalling other investors and creditors about the profitability of the projects. The flows through this channel were estimated at $0.9 billion in 2013 and remained stable around this amount for 2013–2017 (OECD, 2019). There is ambiguity about its leveraging effect. On the one hand, a study claims that the leverage ratio effect can be highest and estimated to be in the range of 1.7–33 (Torvanger et al., 2016). On the contrary to this, it is believed that if financial markets are well-developed, only then can public funds exist; else leveraging effect is low for equity (Lindenberg, 2014). The equity can also help reduce a project's debt to equity ratio, like in clean technology financing by Japan's Green Fund (WWF, 2018). Due to high transaction costs in equity capital markets, it is only applied to finance bigger and fewer projects (Lindenberg, 2014).

4.2.2 Concessional/nonconcessional loan

Concessional loans for climate financing provided by the donors are usually lower than market rates or have extended repayment schedules sometimes spread over three to four decades. It is found that "reducing the cost of debt by 1.8 percentage points and increasing the tenor by eight years will decrease the renewable cost source of energy by approximately 12.7%" compared to projects with no concessions (Shrimali et al., 2014). The loans can be non-concessional too, that are usually provided by the private sector (Meirovich et al., 2013). Loans, like equity, signal the viability of projects and help leverage funds. Nevertheless, since loans are also provided for longer term, loans may have low leverage ratio (Lindenberg, 2014; Torvanger, 2016). The scaling-up and reliability are estimated to be high (Torvanger, 2016). The loans may be preferred over equity by the private sector, as this does not transfer ownership, has reduced transaction costs, and at the same time, it incentivizes the private sector to make profits due to repayment obligation.

As per the OECD (2019) report, amount of loans (both concessional and non-concessional) doubled, from $20.0 billion in 2013 to $40.3 billion in 2017. A majority, i.e., 60% of funds flowing from national DFIs, are flowing through this instrument, amounting to $53.5 billion annually, and its share is not expected to decline in times to come (Meirovich et al., 2013).

The problem is that out of 67 low-income countries, 31 of them are in debt distress or high chance of being, so using loans for financing climate projects could hasten it and, it is likely that cost of servicing the debt will rise in future, which will burden the already vulnerable developing nations (Hirsch et al., 2019). There are challenges in providing concessional loans, like determining the degree of concession so that a balance between climate financing support and optimization of public money is maintained and ensuring that choice of projects is not distortionary or selective (Lindenberg, 2014).

Type of loans

Senior debt: Senior debt provides debt facilities in form of "soft" loans because it provides loans for longer than usually financial markets do and provide range of debt amortization and repayment schedules to customize debt service costs.

Subordinated debt: It covers intermediate funding between senior debt and equity-like *mezzanine or quasi-equity* finance. Mezzanine financing is a hybrid made of debt and equity components that gives the right to the lender to convert it to equity in case of default.

Revolving funds and refinancing schemes: This creates a mechanism to make loans and paid once the project matures and generates income. In case of the project fails, the loan may be partially or fully forgiven.

4.2.3 Credit lines

Credit lines emerged as a solution to the problems associated with loans. In order to save search costs, donors provide debt for on-lending to the local bank, which has a better understanding of local players and local business environment and is given the flexibility to choose interest rates and charges that will apply to the customer (Lindenberg, 2014). However, there are chances of principal-agent problems and divergence in donor and local bank accounting standards which might increase costs for the local bank if it has to comply with the donor's standards.

4.2.4 Grants

Grants are non-repayable funds flowing to mitigation and adaptation projects, and they can be cash or kind. Grants may be given for specific activities and may address different costs like the project's research costs, and to activate a project. Grants are better because they do not further the debt stress in developing nations like loans and provide assistance to developing nations (Hirsch et al., 2019). It is the second-largest source among all flows attributed toward climate finance and has increased by 25% between 2013 and 2017, going from $10.3 billion to $12.8 billion (OECD, 2017). Grants have been estimated to increase the leverage ratio of other projects, have high scaling-up and reliability (Torvanger et al., 2016). Grants have specific features, such as grants are generally provided for nonrevenue generating activities in recipient countries such as knowledge management programs and capacity building. (Meirovich et al., 2013), grants usually flow to public projects (Torvanger et al., 2016), and a greater majority of funds are flowing for adaptation activities (Buchner et al., 2019).

Grant is easily applicable too all instruments but does not incentivize project developers to deliver, and it may have subjectivity in selecting projects (Lindenberg, 2014). However, grants can be made effective to incentivize project owners. The grants may be given based on achieving specific outcomes and making a project viable without grants is commercially nonviable (WWF, 2018).

4.2.5 PPP

Through public-private partnerships models, governments aim at bringing private capital into the projects while taking the contingent liability should the project fail. This instrument can be used effectively as its upper cap for leverage ratio is estimated to be 33 with a high scaling-up (Torvanger et al., 2016). Nevertheless, this also implies an off-the-balance sheet debt liability for the governments and could later translate into debt stress (Hirsch et al., 2019).

4.3 Credit enhancement and risk management instruments

The objective of credit enhancement instruments is to create an attractive environment altering risk/reward profile better that suits different types of investors and ensures smooth flow to projects. Governments can reduce risks by providing guarantees and insurance or by increasing rewards like by providing subsidies.

4.3.1 Guarantees

A guarantee is an effective tool to mitigate the actual or perceived risk that might prevent or obstruct the flow of finance to the project. Guarantors assure in writing to pay a part of the cost involved or fund borrowed in the project if nonperformance or losses accrued. Guarantors may commit to any of these:

- to repay the lender in case the borrower is unable to, in full (credit guarantee) or in part (partial credit guarantee).
- compensate the lender if the project does not deliver as expected or technology underperformance or costs overruns (performance risk guarantee).
- guarantee certain cash flows to the project (revenue guarantee).
- pay in case importers of the shipped goods cannot pay back to the exporters (export credit guarantee).
- guard against the risk of fall of supportive treatment in the host country (regulatory guarantee), says the loss of a supportive tax credit.

The partial credit guarantees can effectively save 10.8% of the delivered cost and A-rated bond; if backed by such guarantee, then its rating may jump up to AAA (Shrimali, 2014). Since guarantees are project-specific, this can help mobilize private financing but at a higher transaction cost (WWF, 2018; Lindenberg, 2014). Guarantees with an estimated leverage ratio between 6 and 10, medium reliability, and scaling-up seems limited due to higher transaction costs (Torvanger et al., 2016; Lindenberg, 2014). Brazil, Colombia, Mexico, and Chile have been the leaders in implementing and designing guarantees (Meirovich et al., 2013).

4.3.2 Insurance

There are two ways to insure projects from the risk of weather-related catastrophes, viz. risk prevention and risk transfer. Risk prevention includes building dams and applying technology for predicting events to reduce vulnerability while risk transfer implies that risk is transferred through capital markets. The former suits events with high frequency and lower level of damage costs; the latter being suitable for events with lower probability but high damage costs in disaster outbreaks (Meirovich et al., 2013). The leverage ratio and scaling-up are estimated to be high while reliability low (Torvanger et al., 2016). The Global Index Insurance Facility (GIIF) was established to "mitigate weather and catastrophic risks in African, Caribbeans and Pacific countries" that guarantees beneficiaries rapid compensation payments once a natural disaster occurs and a related predetermined index hits a predetermined specific value (Schwank et al., 2010).

4.3.3 Interest rate subsidy

The government provides an interest rate subsidy to lower borrowing costs and thus make a project more profitable. The advantage of such an instrument is that it helps in extending climate finance without disturbing competition. An example from India pertaining to solar home systems UNEP and the Shell Foundation paid $900,000 as a subsidy which further mobilized private finance to the tune of $6.7 million (UNEP, 2015), implying a leverage ratio of 7.5, which is very well in the estimated range of 5–12 by Torvanger et al. (2016). Programs like these have been proven to be scalable (UNEP, 2015; Torvanger, 2016).

4.3.4 Securitization

Securitization is a technique that mixes various financial assets like senior debt, subordinated debt, equity, and mezzanine debt, into investment products, and the issuer gets rid of his debt by selling tranches of it to the third-party investor. Various tranches are made out of pooled financial assets to fit varied investors' risk appetites. Usually, senior debt, which has the least risk, is picked up by private investors and mezzanine and junior debt by international investors and public donors. The leverage ratio is estimated to be low at 2.3, while scaling-up can be high (Torvanger et al., 2016).

4.3.5 Technical assistance

Technical assistance is provided to the governments, green projects, and financial intermediaries who partner with these projects. It takes the form of market research, business planning, staff training, setting up technical standards, knowledge dissemination about green infrastructure projects and their benefits. The partner commercial banks are assisted in mitigating risks associated with green financing projects, which are undertaken by some MDBs such as the IADB and the EBRD (Aravamutham et al., 2015). Technical assistance makes banks capable of tackling risks involved and creating additional financing in future. Technical assistance entails huge transaction costs for it is custom-designed, but at the same time is expected to have a high leverage ratio (Lindenberg, 2014).

4.4 Revenue support policy

As the name suggests, the revenue support policy, can ensure minimum revenue from the production or rewards or penalizes for meeting targets. Following are ways how governments can use these:

4.4.1 Feed-in tariff and feed-in premium

Feed-in tariff guarantees the producers of renewable energy sources that they can charge a fixed, predetermined price per unit for some time. It eliminates the uncertainty about market prices and, therefore, about profits. A study of electricity production in Kenya by Nganga et al. (2013) shows how the Kenya government used renewable energy feed-in tariff in 2008 to stimulate electricity production to meet excess demand problems in years preceding 2008. By 2013, the private sector had invested more than $2.8 billion in Kenya's renewable energy industry. The estimated leverage ratio is 5 with high scaling-up (Toravnger et al., 2016). A feed-in premium is like a feed-in tariff in all regards other than defining the guaranteed price; here, the guaranteed price contains a premium over and above the electricity price per unit electricity produced over a period of time (which is equivalent to a higher price in feed-in tariff).

4.4.2 Tradable green certificates

The tradable green certificates as an instrument are expected to promote the production of electricity from green sources. The government imposes a minimum production target, which, if not exactly met, producers, can buy or sell shares in the secondary market. The price of tradables is market-determined. Unlike feed-in tariffs and feed-in premiums, green certificates do not incentivize the use of a specific technology (Torvanger et al., 2016). There is merit in using tradable green certificates since it helps in promoting renewable electricity technologies. Though tradeoffs between keeping customer prices low and bringing about a technical change exists, it has been observed that producers, despite having used tradable green certificates, extracted large rents in Sweden during 2003−2008, contributing marginally to technical change (Bergek and Jacobsson, 2010). By the end of 2015, developing nations like India, Vietnam, Ghana, and Nepal had introduced green certificates in one form or another (REN21, 2015, quoted from Torvanger et al., 2016).

4.4.3 Tendering process

The public authorities announce a quantity of renewable energy produced or some green technology to be built for which private players apply tenders. A competition sets the price, and financial aid is also extended to selected projects. Many developing nations like Indonesia, Morocco, Kenya, the Philippines, and South Africa rely on tendering processes (Torvanger et al., 2016). A recent study by Bento et al. (2020) has demonstrated that tendering has the strongest effect in promoting net renewable capacity compared to feed-in tariffs. The countries that use tendering as an instrument have a higher addition of net capacity of renewables in the order of 1000−2000 MW annually.

5. Flows from developed countries to developing countries

Climate finance forms an important basis for the north-south flows or the flow of funds from developed countries to least developed countries. However, certain issues like funding gaps are prevalent in these flows, along with more focus on adaptation projects leading to a mitigation bias. These issues are discussed in this section.

5.1 The north-south flows and idea of equity

Simms et al. (1999) outlined the concept of "carbon debt," which should form a basis of climate agreements. Projections for sustainable carbon use per person state that the United States currently uses 12 times, and the United Kingdom uses nearly 6 times the allowable amount. However, vulnerable developing countries such as Bangladesh and Tanzania can increase their emission levels up to 10 times and 22 times, respectively. Historically, exploitation and use of natural resources by a few industrialized countries have led to a debate on economic justice, where the use of allowable amounts is supposed to be decided based on of fairness and responsibility. The concept of "contraction and convergence" outlines that developed countries want to use the allowable carbon amount based on proportion to the size of their economies' GDP, but developing countries prefer the size of the population.

As mentioned by Morgan and Waskow (2014), the principle of equity has been at the core of the global climate debates. The principle of "common but differentiated responsibilities and respective capabilities" has been reflected in all discussions and agreements where the transition of developing nations to low-carbon and climate-resilient are to be supported by the developed nations by allowing access and transfer to green technology, policy innovation, and climate finance.

Bond (2010) outlines the climate risks faced by Africa, such as rapid desertification, increasing floods and droughts, worse water shortages, more starvation, increase in climate refugees, and the spread of malaria and other diseases. The author outlines the concept of "ecological debt," a debt accumulated by developed nations toward the developing nations which consist of nutrients in exports such as virtual water, the oil and minerals no longer available, the destruction of biodiversity, sulfur dioxide emitted by copper smelters, the mine tailings, the adverse impact on health from flower exports, water pollution by mining, the commercial use of information and knowledge on genetic resources or biopiracy and agricultural genetic resources. An amount of $1.8 trillion was partially calculated in concrete damages over the years, including factors such as GHG emissions, ozone layer depletion, deforestation, overfishing, and destruction of mangrove swamps for commercial use. Some of the damages which were left out as it was not easy to estimate were excessive freshwater withdrawals, destruction of coral reefs, biodiversity loss, invasive species, and war. Civil society groups have supported the case for developing nations whom the developed nations have denied a fair share of atmospheric space by insisting that these countries are in a "climate debt," which is the sum of "emissions debt." They have further suggested that the "adaptation debt," which is the rise in costs and damage in developing nations, should also be considered during the climate negotiations.

A rise in the financial costs of loans for combating climate risks would make it difficult for developing nations' to tackle climate-related disasters, leading to frequent increase in violent conflicts and migration (ACTAlliance, 2018). Hirsch et al. (2019) state that the developing nations' inability to face climate risks leads to a higher economic inequality and leads to a higher capital costs making it difficult to access loans. (Diffenbaugh & Burke (2019) showed that an increase in economic inequality had slowed economic growth, particularly around the equator where developing nations exist suggesting, that the gap in terms of per capita income between the richest and the poorest countries is 25 percentage points—larger than it would have been without climate change (adjusted for other factors like population growth). There is no doubt in the future that there will be a need to increase support for these nations, and it will require international financial assistance to minimize loss and damage, which is estimated to increase from at least $50 billion in 2022 to $300 billion in 2030 (Bread for the World).

5.2 Funding gap

Leveraging private finance is important to fill in the gap due to inadequate direct government funding in developing nations (Polycarp and Venugopal, 2013). Gianfrate and Peri (2019) have estimated that financial flows which are "consistent with a pathway toward low greenhouse gas emissions and climate-resilient development" will amount to about $2.4 trillion between 2016 and 2035. Increasing focus on the development of the green economy and thereby economic growth and progress (Mundaca et al., 2016) had led to the creation of the Green Climate Fund (GCF) in 2010 to raise $100 billion per year for climate-friendly investment in developing countries by the year 2020 (UNFCCC, 2019). However, only $546 billion has been donated (Climate Policy Initiative, 2019). The economic impact from climate finance can be assessed for donor and recipient countries, but that will vary due to various factors (Román et al., 2018). The authors concluded that through a high-value addition to the development of climate-related industries, there couldbe a benefit from larger shares of the economic impact of climate finance, leading to a further boost to the development and innovation of technologies.

Other examples suggested for covering the funding gap include Foreign Direct Investment (FDI) or through green bonds (Bowen et al., 2015). Green bonds are like conventional bonds issued by corporates, municipalities, and other governmental entities but the proceeds are directed toward climate change mitigation and adaptation projects. Gianfrate and Peri (2019) mention that when a bond is labelled as "green," there is an advantage to the investors regarding lower interest rates thereby reducing debt financing. The rise in their importance and issuance is estimated at $250 billion in 2018 by Moody's and the Climate Bond Initiative, which is expected to reach $1 trillion by 2021.

5.3 Mitigation bias in climate finance

Climate change adaptation refers to actions to reduce vulnerability (IPCC, 2007), whereas mitigation and intervention to reduce the sources or enhance the sinks of GHG gases, meaning both adaptation and mitigation, are complementary and overlapping (IPCC, 2014). The ratio of mitigation to adaptation was found to be 95:5 (Buchner et al., 2011) as the majority of the climate finance is usually aimed at mitigation rather than adaptation projects due to the provision of exclusive benefit for recipient countries, but also generating a public good enjoyed even by the donor nations (Abadie, 2013). More priority given to mitigation than adaptation finance has led to a "Mitigation Bias" in climate financing, which has also been acknowledged in Paris COP 21 Agreement, calling for increased funding for adaptation creating a balance between mitigation and adaptation projects in the GCF. A study, by Abadie (2013), outlines the reasons for a bias towards mitigation. It is because private sources of funds prefer better risk-returns and mitigation has global benefits leading to a preference in investing in mitigation projects, whereas adaptation has greater social benefits and it leads to creation of a local public good. This bias is also due to presence of low-cost mitigation options in developing nations and lack of adaptation finance credit mechanism compared to emission reduction credits in mitigation markets. The author also mentions that investing in mitigation will help save future adaptation costs, and uncertainty regarding climate risks can also discourage private agencies from investing in adaptation projects. Heavy investment in mitigation could lead to a negative impact on the economic growth in developing nations due to "rise in price volatility of resources, a real appreciation of currency as there will demand non-tradables as income from resources rise, and rent-seeking behavior is induced through higher resource rents" (Jakob et al., 2015).

6. Challenges ahead
6.1 Challenge 1: defining and tracking climate finance

The first and foremost challenge is defining the climate finance activities and flows and exploring an operational definition. It is of utmost importance to have a single definition among the international community for clarity (Falconer and Stadelmann, 2014), tracking and biennial assessments submitted by identified (donor) developed nations. It will also help overcome the problem of double-counting in climate finance. In the event of a lack of guidance from UNFCCC on definition, tracking may discourage nations and institutions to improving in this regard. In the of a definition of "mobilised" climate finance, reporting, intervention, and choice of instruments may then become a political decision (Ellis and Caruso, 2013). The 2018 Biennial Assessment and Overview of Climate Finance Flows published by (UNFCCC, 2018a) state that there are data uncertainties (difference in methods applied, and variation in energy efficiency computations) and data gap (lack of clarity between public and private funds, and instruments used).

Since the 2015 Paris Agreement the climate finance includes the mobilized funds too, but it brings us to answer the following: (a) whether an intervention is mobilized or not?, (b) assessing if there is a direct link between intervention and mobilized climate finance, (c) tracking whether financing is public or private, (d) quantifying the amount of financing, and (e) when in the financing chain mobilization is reported (Ellis and Caruso, 2013).

The problem of double counting arises in a situation when funds raised through bonds are deployed to already existing projects, and a total amount raised through green bonds is used to indicate climate finance flows. Alternatively, it could be that the total cost of the green project is attributed to the climate finance flows. Instead, it should be the incremental costs, i.e., additional capital requirement or increased operating costs for mitigation or adaptation when compared to a reference project (Falconer and Stadelmann, 2014).

6.2 Challenge 2: leveraging private capital

It has become imperative that private capital is mobilized in climate change projects. For this to be done successfully, governments and international institutions have to create an environment for leveraging private capital, the choice of instruments have to be made so that they have higher leverage ratios, and at the same time, a successful tool that can be implemented in a nation given its stage of development of financial markets. The public support has to be such that it improves the risk-return calculus for private investors in the low-carbon market (Venugopal and Srivastava, 2012). For this, there has to be the use of innovative instruments. In this regard, there is an emerging concept of "blended finance."

6.3 Challenge 3: blended finance: a potent tool

Blended finance is blending two things: climate funds and private climate. For example, a recent study by Strand (2019) shows that there exists a possibility for blending climate finance and carbon market mechanisms. It is achievable when climate finance is used for subsidizing carbon markets. So, developed countries pay the price, say "p," for carbon credits purchased from least developed countries (LDCs), but LDCs receive a higher price, say "p + s," where s is the subsidy coming from the climate finance funds.

GEF has been using blended finance as their tool for climate financing since its inception. It follows three steps: identify the need for new approaches to finance environmental projects; then the GEF council incubates ideas with different stakeholders and then invests; GEF council has witnessed a leverage ratio as high as 10 in such cases (GEF, 2019). During 2013–2024, GEF provided $1.4 billion through "regular" channels and mobilized $800 million, which is 60 cents mobilized for each dollar spent. In the same year, GEF provided $175 million for blended finance operations and mobilized 6.3 times from the private sector, i.e., US$1.1 billion, making blended finance a potential tool.

6.4 Challenge 4: other challenges

The other challenges of climate finance flows are as follows:

- Projecting mobilized private finance to fulfill the commitments, which requires us to know beforehand the amount of projected public finance, how much of it will be mobilizing private capital, and the leverage ratio if it mobilizes private capital.
- The challenge of enhancing the readiness of developing nations remains. It includes building the capacity of developing countries to plan for, access, deliver and monitor and report on climate finance.
- To effectively monitor the ways deploying funds in developing countries that do not put them in debt distress as it does when "debt" is used as an instrument for financing.
- Better classification of projects into mitigation, adaptation, or both so that there does not appear to be a bias, which in reality may not exist.
- Climate finance can be prone to corruption which needs attention, which requires effective governance.
- Equity or social justice should prevail at all times in climate financing and should not burden the already vulnerable.

References

Abadie, L., Galarraga, I., Rübbelke, D., 2013. An analysis of the causes of the mitigation bias in international climate finance. Mitig. Adapt. Strategies Glob. Change 18 (7), 943–955.

Abeygunawardena, P., et al., 2009. Poverty and Climate Change: Reducing the Vulnerability of the Poor through Adaptation. World Bank Group, Washington, D.C. http://documents.worldbank.org/curated/en/534871468155709473/Poverty-and-climate-change-reducing-the-vulnerability-of-the-poor-through-adaptation.

ACT Alliance, 2018. Submission on type and nature of actions to address loss and damage for which finance may be required. ACT Alliance. https://unfccc.int/files/adaptation/workstreams/loss_and_damage/application/pdf/act_alliance.pdf.

Adaptation Fund, 2019. About the Adaptation Fund. https://www.adaptation-fund.org/about/.

Advances in Blended Finance: GEF's Solutions to Protect the Global Environment, 2019. Global Environment Facility. https://www.thegef.org/sites/default/files/publications/gef_advances_blended_finance_201911_0.pdf.

Aravamuthan, M., Ruete, M., Dominguez, C., May 2015. Credit Enhancement for Green Projects: Promoting Credit-Enhanced Financing from Multilateral Development Banks for Green Infrastructure Financing. International Institute for Sustainable Development. https://www.iisd.org/sites/default/files/publications/credit-enhancement-green-projects.pdf.

Asiedu-Akrofi, D., 1991. Debt-for-nature swaps: extending the frontiers of innovative financing in support of the global environment. Int. Lawyer 25 (3), 557–586.

Baker, M., Bergstresser, D., Serafeim, G., Wurgler, J., 2018. Financing the response to climate change: the pricing and ownership of U.S. green bonds. In: IDEAS Working Paper Series from RePEc.

Bento, Nuno, et al., 2020. Market-pull policies to promote renewable energy: A quantitative assessment of tendering implementation. Journal of Cleaner Production 248.

Bergek, A., Jacobsson, S., 2010. Are tradable green certificates a cost-efficient policy driving technical change or a rent-generating machine? Lessons from Sweden 2003–2008. Energy Pol. 38 (3), 1255–1271.

Bond, P., 2010. Climate debt owed to Africa: What to demand and how to collect? Links International Journal of Socialist Renewal. http://links.org.au/node/1675.

Bowen, A., et al., 2015. The 'optimal and equitable' climate finance gap. Centre for Climate Change Economics and Policy, Working Paper No. 209. Grantham Research Institute on Climate Change and Environment. https://www.cccep.ac.uk/wp-content/uploads/2015/10/Working-Paper-184-Bowen-et-al.pdf.

Brown, J., Bird, N., Brown, J., 2010. Climate Finance Additionality: Emerging Definitions and Their Implications. Climate Finance Policy Brief No.2. Overseas Development Institute.

Buchner, B, et al., 2011. The Landscape of Climate Finance. Climate Policy Initiative.

Buchner, B., Clark, A., Falconer, A., Macquarie, R., Meattle, C., Tolentino, R., Wetherbee, C., 2019. Global landscape of climate finance 2019. Climate Policy Initiative. https://climatepolicyinitiative.org/wp-content/uploads/2019/11/2019-Global-Landscape-of-Climate-Finance.pdf.

Burke, M., Dykema, J., Lobell, D., Miguel, E., Satyanath, S., 2015. Incorporating climate uncertainty into estimates of climate change impacts. Rev. Econ. Stat. 97 (2), 461.

CARE, 2017. G20 and Climate Change: Time to Lead for a Safer Future. https://careclimatechange.org/wp-content/uploads/2017/06/G20-REPORT-.pdf.

Clapp, C., Ellis, J., Benn, J., Coffee-Morlot, J., 2012. Tracking Climate Finance.

Climate Bonds Initiative, 2017. Bonds and Climate Change: The State of the Market 2017. https://www.climatebonds.net/resources/reports/bonds-and-climatechange-state-market-2017.

Climate Bonds Initiative, 2019. Green Bonds: The State of the Market 2018. https://www.climatebonds.net/resources/reports/green-bonds-state-market-2018.

Climate Investment Fund (CIF), 2018. Donors & MDBS. https://www.climateinvestmentfunds.org/finances.

Diffenbaugh, N.S., Burke, M., 2019. Global Warming Has Increased Global Economic Inequality. Proceedings of the National Academy of Sciences of the United States of America 116 (20), 9808–9813. https://www.pnas.org/content/pnas/116/20/9808.full.pdf.

Earth Journalism Network, June 09, 2016. Clean Development Mechanism. https://earthjournalism.net/resources/clean-development-mechanism.

Eboli, F., Parrado, R., Roson, R., 2010. Climate-change feedback on economic growth: explorations with a dynamic general equilibrium model. Environ. Dev. Econ. 15 (5), 515–533.

Ellis, J., Caruso, C., 2013. Comparing Definitions and Methods to Estimate Mobilised Climate Finance. Climate Change Expert Group Paper No. 2013(2). OECD. https://www.oecd-ilibrary.org/docserver/5k44wj0s6fq2-en.pdf?expires=1594968343&id=id&accname=guest&checksum=852A13440BB00D74C9A2F13411E6977E.

European Commission, n. d. Kyoto 1st Commitment Period (200–12). https://ec.europa.eu/clima/policies/strategies/progress/kyoto_1_en.

Falconer, A., Stadelmann, M., 2014. What is climate finance? Definitions to improve tracking and scale up climate finance. Climate Policy Initiative. https://climatepolicyinitiative.org/wp-content/uploads/2014/09/Climate-Finance-Brief-Definitions-to-Improve-Tracking-and-Scale-Up.pdf.

Farid, M., Keen, M., Papaioannou, M., Parry, I., Pattillo, C., Ter-Martirosyan, A., Other IMF Staff., 2016. After Paris: Fiscal, Macroeconomic, and Financial Implications of Climate Change. SDN/16/01. https://www.imf.org/external/pubs/ft/sdn/2016/sdn1601.pdf.

Gerard, D., Wilson, E., 2009. Environmental bonds and the challenge of long-term carbon sequestration. J. Environ. Manag. 90 (2), 1097–1105.

Gianfrate, G., Peri, M., 2019. The green advantage: exploring the convenience of issuing green bonds. J. Clean. Prod. 219, 127–135.

Green Climate Fund (GCF), n.d. https://www.greenclimate.fund/about.

Green Facts, n.d. Parties & Observers of the UNFCCC. https://www.greenfacts.org/en/climate-change-ar4/toolboxes/6.htm.

Hirsch, T., et al., 2019. Climate finance for addressing loss and damage: how to mobilize support for developing countries to tackle loss and damage. In: Cedillo, E., et al. (Eds.), Brot für die Welt.

IMF, 2019. Macroeconomic Developments and Prospects in Low-Income Developing Countries—2019. Policy Paper No. 19/039. https://www.imf.org/en/Publications/Policy-Papers/Issues/2019/12/11/Macroeconomic-Developments-and-Prospects-in-Low-Income-Developing-Countries-2019-48872.

IPCC, 2014. In: Core Writing Team, Pachauri, R.K., Meyer, L.A. (Eds.), Climate Change 2014: Synthesis Report. Contribution of Working Groups I, II and III to the Fifth Assessment Report of the Intergovernmental Panel on Climate Change. IPCC, Geneva, Switzerland, p. 151.

IPCC, 2007. Fourth Assessment Report. https://www.ipcc.ch/assessment-report/ar4/.

Jakob, M., Steckel, J., Flachsland, C., Baumstark, L., 2015. Climate finance for developing country mitigation: blessing or curse? Clim. Dev. 7 (1), 1—15.

Kato, T., Ellis, J., Clapp, C., 2014. The Role of the 2015 Agreement in Mobilising Climate Finance. OECD/IEA Climate Change Expert Group Papers. No. 2014/07. OECD Publishing, Paris.

Labatt, S., White, R.R., 2007. Carbon Finance the Financial Implications of Climate Change (Wiley Finance Series). John Wiley & Sons, Hoboken, N.J.

Lindenberg, N., 2014. Public instruments to leverage private capital for green investments in developing countries. In: German Development Institute Discussion Paper 4/2014.

Meirovich, H., Peters, S., Rios, A., 2013. Financial Instruments and Mechanisms for Climate Change Programs in Latin America and the Caribbean. Inter-American Development Bank Climate Change and Sustainability Division. No. IDB-PB-212. https://publications.iadb.org/publications/english/document/Financial-Instruments-and-Mechanisms-for-Climate-Change-Programs-in-Latin-America-and-the-Caribbean-A-Guide-for-Ministries-of-Finance.pdf.

Mendelsohn, R., 2008. The impact of climate change on agriculture in developing countries. J. Nat. Resour. Pol. Res. 1 (1), 5—19.

Mersmann, F., Wehnert, T., Göpel, M., Arens, S., Ujj, O., 2014. Shifting Paradigms Unpacking Transformation for Climate Action A Guidebook for Climate Finance & Development Practitioners. Wuppertal Institute for Climate, Environment and Energy GmbH.

Michaelowa, A., Bouzidi, A., Friedmann, V., 2016. Boosting climate action through innovative debt instruments. Perspect. Climate Res. https://www.perspectives.cc/fileadmin/Publications/Boosting_climate_action_through_innovative_debt_instruments_Michaelowa_Axel__Friedmann_Valentin_22016.pdf.

Mitchelli, T., 2015. Debt swaps for climate change adaptation and mitigation: a commonwealth proposal. In: Commonwealth Secretariat Discussion Paper No. 19.

Morgan, J., Waskow, D., 2014. A new look at climate equity in the UNFCCC. Equity, Sustainable Development and Climate Policy 14 (1), 17—22.

Mullins, F., 2002. Joint Implementation Institutions: Implementing JI at the National Level. OECD. https://www.oecd.org/czech/2766355.pdf.

Mundaca, L., Markandya, A., 2016. Assessing regional progress towards a 'Green Energy Economy'. Applied Energy 179, 1372—1394. https://www.researchgate.net/profile/Luis-Mundaca/publication/284130989_Assessing_regional_progress_towards_a_%27Green_Energy_Economy%27/links/5de6ca7d92851c83645fc495/Assessing-regional-progress-towards-a-Green-Energy-Economy.pdf.

Nganga, J., Wohlert, M., Woods, M., Becker-Birk, C., Rickerson, W., 2013. Powering Africa through Feed-In-Tariffs. Renewables Energy Ventures (K) Ltd. https://www.sipua-consulting.com/app/download/9923416995/PoweringAfricathroughFeedinTariffs.pdf?t=1410365773.

OECD, 2019. Climate Finance Provided and Mobilised by Developed Countries in 2013—17. OECD Publishing, Paris. https://www.oecd-ilibrary.org/docserver/39faf4a7-en.pdf?expires=1595035624&id=id&accname=guest&checksum=C94B00D2A95FF4DACA012D9965C39147.

Park, S., 2018. Investors as regulators: green bonds and the governance challenges of the sustainable finance revolution. Stanford J. Int. Law 54 (1). University of Connecticut School of Business Research Paper No. 18−12.

Partridge, C., Medda, F., 2020. The evolution of pricing performance of green municipal bonds. J. Sustain. Finance Invest. 10 (1), 44−64.

Polycarp, C., Venugopal, S., 2013. 3 Ways to Unlock Climate Finance. World Resources Institute. https://www.wri.org/insights/3-ways-unlock-climate-finance.

Román, M.V., et al., 2018. Why do some economies benefit more from climate finance than others? A case study on North-to-South financial flows. Economic Systems Research 30 (1), 37−60. https://www.tandfonline.com/doi/abs/10.1080/09535314.2017.1334629.

Schwank, O., Steinemann, M., Bhojwani, H., Holthaus, E., Norton, M., Osgood, D., Sharoff, J., Bresch, D., Spiegel, A., 2010. Insurance as an Adaptation Under the UNFCCC: Background Paper. Swiss Federal Office for the Environment.

Shishlov, I., Morel, R., Cochran, I., June 2016. Beyond Transparency: Unlocking the Full Potential of Green Bonds. Institute for Climate Economics. https://www.cbd.int/financial/greenbonds/i4ce-greenbond2016.pdf.

Shrimali, G., Konda, C., Srinivasan, S., 2014. Solving India's Renewable Energy Financing Challenge: Instruments to Provide Low-Cost, Long-Term Debt. A CPI-ISB Series.

Simms, et al., 1999. Who owes who? Climate change, debt, equity and survival. London: Christian Aid. http://www.ecologicaldebt.org/Who-owesWho/Who-owes-who-Climate-change-dept-equity-and-survival.html.

Steckel, J., Jakob, M., Flachsland, C., Kornek, U., Lessmann, K., Edenhofer, O., 2017. From climate finance toward sustainable development finance. Wiley Interdiscip. Rev. Climate Change 8 (1).

Stern, N., 2008. The economics of climate change. Am. Econ. Rev. 98 (2), 1−37.

Strand, J., 2019. Climate finance, carbon market mechanisms and finance "blending" as instruments to support NDC achievement under the Paris Agreement. In: Policy Research Working Paper 8914. World Bank Group 2019.

The World Bank, 2020. Climate Change. https://www.worldbank.org/en/topic/climatechange/overview#1.

Tol, R., 2002. Estimates of the damage costs of climate change. Part 1: benchmark estimates. Environ. Resour. Econ. 21 (1), 47−73.

Torvanger, A., Narbel, P., Pillay, K., Clapp, C., 2016. Instruments to Incentivize Private Climate Finance for Developing Countries. CICERO. https://pub.cicero.oslo.no/cicero-xmlui/bitstream/handle/11250/2420794/CICERO%20Report%202016%2008%20web%20rettet.pdf?sequence=8&isAllowed=y.

UNECE, 2016. Summary of the Paris Agreement. https://www.unece.org/fileadmin/DAM/stats/documents/ece/ces/2016/mtg/Session_1_Bigger_picture_of_COP21.pdf.

UNEP, 2020. Climate Finance. https://www.unenvironment.org/explore-topics/climate-change/what-we-do/climate-finance.

UNEP, 2015. Indian Solar Loan Programme Overview. United Nations Environment Programme. https://siteresources.worldbank.org/EXTRENENERGYTK/Resources/5138246-1238175210723/India0consumer0credit0program0for0PV.pdf.

UNFCCC, 1992. United Nations Framework Convention on Climate Change. New York: United Nations, General Assembly, 1992. https://unfccc.int/resource/docs/convkp/conveng.pdf.

UNFCCC, 2020a. Climate Finance in the Negotiations. https://unfccc.int/topics/climate-finance/the-big-pcture/climate-finance-in-the-negotiations.

UNFCCC, 2020b. International Emissions Trading. https://unfccc.int/international-emissions-trading.

UNFCCC, 2020c. Joint Implementation. https://ji.unfccc.int/index.html.

UNFCCC, 2020d. UNFCCC. Process-And-Meetings. In: https://unfccc.int/process-and-meetings#:2cf7f3b8-5c04-4d8a-95e2-f91ee4e4e85d.

UNFCCC, March 4, 2019. Climate Finance Decision Booklet 2020. https://unfccc.int/sites/default/files/resource/2019%20CF%20Decisons%20Booklet.pdf.

UNFCCC, n.d. United Nations Framwork Convention on Climate Change. UNFCCC. https://unfccc.int/.

UNFCCC Standing Committee on Finance, 2018a. Biennial Assessment and Overview of Climate Finance Flows Technical Report. https://www.mainstreamingclimate.org/publication/unfccc-2018-biennial-assessment-and-overview-of-climate-finance-flows/.

UNFCCC Standing Committee on Finance, 2018b. In: Report of the Conference of the Parties Serving as the Meeting of the Parties to the Kyoto Protocol on its Thirteenth Session, Held in Bonn from 6 to 18 November 2017. https://www.informea.org/sites/default/files/decisions/FCCC_CMP_2017_7_Add.1_1.pdf.

Union of Concerned Scientists, December 23, 2009. The Copenhagen Accord: Not Everything We Wanted, but Something to Build on. https://www.ucsusa.org/resources/copenhagen-accord.

Venugopal, S., Srivastava, A., 2012. Moving the Fulcrum: A Primer on Public Climate Financing Instruments Used to Leverage Private Capital. WRI Working Paper. World Resources Institute, Washington DC. https://pdf.wri.org/moving_the_fulcrum.pdf.

Warland, L., Michaelo, A., 2015. Can debt for climate swaps be a promising climate finance instrument? Lessons from the past and recommendations for the future. Perspect. Climate Change. https://www.perspectives.cc/fileadmin/Publications/Can_debt_for_climate_swaps_be_a_promising_climate_finance_instrument_Warland_Linde__Michaelowa_Axel_2015.pdf.

World Wildlife Fund, 2018. Financial Instruments Used by Governments for Climate Change Mitigation. https://wwfafrica.awsassets.panda.org/downloads/wwf_2018_financial_instruments_used_by_governments_for_climate_change_mitigation.pdf.

Further reading

Black to Green Consulting, June 23, 2017. Climate Finance: Who Are the Main Players? https://blacktogreen.com/2017/06/climate-finance-main-players/.

Global Environment Facility (GEF), 2020. About Us. https://www.thegef.org/about-us.

OECD, n.d. About the OECD. https://www.oecd.org/about/.

UNDP, 2020. Global Environmental Finance. https://www.undp.org/content/undp/en/home/2030-agenda-for-sustainable-development/planet/environment-and-natural-capital/global-environmental-finance.html.

Green economy and sustainable development: a macroeconomic perspective

16

Saumya Verma[1], Deepika Kandpal[2]

[1]*Lady Shri Ram College, University of Delhi, New Delhi, Delhi, India;* [2]*Department of Economics, University of Delhi, New Delhi, Delhi, India*

Chapter outline

1. Introduction

Globally, there has been a significant deterioration in environmental quality over the years, which can be attributed to climate change. An increase in global average surface temperature above 1.5°C is predicted to cause more frequent and intense climate and weather extremes, such as floods, droughts, heat waves, along with other effects, such as sea-level rise (IPCC Special Report 2018). Throughout the history of mankind, natural and manmade catastrophes have plagued human well-being. Be it the 2008 recession or the health pandemic in 2019, the crisis initially reduces the burden on the environment through lower economic activity. However, economic recovery from the crisis without environmental concern may worsen climate change beyond repair (Hepburn et al., 2020). Consequently, it requires a policy response that drives economies along the path of recovery, without renouncing the climate change goals.

Environmental Sustainability and Economy. https://doi.org/10.1016/B978-0-12-822188-4.00016-6

The links between economy and the environment are multifaceted. While environmental degradation cripples economic growth by damaging natural capital, a higher level of economic activity also leads to environmental degradation (Ocampo, 2011). Nevertheless, an increase in per capita GDP beyond a threshold provides opportunities for technological innovation, which may improve environmental quality (Stern et al., 1996). The tradeoff between higher economic growth and environmental quality is known as Environmental Kuznets Curve (EKC) hypothesis which postulates an inverted U-shaped relationship between economic growth and environmental degradation (Shafik, 1994). However, it should be noted that empirical evidence on the EKC hypothesis is mixed. For example, Lau et al. (2019) provide evidence in favor of EKC hypothesis for OECD countries during 1995−2015, whereas Kisswani et al. (2019) find that the EKC hypothesis does not hold good for a sample of five countries of ASEAN during 1971−2013.

The economy-environment interlinkage is incomplete without an understanding of the impact of climate change policies on present and future economic growth. Mitigating climate change requires incurring abatement expenditure in the current period, the benefits of which are enjoyed by both current and future generations (Padilla, 2002). However, such policies warrant short and long-term economic growth by stimulating demand, reducing future losses in output, and increasing technological innovation (Dechezlepretre et al., 2019). The decision of whether to undertake a mitigation project is based on a cost-benefit analysis. The social discount factor used for discounting benefits and costs of the project plays a crucial role in determining if it will be viable to undertake the project. The discount factor is to be chosen considering both efficiency and equity (Lind, 1995).

However, it should be noted that there are market failures associated with technological innovations required to accomplish green growth (Kemp and Never, 2017). This suggests the need for government intervention. An assessment of the impact of green policies on employment has attracted a great deal of attention among policymakers and environmental economists. Transformation toward a green economy involves displacement of labor from low energy to high energy-efficient firms (Fankhaeser et al., 2008). It is, therefore, imperative to build a policy framework aimed at labor market reforms to avoid any inadvertent effects during the transition process.

In the wake of the quandary of climate change and fluctuating economic growth, it is crucial to develop green growth policies that focus on the environment without compromising on economic growth. In this chapter, we discuss some of the issues related to sustainable development and economic growth. In Sections 2 and 3, we discuss the widely used notions in environmental economics literature while focusing on intergenerational tradeoffs. Issues pertaining to ecological economics in a macroeconomic framework are analyzed in Section 4. Section 5 presents the intersection between sustainable development and green growth. Section 6 discusses the need for green accounting and commonly used measures of green accounting. In Section 7 policies aimed at attaining green growth and sustainable development are examined. We conclude the chapter in Section 8.

2. Introduction to sustainable development

Sustainability is an ethical issue that involves intergenerational tradeoffs. According to the Brundtland Report (1987), "Sustainable development is development that meets the needs of the present without compromising the ability of future generations to meet their needs." A total of 17 Sustainable Development Goals (SDGs) were formulated at the United Nations General Assembly in September 2015. These goals aim at provision of a decent standard of living for all, with significant improvements, particularly in education and health sectors.

There are significant interlinkages both among the natural ecosystems and higher economic growth and environment which is the essence of sustainability. Sustainability was initially understood as a path of nondeclining consumption over time, which will be satisfied if the Hartwick rule of constant consumption holds (Perman et al., 2003, p. 89). The notion of sustainability mainly depends on the extent of substitution between natural and other forms of capital (Perman et al., 2003, pp. 90−91). Natural capital includes all natural ecosystems and resources used as a life support system for all living species, in addition to other uses such as direct inputs in production processes and/or for recreation purposes, also known as amenity value of environmental resources. Economists who believe in weak sustainability consider limited substitution between these two forms of capital. On the other hand, strong sustainability requires a minimum level of natural capital with a nondeclining path of natural capital over time (Perman et al., 2003, pp. 90−91). Strong sustainability considers perfect complementarity between natural and other forms of capital, with zero substitution between the two (Perman et al., 2003, pp. 90−91).

Economic growth is dependent both on the availability and quality of environmental resources. Environmental resources are directly used as an input in the production process for major sectors of the economy (such as agriculture and industry) or as recreation or amenity value in consumption. However, it should be noted that while we use natural resources for production, wastes generated from production are discharged back into the environment. This in turn requires that all production processes should be undertaken considering the assimilative capacity of the environment. Globally, this has resulted in the negative global externality of climate change, the adverse effects of which are evident in our day-to-day lives.

The intergenerational tradeoff involved in designing policies to combat climate change is extensively discussed in the arena of environmental economics. The stringency of any policy aimed at attaining sustainability clearly depends on the relative importance of future reduction in welfare from higher greenhouse gas emissions today. This has been studied in detail by three novel economists— Nicholas Stern, Martin Weitzman, and William Nordhaus, among others. Both Nicholas Stern and Martin Weitzman are in favor of adopting stringent emission reduction policies in the present. Nordhaus (2014), on the other hand, proposes a policy ramp, i.e., policies aimed at modest reduction in greenhouse gas emissions in the near term and aggressive abatement policies in future. One of the reasons for differences in policy prescription by these economists is variation in the weight attached to welfare losses of future generations.

Though there has been a continued debate regarding the appropriate discount rate to be used in climate change policy, the issue of sustainability is well understood in the economic literature, and seems to be a desirable objective from the perspective of both equity and efficiency. In the next section, we discuss the meaning of a green economy.

3. Definition of a green economy

The notion of *Green Economy* originated in 1989 in an advisory report submitted by leading economists in this field to the UK Government. It regained its popularity after the economic crisis of 2008 and the Rio+20 Conference in 2012. According to the UNEP (United Nations Environment Program) (2011), a green economy is "one that results in improved human well being and social equity, while significantly reducing environmental risks and ecological scarcities. It is low-carbon, resource efficient and socially inclusive" (UNEP, 2011). Its objective is to enhance economic opportunities in a manner that improves social welfare and equity, without deterioration of the natural environment.

Green growth serves the dual objective of job creation and economic growth with lesser impact on the environment. Strategies of green growth are meant to encourage ecofriendly innovations leading to a reallocation of inputs used in production, toward environmental friendly projects. In the next section, we discuss the importance of a green economy from a macroeconomic perspective.

4. Green economy from a macroeconomic perspective

Both sustainability and green growth involve links between economy and the environment. In addition to microeconomic effects, environmental policies are affected by, and in turn affect, the economy at large. In this section, we discuss some aspects of the nexus between macroeconomy and the environment.

4.1 Intergenerational equity and sustainable development

As discussed in Section 2, investing in projects targeted at reducing environmental degradation involves intergenerational tradeoffs. The size of the discount rate is critical in cost-benefit analysis because it determines whether an environmental project produces net benefits or losses. An extremely high discount rate can result in rejection of a desirable environmental project whereas a very low discount rate may result in accepting of an undesirable project (Harrison, 2010). One of the main reasons for considering the welfare of future generations while implementing policies in the present is the low probability of survival of future generations, since there is a high probability of low-frequency high-impact catastrophic events, which in future may even lead to loss of a civilization (for example, an asteroid hitting the earth).

In a report submitted to the government of United Kingdom in 2007, Nicholas Stern used a very low discount rate (1.4%), attaching a higher weight to even minor reductions in future welfare (Stern, 2007). However, it is also argued that a high discount factor (or, a low discount rate) would divert most of the resources used for the present generation toward avoidance expenditure, which is unfair from an equity perspective (Weisbach and Sunstein, 2009). The Stern Review concludes that mitigation costs amounting to 1% of global GDP can reduce damage costs by 5% of GDP by 2050. Van der Ploeg (2011) suggests variation in discount rate based on economic capability. The authors believe that developed countries should bear a greater share of the burden of reduction in emissions due to higher abatement potential. They recommend using a lower social discount rate for rich countries because of their ability to afford investments which would translate into an aggressive abatement policy. Moreover, owing to uncertainty about future economic growth, studies propose to use a social discount rate that declines over time (Freeman and Groom, 2016). The correct discount rate to be used is thus a debatable issue in environmental economics. In the next section, we discuss in detail the links between economy and environment, which form the basis of both sustainability and green growth.

4.2 Ecosystem and macroeconomy

Another macroeconomic dimension in environmental economics is the complex linkages between the ecosystem and macroeconomic indicators. There is a bidirectional relationship between economic activity and the environment. Higher economic activity results in environmental degradation whereas policies in favor of sustainable development affect the macroeconomy through changes in both aggregate supply and aggregate demand (Ocampo, 2011). The critical nature of natural resources is evident from the following facts: (1) natural resources are essential for manufacturing goods and services, (2) consumption of natural resources cannot be reversed, and (3) higher consumption in future, as a result of higher economic activity, cannot substitute for any loss in natural resources in the current period (Pelenc et al., 2015).

Neoclassical and ecological economists offer varied perspectives on the association between macroeconomic factors and sustainable development. Neoclassical economists treat the ecosystem as a subset of the economy whereas ecological economists consider the economy as a part of the ecosystem (Pollitt et al., 2010). A major drawback of the former models is that they imply weak sustainability (Daly and Farley, 2011). Moreover, these models assume a high discount rate, which washes away the detrimental effects of environmental damages (Ackerman, 2008). Nevertheless, the notion of strong sustainability asserts that degradation of natural capital is critical to human welfare. They propound that the two forms of capital are not substitutable (Dietz and Neumayer, 2007; Pollitt et al., 2010).

The paradigm of strong sustainability provides an economic rationale behind the negative impact of climate change on aggregate supply. It indicates that climate change and depletion of natural resources, if not controlled, pose a serious threat to aggregate supply and limit economic growth. Burke et al. (2015) estimate that global climate change is likely to reduce average global output by 23% by 2100. Low-income developing countries are expected to be the worst affected due to their geographic location and limited adaptation potential. Furthermore, greenhouse gas emissions by any country are found to have negative spillover effects for other countries. These arguments and empirical evidence indicate that environmental degradation leads to higher damages globally.

Investment in policies promoting green growth affects the economy through a reduction in aggregate supply and expansion in aggregate demand. The government incurs abatement expenditure while implementing climate change policies. Also, according to the Keynesian approach, green investments are expected to stimulate the growth process by boosting aggregate demand in an economy. The multiplier effect of green investment spending on aggregate output and employment provides ground for the government to offer the green stimulus package (Jacobs, 2012). Investments in climate change policy would induce technological change along with the accumulation of assets, which stimulates growth in the long run (Kaldor, 1978). Green stimulus packages were offered by several countries as expansionary policies in response to the economic recession during 2008–09. The policies involved public investment in environmental protection, subsidizing renewable energy, increasing energy efficiency, carbon laws, among others (Barbier, 2010). The crucial role of technological change in attaining green growth is explained in detail in the next section.

4.3 Green growth, technological innovation, and economic growth

Green investment, also a fundamental driver of economic growth, brings us to the next macroeconomic dimension of a green economy, i.e., technological change. Green growth requires undertaking policy initiatives to tackle the three crises of climate change, economic growth, and food security. Technological change offers a mechanism to achieve higher economic growth through increased adoption of green technologies. Hence, increased adoption of mitigation technology coupled with limited use of exhaustible resources is a prerequisite for green growth (Rodrik, 2014).

Existing growth theories and empirical evidence suggest that inequalities in economic growth across nations cannot be explained solely by disparities in physical capital (Broughel and Thierer, 2019).

To a larger extent, differences in technological innovation and diffusion contribute to variations in level of productivity across countries (Acemoglu, 2012). According to the structuralist view, technological innovations in green technology are imperative for green growth. Technological change drives advancement in production structure, which results in economic growth and development (Gabardo et al., 2017). Thus, investment in green technologies is not only crucial for reducing environmental damages but also helps create favorable conditions for economic growth.

Embracing green policies is a conscious decision to promote sustainable development. However, there are various impediments in the transition to a green economy. Firstly, the innovation of such technologies is susceptible to market failure because of positive spillover effects of R&D activities by a firm on other firms, industries, and countries, and due to underpricing of greenhouse gas emissions (Rodrik, 2014). The innovating firms receive lower returns because knowledge generated in this process can be easily transmitted to other producers. Consumers, on the other hand, benefit from increased innovation because of competitive pricing. Hence, returns to the innovating firm are far lower than benefits to society. Second, the rate of diffusion of technology has remained low because of institutional and behavioral issues, in addition to market failure owing to externalities with technology adoption (Jaffe et al., 2005). The adoption of new technology requires prior information related to its success rate. The use of technology by a firm helps other firms in bridging this information gap. This positive externality in adoption is another hindrance in the transition toward a greener economy. Third, there is asymmetric information between investors and firms regarding returns from investment in green technology (Jaffe et al., 2005). Due to these reasons, investment in technologies is suboptimal, in the absence of government intervention. Government intervention in innovation and the adoption of green technologies is essential to guarantee a steady shift toward a green economy. Adoption of green technologies also affects employment, which is discussed below.

4.4 Green policy and employment

Green policies refer to investments that reduce carbon emissions, increase energy efficiency, and preserve biodiversity (UNEP, 2011). These policies lead to a shift in demand for labor from unskilled to skilled workers. Looking at the supply side of the market, green firms attract better employees (Lanfranchi and Pekovic, 2014; Jones et al., 2016). In their analysis of the impact of corporate social responsibility practices on job attractiveness, Jones et al. (2016) show that being green provides positive signals about organizational values of the employer, job reputation in the market, and prosocial orientation. Potential job seekers expect a good treatment upon employment and consider applying in these firms. Similarly, being green also attracts higher capital investment, increasing profitability. It has been seen that green companies are high-profit companies that consider these sustainability practices as their social responsibility in addition to earning higher profits.

There are contrasting arguments on the impact of green policies on employment. In an analysis of the effect of energy-efficient technologies on employment, results from Fankhaeser et al. (2008) indicate that switching to low-carbon technologies results in destruction of employment in high-carbon industries and creation of employment opportunities in industries using cleaner technologies. They argue that the net effect would be an increase in jobs in the short and medium run because green industries are labor-intensive. However, benefits disappear in the long run as technology becomes efficient.

Empirical findings on the influence of green policies on employment are also polarized (Jacobs, 2012; Liu et al., 2020). Some studies find that environmental policies increase employment because green policies (such as making energy-efficient buildings) are highly labor-intensive, and the resulting job losses are compensated by new employment opportunities. In contrast, the shift to energy-efficient and renewable energy technologies has also been found to hurt employment, especially among unskilled workers (Marin and Vona, 2019).

The consequences of adopting a green fiscal policy for employment are also highly disputed. Keynesian supporters maintain that green stimuli investments would trigger gains in employment, while skeptics argue that green investments are not effective in generating long-term growth because it crowds out private investment (Jacobs, 2012). Empirical evidence on the effect of the US green stimulus package on economic recovery and employment is mixed. Official estimates suggest a positive multiplier effect whereas other studies find adverse effects of the package on employment (Jacobs, 2012).

Thus we see that both sustainable development and green economy are closely linked with macroeconomic growth and development of a country. The two concepts however differ in some respects. In the next section, we explain the parallels and dissimilarities between sustainable development and green economy.

5. Green economy and sustainable development

The terms *green economy* and *sustainable development* are often used interchangeably with a common objective of preservation of environmental resources for future generations. Moreover, both green growth and sustainable development focus on reducing social inequality. Despite a common underlying objective, both concepts address different dimensions of human well-being.

Sustainable development is a much broader concept than green growth, which mainly considers economy-ecology link. Sustainable development incorporates all forms of investments and technological innovations that are significant for the economy. This includes investments directed toward increases in both natural and human-made capital. Green growth, on the other hand, stresses on attaining economic growth only through investments and innovations which lead to better environmental quality.

Sustainable development aims at balancing economic development with environmental degradation. It requires implementing a much broader set of policies with an objective of significant advances in the quality of life, not only of human beings but of all living species on our planet. In addition to undertaking policies to provide enough opportunities for entrepreneurship and job creation, sustainable development goals also aim at providing good quality education and health facilities and raising the living standards for all species. The goals are formulated to reduce inequality and ensure a good quality of life for all. Green economy, as defined by UNEP (2011), focuses on social inclusiveness, with an objective of poverty reduction through economic growth and green employment opportunities. Therefore, "green growth should be viewed as a means to attain sustainable development rather than being a substitute for the latter" (World Bank, 2012).

To achieve the dual objectives of sustainable development and green growth, however, it is essential to account for changes in the environment with economic growth in the official national accounting procedures, also known as Green Accounting, discussed in detail in the next section.

6. Green accounting

Encapsulating environmental quality in national income accounting is of utmost importance to formulate policies for green growth. This can be done by accounting for degradation of natural capital while measuring income (or wealth) of the economy, also known as Green Accounting. It is a common saying in economics—*What gets measured, gets managed.* Accounting for the environment is thus a prerequisite for the conservation of environmental resources. When environmental accounting (also known as natural resource accounting or green accounting) was introduced in the 1950s, there was a limited understanding of the implications of higher economic growth for environmental degradation. Norway was among the first countries to collect data on the state of natural resources in the 1970s (Perman et al., 2003, p. 628). It was followed by the development of a "pressure state response" model by OECD in 1994, which aims at measuring increased environmental pressures due to higher economic activity (Perman et al., 2003, p. 629).

The United Nations Statistical Division (UNSTAT) proposed a system of satellite accounting in 2008 to account for the impact of higher economic activity on natural capital base. They suggest calculating the environmental cost of economic activity, by observing trends in the stock of environmental assets. The conventionally computed net domestic product (NDP) (which includes depreciation of physical capital) is adjusted by subtracting the environmental cost to yield Environmentally Adjusted NDP (EDP). They suggest that conventional accounts should include information on natural assets. However, there are issues with computing a measure of the stock of natural assets, particularly for fossil fuels and minerals, as there are concerns related to property rights for these environmental assets. Indonesia and Australia were among the first countries to report values of GDP and EDP during 1970s and 1980s. Indonesia accounted for depreciation of oil deposits, timber, and soil in its GDP measures to find a growth rate of EDP of 4.1%, with a much higher expected growth rate of GDP of 7.1%. Data reported for 1971−84 indicates that with an annual growth rate of 51% EDP grows at a much slower rate than GDP (Perman et al., 2003, p. 647). Similar results are obtained from trends in EDP reported for Australia during 1980−88 (Perman et al., 2003, p. 647).

Conventionally used accounting measures are based on the assumption of perfect substitutability between increases in GDP and degradation of natural capital. These measures do not account for sustainable income (i.e., income net of negative impact on the environment which includes any environmental degradation or reduction in natural assets such as a reduction in forest cover to clear up land area for industrial activity). Given the limitations of GDP as an indicator of welfare, other alternative indicators of sustainability such as Gross National Happiness (GNH), Adjusted Net Savings, and Genuine Savings are used. Genuine savings rate is based on net natural resources in the economy, i.e., investment in all forms of capital net of capital depreciation (Neumayer, 2004). As measured by the World Bank, genuine savings take into account depreciation of mainly nonrenewable sources of energy. Countrywise performance in Human Development Index (HDI) and genuine savings rate are interrelated. It has been observed that countries with higher HDI invest more in human and manmade capital, which compensates for the degradation of natural capital (Neumayer, 2004). Genuine saving rates are particularly low for countries with low or medium human development (Neumayer, 2004).

Adjusted Net Saving (ANS) differs from genuine savings rate, as it includes government expenditure on education and also accounts for degradation of the environment through higher CO_2 emissions along with depletion of natural capital. As defined by World Bank (2018),

$$ANS = GNS - CFC + EDU - NRD - GHG - POL$$

where ANS: Adjusted Net Saving rate

GNS: Gross National Saving
CFC: Depreciation of fixed capital
EDU: Education expenditure in current period
NRD: Depletion of natural resources, which includes forests, fossil fuels, metals, and minerals. Energy generated from fossil fuels includes nonrenewable resources. Minerals include major minerals such as gold, lead, zinc, among others.
GHG: All damages from CO_2 emissions from use of fossil fuels and cement manufacturing.
POL: Cumulative damages due to higher exposure to air pollution, mainly ambient $PM_{2.5}$ concentration and higher indoor concentration of air pollutants.

Malaysia has witnessed consistently positive ANS rates from 1970 to 2012, with cyclical fluctuations during this period (Pardi et al., 2015). It has been observed that increases in agricultural productivity, particularly use of natural agro production technologies, increases ANS rates significantly (Katan et al., 2018). This is particularly important for agrarian countries like India. A rising trend in ANS rates has been observed for Latvia, Czech Republic, Kazakhstan, Poland, and Moldova during 2009−17. For others, such as Slovak Republic, Kyrgz Republic, and Bulgaria, ANS exhibits a downward trend during the same time period (Katan et al., 2018). Among the countries of European Union (EU), ANS values are the least for Romania, Greece, and Bulgaria during 1990−2011. Countries with high extraction of minerals exhibit negative values of ANS (for e.g., Sweden) whereas Luxemborg had high rates of ANS during this period (Drastichova, 2014). Hanley et al. (2016) confirm the use of genuine savings as a reliable indicator of changes in consumption in future in Britain, Germany, and the USA. In addition to economic measures for green accounting, there are other noneconomic indicators, discussed below.

Increases in GDP growth not only result in depletion of natural capital but also increases income inequality (both within and across a nation). Beyond a point, "increases in GDP reduce social welfare through reduced community cohesion, healthy relationships, knowledge, connection with nature and other dimensions of human happiness" (Costanza et al., 2016). This holistic approach to development is being practiced in Bhutan, where Gross National Happiness (GNH) is used as an indicator of well-being. The indicator is mainly based on Meadows principle of "Limits to Growth" which is against the growth-centric view of development. The index is based on several dimensions determining human welfare—social indicators such as education and health along with environmental indicators such as ecological and cultural diversity and overall psychological well-being (Heal, 2012). The index is based on the principle that mere ownership of assets is not enough to lead a healthy and fulfilling life. There are issues with arriving at an accurate measure of GNH, as happiness is subjective and this index is usually computed from data obtained from surveys, which is difficult to obtain for countries with higher population, such as China. Alternative indices measuring the quality of human well-being include HDI, Happy Planet Index, Social Progress Index, Policy Effectiveness Index, and Physical Quality of Life Index (PQLI). Apart from income, education, and health, there are other social factors determining happiness such as willingness to help, tolerance, and security. In the next section, we discuss trends in adjusted net saving rates for selected countries during 1990–2018.

6.1 Data analysis

World Bank provides data on Adjusted Net Savings (ANS) (measured as a proportion of Gross National Income (GNI)) for 268 countries from 1990 to 2018, following the methodology explained previously. We compare the ANS rates for India and developed countries namely China, United States (US), and European Union (EU) in Fig. 16.1A. It shows a clear rising trend of adjusted net savings for India, with a sharp increase from 14% in 2001 to 24% in 2007. However, ANS decreased considerably after 2011 from 24% in 2011 to 17% in 2018. Overall, the ANS rate for India increased from 9% in 1990 to 17% in 2018. ANS rate in EU and US is consistently lower than that for India, which can be explained by increased exhaustion of natural resources and damages due to CO_2 emissions. At 16%, ANS rate for China exceeds that of India in 1990, with both countries experiencing an ANS rate close to 20% in 2018. Globally averaged adjusted net saving rate is close to 8% during this period. Among these four countries, the United States is the only country with adjusted net savings rate consistently below the global average during this period.

A comparison of ANS rates for countries exhibiting higher adjusted net saving rates (relative to the global average) in Fig. 16.1B indicates a clear rising trend for Bangladesh and Nepal. Nepal has witnessed the highest increase in ANS rate, from 5% in 1990 to 38% in 2018. It can be seen from Fig. 16.1C that Saudi Arabia has witnessed significant improvements in ANS rate with negative saving rates before 1992 to 20% in 2018. Bulgaria also exhibits a similar trend, with both countries

FIGURE 16.1A

Trends in adjusted net saving rate (ANS) (proportion of gross national income (GNI)).

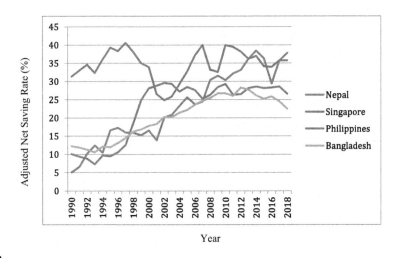

FIGURE 16.1B

Trends in adjusted net saving rates for countries with high ANS (% of GNI).

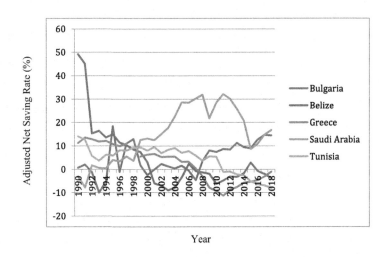

FIGURE 16.1C

Trends in adjusted net saving rates for countries with negative ANS (% of GNI).

converging to a 20% ANS rate in 2018. There have been significant reductions in ANS rate for Belize, Greece, and Tunisia. For Belize, ANS rates have fallen drastically from 50% in 1990 to close to 0% in 2018, with a negative saving rate for most of the period. Tunisia and Greece also exhibit a decreasing trend in ANS rate during this period. One of the possible reasons for this decline could be increased environmental degradation due to increases in growth rate. Moreover, some countries such as Uganda, Togo, Sierra Leone, Romania, Oman, and Republic of Congo have consistently witnessed negative adjusted net saving rates.

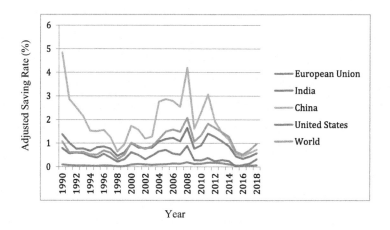

FIGURE 16.2

Trends in adjusted savings from energy depletion (% of GNI).

We also examine the trends in adjusted savings from energy depletion[1] (measured as a proportion of Gross National Income (GNI)) for these countries in Fig. 16.2. Analysis reveals highest rates of adjusted savings from energy depletion for China, with close to zero adjusted savings for European Union. Both United States and India exhibit close to 1% adjusted savings from energy depletion. Similarly, graphs for adjusted savings from depletion of minerals[2] (measured as a percentage of GNI) for these countries (Fig. 16.3) shows high adjusted savings for China and India with a steep increase in saving rates in 2006 and 2008. This was followed by a decrease in adjusted savings thereafter. It should be noted that globally adjusted savings from depletion of minerals are generally low, increasing gradually from 2004. The depletion rate is close to zero for both European Union and the United States.

Trends in adjusted savings from net forest depletion[3] (Fig. 16.4) have been particularly low with an adjusted saving rate of 0% for most countries, including China, United States, and European Union. Globally, adjusted savings are low and less than 0.1%. For India, adjusted saving rates are low and decreasing from 0.5% in 1990 to 0.1% in 2018. Excluding damages from particulate emissions does not alter the relative ranking in adjusted saving rates for these countries (Fig. 16.5) indicating that damages from higher particulate emissions are not a major source of variation in adjusted net savings rate. It should be noted that these observed differences in trends in adjusted net saving rates (and its individual components) can be attributed to country-specific policies to curb environmental degradation and trends in economic growth in these countries, which is left for future research. In the next section we discuss policies followed in other countries to attain the dual objectives of green

---

[1]Energy depletion is the ratio of value of the stock of energy resources to the remaining reserve lifetime. It covers coal, crude oil and natural gas (World Bank, 2018).

[2]Mineral depletion is the ratio of value of stock of mineral resources to the remaining reserve lifetime, It covers tin, gold, lead, zinc, copper, nickel, silver, bauxite and phosphate (World Bank, 2018).

[3]Net forest depletion is calculated as the product of unit resource rents and excess of round wood harvest over natural growth. If growth exceeds harvest, this figure is zero (World Bank, 2018).

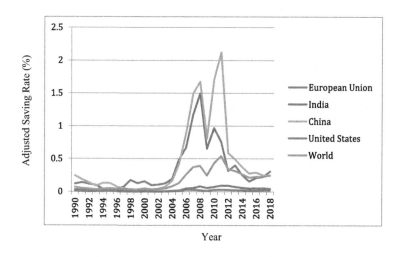

FIGURE 16.3

Trends in adjusted savings from depletion of minerals (% of GNI).

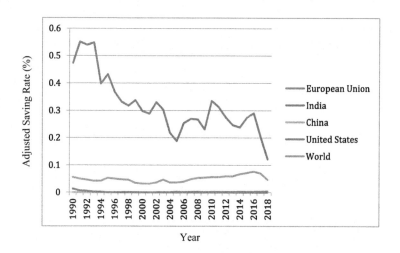

FIGURE 16.4

Trends in adjusted savings from net forest depletion (% of GNI).

economy and sustainable development. We also elaborate the policy challenges faced in climate change mitigation.

7. Toward a green economy—cross-country experience

The adoption of environment-friendly technologies leads to a low-carbon economy path along with providing opportunities for increased innovation. For this, we need to undertake country-level policies

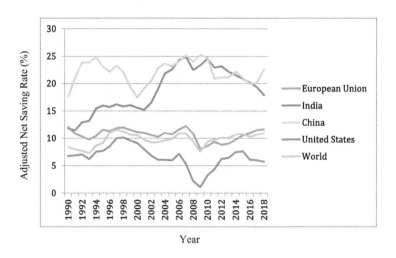

FIGURE 16.5

Trends in adjusted net savings (excluding particulate emission damages) (% of GNI).

to combat the problem of climate change. One way to achieve green growth is through improvements in energy efficiency. There is sufficient empirical evidence of significant reduction in carbon emissions due to successful implementation of programs such as US Energy Star Label, which reduced carbon emissions by 37.6 million metric tons (US EPA, 2008). Consumers in the United States also find it beneficial to purchase green goods (for example, green electricity), since it lowers energy consumption and in turn household consumption expenditure.

As stated earlier, the economy-environment link forms the basis of formulating any strategy targeted toward a green economy. The stringency (or leniency) of the policy mainly depends on the current climate system. The global climate system, which forms the basis of survival of all living species, is determined by the net energy balance of earth, i.e., net inflow and outflow of energy into the surface of the earth. Any policy which meets the dual objectives of green economy and sustainable development is incomplete without implementing appropriate policy to resist climate change.

Climate change has been extensively dealt with by the United Nations Framework Convention on Climate Change (UNFCCC) which organizes the Conference of Parties (COP), the last one being the 25th Conference held in Madrid, Spain. However, not much progress has been achieved in stabilizing the greenhouse gas emissions globally. There is a limited global consensus as to what appropriate policies should be undertaken, which provides room for more initiatives at the local level, such as voluntary environmental agreements. In one of the seminal reports on addressing this issue, the UN secretary general highlights the policy framework required to attain Sustainable Development Goals (SDG). It involves an expansion of public support for sustainable, energy-efficient infrastructural capital, preserving the existing base of natural capital and increase in public investment to increase ecological development. These can be achieved through policies promoting the greening of business and markets (UNEP, 2011). The government can also use fiscal policy instruments such as providing tax benefits to firms which invest in developing energy-efficient alternatives.

The Indian government has taken several initiatives in this direction, the first one being setting up of a National Action Plan on Climate Change (NAPCC) in 2008. The NAPCC emphasizes the growing need for harnessing the full potential of renewable sources of energy and using them as a substitute for nonrenewable energy sources. This was implemented by creating eight National Missions focusing on key aspects of energy conservation and sustainable development. The Indian government submitted India's Intended Nationally Determined Contribution (INDC) to the United Nations Framework Convention on Climate Change (UNFCCC) after the Paris Agreement in 2015. India's INDC goals focus on energy conservation to reduce the extent of climate change. Its goals include reduction of energy intensity of growth by 33%–35% by 2030 (relative to 2005) and generation of 175 GW of renewable energy by 2022, by increasing the share of power generated by use of non-fossil-fuel-based sources of energy to 40% by 2030. The goals also stipulate afforestation to create a carbon sink of 2.5 to 3 billion tons of CO_2.

Highlighting the role of technological innovations in improving energy efficiency, Walz et al. (2017) find Finland, Denmark, Germany, and Austria to be the most capable countries in green technological innovation. Moreover, Korea, China, Germany, and Japan exhibit high innovation potential. With a clear dominance of countries of the South in green technology innovations, there is a scope for technology transfer from developed to developing countries. China has been formulating policies to attain higher economic growth with least possible environmental degradation. It has channelized its financial resources toward projects aimed at reducing dependence on fossil fuel by substituting them with renewable energy, particularly solar and wind. The country has financed a number of power projects in sub-Saharan Africa, for example, "Aysha Wind Farm Project" (Gu et al., 2018). Malaysia has undertaken several programs (such as Small Renewable Energy Power Program) targeted toward generating electricity from renewable sources of energy. In the process of transformation to a green developing economy, the government of Malaysia has undertaken several programs to create awareness among the consumers about the use of renewable sources of energy (Mekhilef et al., 2014). The success of Small Renewable Energy Power program can be seen from the fact that the program was successful in generating electricity from major sources of renewable energy.

Germany has also contributed significantly toward meeting global solar energy demand through a significant reduction in the price of photovoltaic modules (Buchholz et al., 2019). Germany implemented the Energy Transition program and has emerged as the leading exporter of wind energy converters during 2004–12 in addition to being a potential competitor in the global solar photovoltaic market (Pegels and Lütkenhorst, 2014). The country has undertaken several other programs to promote technological innovation in this direction.

Pneuli and Zussman (1997) highlight the need to reduce toxic wastes generated in production processes by taking steps toward recycling of waste materials and adopting cleaner production technology. This can be achieved through appropriate product design, which determines the bulk of recycling costs (Pneuli and Zussman, 1997). Numerous policies have been implemented in this direction in advanced economies, such as European Union. Cleaner production processes have led to a significant reduction in industrial pollution in China (Hang et al., 2019). The Indian government should consider implementing similar policies in our country. There is scope for increased public-private partnerships and stringent regulation to attain the dual goals of sustainable development and green growth. This will have to be accompanied by higher investments in R&D to tap the full potential of renewable resources in India. There is sufficient empirical evidence indicating high and positive income elasticity of demand for environmental goods, particularly in developed countries (McConnell, 1997). Thus, there is a growing need to increase awareness and encourage consumption of green goods among consumers in developing countries as well.

8. Concluding remarks

This chapter outlines the issues of sustainability and green economy, examining its implications for the macroeconomy. Though data on economic measures of green accounting indicate sustainable growth, there is a need for concerted efforts to achieve the twin objectives of sustainable development and green growth. The transformation process toward attaining these two goals requires macroeconomic changes in the economy. Green investments are expected to drive long-term growth and increase employment opportunities for the skilled labor force. However, state intervention is required to correct for market failures in green investments and to develop the human capital base of the economy. Countries with higher government intervention in the form of green stimulus packages are more likely to excel in exporting green products and reducing greenhouse gas emissions (Mealy and Teytelboym, 2020). Moreover, there is emerging evidence in favor of path dependency[4] in green innovation (Aghion et al., 2019; Fischer and Heutel, 2013). The countries, therefore, must be directed to encourage green innovation, to reap the benefits of higher production possibilities while attaining green growth.

Both green growth and sustainable development cannot be achieved without appropriate measures for climate change mitigation. Several initiatives have been taken in different countries to mitigate climate change. The Government of Cape Verde, for instance, aimed to produce all the electricity needed for the country through renewable energy sources by 2020. The Middlesbrough's Sustainability Action Plan aims at a 90% reduction in emissions from buildings and reducing waste (both domestic and municipal) significantly by 2025. The plan also includes measures for creating awareness among consumers and encouraging emission-free modes of transport. This is encouraging, since globally road vehicles account for 72% of transport emissions. To promote consumption of green products, various mechanisms of certification of green products have been introduced, such as Emblem of Guarantee of Environmental Quality, an ecological labeling system (EMAS). Global initiatives also include the G20 Energy Efficiency Action Plan adopted in 2014, which aims at formulating policies to reduce emissions from vehicles, buildings, industrial processes and lower energy consumption from all network-connected devices. The plan also aims at increased public-private partnerships in programs aimed at improving energy efficiency.

The government of India has also embarked upon policies to combat climate change. Some of the initiatives taken in this direction include National and State level Action Plans on Climate Change (NAPCC), National Adaptation Fund on Climate Change, Climate Change Action Program, International Solar Alliance, and Atal Mission for Rejuvenation and Urban Transformation. The National Energy Policy concentrates on adopting energy-efficient methods in production. This is coupled with the introduction of Perform, Achieve and Trade (PAT) scheme targeting reduction in industrial emissions. In addition to formulating national-level policies, setting a price for carbon emissions would also encourage the adoption of low-carbon and energy-efficient technology. A carbon tax of 35\$ per ton of CO_2 would reduce emission intensity to 22% by 2030 (Ernst and Young, 2018). The government aims to resolve the climate change issue through the principle of "common but differentiated responsibilities," and encourages developed countries to undertake development projects in

---

[4]Path dependence implies that firms or countries capable of green production are more likely to diversify and innovate in cleaner technologies in future (Aghion et al., 2019)

India. The progress made by the government in this regard is however insufficient, and there is scope for adopting stringent mitigation strategies both at the local and national level.

Similarly, India has experienced notable success in promoting policies toward attaining sustainable development goals (GoI, 2020). This was accomplished through the introduction of antipoverty programs, provision of clean energy fuels in rural areas, and promotion of gender equality, among other related policies. Significant improvements have been made in the health sector, with reduction in infant mortality rate (measured as deaths per 1000 live births) from 129 in 1971 to 32 in 2018. Moreover, the Indian government acknowledges the need for a global partnership to attain sustainable development, which includes provision of necessary infrastructure to meet these goals. However, there is a growing need for formulating policies that enable harnessing the complete potential of renewable energy resources in India. This needs to be coupled with stringent emission reduction strategies particularly in the industrial sector and adopting energy-efficient strategies.

References

Acemoglu, D., 2012. Introduction to economic growth. J. Econ. Theor. 147 (2), 545−550.

Ackerman, F., 2008. Climate economics in four easy pieces. Development 51 (3), 325−331.

Aghion, P., Hepburn, C., Teytelboym, A., Zenghelis, D., 2019. Path dependence, innovation and the economics of climate change. In: Handbook on Green Growth. Edward Elgar Publishing.

Barbier, E.B., 2010. Green stimulus, green recovery and global imbalances. World Econ. 11 (2), 149−177.

Broughel, J., Thierer, A.D., 2019. Technological Innovation and Economic Growth: A Brief Report on the Evidence. In: Mercatus Research Paper (forthcoming).

Brundtland, G.H., Khalid, M., Agnelli, S., et al., 1987. Our Common Future. New York. Retrieved from: https://sustainabledevelopment.un.org/content/documents/5987our-common-future.pdf.

Buchholz, W., Dippl, L., Eichenseer, M., 2019. Subsidizing renewables as part of taking leadership in international climate policy: the German case. Energy Pol. 129, 765−773.

Burke, M., Hsiang, S.M., Miguel, E., 2015. Climate and conflict. Ann. Rev. Econ. 7 (1), 577−617.

Costanza, R., Daly, L., Fioramonti, L., et al., 2016. Modelling and measuring sustainable wellbeing in connection with the UN Sustainable Development Goals. Ecol. Econ. 130, 350−355.

Daly, H.E., Farley, J., 2011. Ecological Economics: Principles and Applications. Island Press.

Dechezleprêtre, A., Martin, R., Bassi, S., 2019. Climate change policy, innovation and growth. In: Handbook on Green Growth. Edward Elgar Publishing.

Dietz, S., Neumayer, E., 2007. Weak and strong sustainability in the SEEA: concepts and measurement. Ecol. Econ. 61 (4), 617−626.

Drastichová, M., 2014. Measuring sustainable development in the European union using the adjusted net saving. In: Proceedings of the 2nd International Conference on European Integration, pp. 87−101.

Ernst and Young, 2018. Discussion Paper on Carbon Tax Structure for India. Retrieved from: https://shaktifoundation.in/wp-content/uploads/2018/07/Discussion-Paper-on-Carbon-Tax-Structure-for-India-Full-Report.pdf.

Fankhaeser, S., Sehlleier, F., Stern, N., 2008. Climate change, innovation and jobs. Clim. Pol. 8 (4), 421−429.

Fischer, C., Heutel, G., 2013. Environmental macroeconomics: environmental policy, business cycles, and directed technical change. Ann. Rev. Res. Econ. 5 (1), 197−210.

Freeman, M.C., Groom, B., 2016. How certain are we about the certainty-equivalent long term social discount rate? J. Environ. Econ. Manag. 79, 152−168.

Gabardo, F.A., Pereima, J.B., Einloft, P., 2017. The incorporation of structural change into growth theory: a historical appraisal. EconomiA 18 (3), 392−410.

GoI (Government of India), 2020. Sustainable Development Goals National Indicator Framework Progress Report 2020. Ministry of Statistics and Programme Implementation, National Statistical Office.

Gu, J., Renwick, N., Xue, L., 2018. The BRICS and Africa's search for green growth, clean energy and sustainable development. Energy Pol. 120, 675−683.

Hang, Y., Wang, Q., Wang, Y., Su, B., Zhou, D., 2019. Industrial SO2 emissions treatment in China: a temporal-spatial whole process decomposition analysis. J. Environ. Manag. 243, 419−434.

Hanley, N., Oxley, L., Greasley, D., et al., 2016. Empirical testing of genuine savings as an indicator of weak sustainability: a three-country analysis of long-run trends. Environ. Resour. Econ. 63 (2), 313−338.

Harrison, M., 2010. Valuing the Future: The Social Discount Rate in Cost-Benefit Analysis. Retrieved from: https://ssrn.com/abstract=1599963.

Heal, G., 2012. Reflections - defining and measuring sustainability. Rev. Environ. Econ. Pol. 6 (1), 147−163.

Hepburn, C., O'Callaghan, B., Stern, N., et al., 2020. Will COVID-19 fiscal recovery packages accelerate or retard progress on climate change? Oxf. Rev. Econ. Pol. 36.

IPCC 2018. https://www.ipcc.ch/2018/10/08/summary-for-policymakers-of-ipcc-special-report-on-global-warming-of-1-5c-approved-by-governments/.

Jacobs, M., 2012. Green Growth: Economic Theory and Political Discourse. Working Paper No. 92. Grantham Research Institute on Climate Change and the Environment.

Jaffe, A.B., Newell, R.G., Stavins, R.N., 2005. A tale of two market failures: technology and environmental policy. Ecol. Econ. 54 (2−3), 164−174.

Jones, D.A., Willness, C.R., Heller, K.W., 2016. Illuminating the signals job seekers receive from an employer's community involvement and environmental sustainability practices: insights into why most job seekers are attracted, others are indifferent, and a few are repelled. Front. Psychol. 7, 1−16.

Kaldor, N., 1978. Further Essays on Economic Theory (No. 04; HB171, K3.).

Katan, L., Dobrovolska, O., Espejo, R.J.M., 2018. Economic growth and environmental health: a dual interaction. Probl. Perspect. Manag. 16 (3), 219−228.

Kemp, R., Never, B., 2017. Green transition, industrial policy, and economic development. Oxf. Rev. Econ. Pol. 33 (1), 66−84.

Kisswani, K.M., Harraf, A., Kisswani, A.M., 2019. Revisiting the environmental kuznets curve hypothesis: evidence from the ASEAN-5 countries with structural breaks. Appl. Econ. 51 (17), 1855−1868.

Lanfranchi, J., Pekovic, S., 2014. How green is my firm? Workers' attitudes and behaviors towards job in environmentally-related firms. Ecol. Econ. 100, 16−29.

Lau, L.S., et al., 2019. Is nuclear energy clean? Revisit of Environmental Kuznets Curve hypothesis in OECD countries. Econ. Modell. 77, 12−20.

Lind, R.C., 1995. Intergenerational equity, discounting, and the role of cost-benefit analysis in evaluating global climate policy. Energy Pol. 23 (4−5), 379−389.

Liu, Y., Park, S., Yi, H., Feiock, R., 2020. Evaluating the employment impact of recycling performance in Florida. Waste Manag. 101, 283−290.

Marin, G., Vona, F., 2019. Climate policies and skill-biased employment dynamics: evidence from EU countries. J. Environ. Econ. Manag. 98, 102253.

McConnell, K.E., 1997. Income and the Demand for Environmental Quality. Environment and Development Economics, pp. 383−399.

Mealy, P., Teytelboym, A., 2020. Economic Complexity and the Green Economy. Research Policy, pp. 1−24.

Mekhilef, S., Barimani, M., Safari, A., Salam, Z., 2014. Malaysia's renewable energy policies and programs with green aspects. Renew. Sustain. Energy Rev. 40, 497−504.

Neumayer, E., 2004. Sustainability and Well Being Indicators. WIDER Research Papers, 2004/23, UNU-WIDER.

Nordhaus, W.D., 2014. A Question of Balance: Weighing the Options on Global Warming Policies. Yale University Press.

Ocampo, J.A., 2011. The macroeconomics of the green economy. Transit. Green Econ. 16.

Padilla, E., 2002. Intergenerational equity and sustainability. Ecol. Econ. 41 (1), 69–83.

Pardi, F., Salleh, A.M., Nawi, A.S., 2015. A conceptual framework on adjusted net saving rate as the indicator for measuring framework on adjusted net saving rate as the indicator for measuring sustainable development in Malaysia. J. Technol. Manag. Bus. 2 (2), 1–10.

Pegels, A., Lütkenhorst, W., 2014. Is Germany's energy transition a case of successful green industrial policy? Contrasting wind and solar PV. Energy Pol. 74, 522–534.

Pelenc, J., Ballet, J., Dedeurwaerdere, T., 2015. Weak Sustainability Versus Strong Sustainability. Brief for GSDR United Nations. Retrieved from: https://sustainabledevelopment.un.org/index.php?page=view&type=111&nr=6569&menu=35.

Perman, R., Ma, Y., Mc Gilvray, J., Common, M., 2003. Natural Resource and Environmental Economics. Pearson Education.

Pnueli, Y., Zussman, E., 1997. Evaluating the end-of-life value of a product and improving it by redesign. Int. J. Prod. Res. 35 (4), 921–942.

Pollitt, H., Barker, A., Barton, J., et al., 2010. A Scoping Study on the Macroeconomic View of Sustainability. Final Report for the European Commission, DG Environment. Cambridge Econometrics, Cambridge.

Rodrik, D., 2014. Green industrial policy. Oxf. Rev. Econ. Pol. 30 (3), 469–491.

Shafik, N., 1994. Economic development and environmental quality: an econometric analysis. Oxf. Econ. Pap. 46 (October), 757–773.

Stern, D.I., Common, M.S., Barbier, E.B., 1996. Economic growth and environmental degradation: the environmental Kuznets curve and sustainable development. World Dev. 24 (7), 1151–1160.

Stern, N., 2007. The Economics of Climate Change: The Stern Review. Cambridge University Press.

UNEP (United Nations Environmental Programme), 2011. Towards a Green Economy: Pathways to Sustainable Development and Poverty Eradication. UNEP, Nairobi, Kenya.

US EPA (United States Environmental Protection Agency), 2008. 2008 Report on the Environment. Retrieved from: http://wedocs.unep.org/handle/20.500.11822/9043.

Van der Ploeg, F., 2011. Macroeconomics of sustainability transitions: second-best climate policy, green paradox, and renewables subsidies. Environ. Innov. Soc. Transit. 1 (1), 130–134.

Walz, R., Pfaff, M., Marscheider-Weidemann, F., et al., 2017. Innovations for reaching the green sustainable development goals—where will they come from? Int. Econ. Econ. Pol. 14 (3), 449–480.

Weisbach, D., Sunstein, C.R., 2009. Climate change and discounting the future: a guide for the perplexed. Yale Law Pol. Rev. 27 (2), 433–457.

World Bank, 2012. Inclusive Green Growth: The Pathway to Sustainable Development. World Bank Publications.

World Bank, 2018. Estimating the World Bank's Adjusted Net Saving: Methods and Data. Retrieved from: https://development-data-hub-s3-public.s3.amazonaws.com/ddhfiles/143151/ans-methodology-january-30-2018_2_0_0.pdf.

CHAPTER

Challenges and opportunities at the crossroads of *Environmental Sustainability and Economy* research

17

Daniela Perrotti[1], Pramit Verma[2], K.K. Srivastava[3], Pardeep Singh[4]

[1]*Research Institute for Landscape, Architecture and Built Environment, University of Louvain UCLouvain, Ottignies-Louvain-la-Neuve, Belgium;* [2]*Integrative Ecology Laboratory (IEL), Institute of Environment & Sustainable Development (IESD), Banaras Hindu University Varanasi, U.P., India;* [3]*PGDAV College, University of Delhi, New Delhi, India;* [4]*Department of Environmental Studies, PGDAV College, University of Delhi, New Delhi, India*

Chapter outline

Environmental Sustainability and Economy. https://doi.org/10.1016/B978-0-12-822188-4.00013-0

1. Introduction

In his masterpiece *Faust* the poet and writer Johann Wolfgang von Goethe examines the pathology of human society running in a ceaseless quest after fastest means of transport, the quickest and most powerful weapon, and the largest and easily earned wealth, all of which are utilized to fulfill a succession of human desires but lead ultimately to the society's downfall.

Most global societies today are also suffering from the same pathology running after growth and development in a ceaseless quest, at the cost of most marginalized communities, unabated ecological destruction, and worldwide food crisis.

While urban populations are expected to reach 5.6—7.1 billion by 2050, the majority of urban population growth is expected to take place in small- to medium-sized urban areas with lower levels of economic development (Seto et al., 2014). Economic development has the potential to drive urban growth, innovation, socioeconomic balance, healthcare, and employment, as well as, air pollution, greenhouse gas emissions, waste production and socioeconomic disparity, accessibility and availability of cleaner sources of energy (Fig. 17.1). This is not an exhaustive list, but it indicates the potential for economic development to be used in response to local priorities and needs. Prioritizing the application of sustainable measures lies at the core of the green economy concept.

1.1 Environmental sustainability and economy

According to the United Nations Environment Programme (UNEP, 2011), green economy is defined as "one that results in improved human well-being and social equity, while significantly reducing environmental risks and ecological scarcities. In its simplest expression, a green economy is

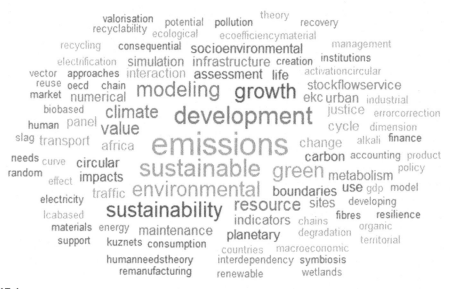

FIGURE 17.1

Word cloud image of author's keywords from chapters in the book.

low-carbon, resource-efficient, and socially inclusive. In a green economy, growth in income and employment are driven by public and private investments that reduce carbon emissions and pollution, enhance energy and resource efficiency, and prevent the loss of biodiversity and ecosystem services."

Traditional economic practices are generally resource-intensive and do not give much regard to the environmental fallout in the form of degradation of ecosystems, disruption of biogeochemical cycles, waste production, and social equality, among many others. As opposed to the traditional practices in the economy, the green economy aims to extend the life cycle of natural resources by considering their sustainability over long periods of time in a holistic manner. The system working of natural fluxes of mass and energy is used to create economic benefits, which should cause the least harm to environmental and economic systems, and potentially increase their resilience. Knowledge of systems, their interactions, and feedback loops, the impact of sustainable policies and socioeconomic equality are essential in creating green economic models. However, often the socioeconomic equality takes a backseat while designing green economy functions.

In a study by Glazyrina et al. (2015), it was reported that in forest-use policy, the use of budgetary and socioeconomic indicators was more effective in creating green practices than the traditional indicators of "felling, export and processing volumes" Political, socioeconomic, and cultural constraints should be considered (Brand, 2012). A diversification in economic activities and priority on socioeconomic indicators would help create green economic policies (Zabortseva et al., 2017). The green economy requires a shift in assumptions about traditional economic principles and a restructuring of priorities, where accountability is included from the conception (Najam and Halle, 2010). At present, most of the countries worldwide follow the economic models set by the predominant economic powers. However, such development has happened at the cost of significant, and in some cases, irreparable environmental damage. As exposed by Rockström et al. (2009), the scale of human activities has influenced the global cycles to the point that the chances of global-scale environmental disruption are increasing. The planetary boundary concept is crucial for creating sustainable "growth within limits" (Crépin and Folke, 2015). It signifies the biophysical limits of growth beyond which the global balance might tip permanently toward chaos, causing shifts in global biophysical processes. For example, four of the nine planetary boundaries had already been crossed in 2015, as reported by Stephen et al. (2015). These were climate change, loss of biosphere integrity, land system change, altered biogeochemical cycles (phosphorus and nitrogen). Out of these, climate change and biophysical integrity are considered core boundaries which have effects on multiple scales. It is essential to mention here that the same research drew on these results to conclude that such a change would have not only an impact on the environment but also on the economic status and human well-being (Stephen et al., 2015). Planetary boundaries may be considered as warning signs indicating a need to shift the current economic policies from GDP growth toward human well-being (Crépin and Folke 2015). A more recent analysis of tipping cascades suggested that a potential planetary threshold could occur at a $\sim 2°C$ temperature rise above preindustrial level (Steffen et al., 2018). A "Stabilised Earth Pathway" to maintain the earth system within such limits would require fundamental change in humanity's relationship with earth based on reorientation of human values, equity, behavior, institutions, economies, and technologies. This evidence points to the need for an integrated approach of environmental sustainability and socioeconomic development.

Environmental sustainability is seldom practiced in the supply chain of products and services, and the process toward integrating sustainability into the economic process is not always clear-cut, especially for developing countries (Suhi et al., 2019). Further, people directly involved in product chains are unsure about how to measure sustainability; there is a lack of indicators (Verma and Raghubanshi, 2018; Suhi et al., 2019). This also affects the efficiency of economic processes. Certain indicators focus on the economic state only, while others might deal with environmental impact. The concept of the circular economy is also essential in this regard, as it has the potential to open pathways for reduced waste generation and increased efficiency of production processes. The recycling rate is one such indicator used to measure the circular economy; however, it fails to capture the environmental dimensions clearly (Haupt and Hellweg, 2019). This chapter essentially focuses on the current state of research into environmental sustainability and economic practices. It gives an overview of the subjects addressed within this book across four themes (*environment and economy, indicators and sustainability, circular economy and urban metabolism,* and *market and sustainability*), giving a background for each as well as proposing future directions for research in these fields.

2. Environment and economy

2.1 State of the art

Economic activities drive urbanization and its associated impacts in the form of greenhouse gas emissions, land-use and land cover change, urban heat island effect, urban design and architecture, social structure and human health, resource management and sustainability, climate change, and loss of ecosystem services and biodiversity (Verma et al., 2020a, 2020b). Economic, geographical drivers refer to the scale at which economic activities take place and their boundary in the global hierarchical structure of human settlements. The resulting trade and commerce, and the flows of material, energy, and services that ensue, generate income which is generally measured as per capita GDP or Gross Regional Product (GRP) (i.e., the GDP normalized at the scale of human settlements), calculated either as an urban total, or normalized on a per capita basis (Seto et al., 2014). Urban areas have been called economic centers of energy consumption (Xia et al., 2015). Sustainability is inherently intertwined with urban growth, since human activities converge in urban areas of high resource consumption. Sustainability needs to be analyzed, since the growth and development associated with human activities is set on a different course than the natural functioning of ecosystems. However, with the right policy measures, urban areas might embrace resilient and resource-efficient pathways for development, including a green economy model based on an understanding of different components of the environment (Verma et al., 2020a, 2020b). In the next section, we discuss in brief the various components of this book aimed at describing the relationship between environmental sustainability and economy.

2.2 Highlights from chapters relevant to the environment and economy theme

The transportation sector is closely associated with economic growth, as it serves one of the essential functions of transporting resources and provides employment to a large number of people. However, the transport sector is also responsible for a significant impact on the environment. It was reported by the International Energy Agency (IEA) that in Organization for Economic Co-operation and Development (OECD) countries transport is responsible for more than half of their energy consumption (Chapter 1). Air and rail transports contribute to higher GHG emissions; however, this impact is not uniform for developed and developing countries. This impact of the transport sector is a major concern for sustainability, as it is an essential part of economic development as well as responsible for environmental impacts (Chapter 2). A major shift in transport technology and resource movement policies is needed to offset its environmental costs. The policy interventions to achieve energy sustainability along with economic growth requires the involvement of economic institutions. More consumption of resources leads to higher energy-related emissions necessitating decoupling of economic growth from energy consumption. The role of economic institutions in this respect needs more attention (Chapter 3).

Chapter 4 discussed the role of pollution, energy consumption, and economic growth. Emissions from urban activities, transport sector and industries, energy generation, and waste management constitute a number of pollutants and GHGs like NOx and SOx, methane, and others. According to the Environmental Kuznets Curve (EKC) theory, during the initial stages of economic growth, the environmental pollution rises rapidly, reaches a stable state for some time, and then decreases when more and more people start getting richer, forming an inverted U-shaped curve (Dasgupta et al., 2002). However, there are several concerns regarding this type of growth. First, the per capita pollutants may fall, but the total amount of pollution increases. Some developed countries have shown an N-shaped curve where, after reaching the U-shape, economic growth leads to much higher emissions. The role of particular pollutants was investigated in Chapter 5. It remains an emerging challenge to assess the effect of economic growth on different types of pollutants. Chapter 19 deals with the various aspects of using EKC hypothesis in environmental research.

The impact of economic growth and human activities is now apparent at the global scale in the form of climate change. The Paris Agreement toward reducing carbon emissions by 2030 to keep global average temperature rise below 2°C introduced the concept of Nationally Determined Commitments (Rogelj et al., 2016). This has given governments the opportunity to take action to mitigate the effects of climate change. As discussed in Chapter 6, the role of governmental institutions toward formulating new strategies for mitigating the impact on the environment and, at the same time, sustaining economic growth is crucial in this regard. However, it has been observed that it is becoming increasingly challenging to achieve sustainable development goals in many parts of the world. This problem was analyzed in Chapter 7 for the sub-Saharan Africa region. The major problem involves creating sustainable energy policies, which can ensure environmental sustainability as well as economic growth in underdeveloped and developing countries.

Evidence-based policy decision is important to develop effective plans of action toward the uptake of environmental and socioeconomic sustainability principles. Integration of scientific knowledge into policy decision is a major challenge, since both are traditionally considered separately (Moldan and Dahl, 2007; Verma and Raghubanshi, 2018). Integration of knowledge and knowledge coproduction requires removal of the ambiguity of assumptions, which may not be applicable uniformly across scientists, policymakers, and other stakeholder groups (Perrotti, 2019). This requires a situation-specific production of knowledge (Brugnach and Ingram, 2012). Apart from policy initiatives, other critical aspects of economic development are the resilience of the environment in the changing economic circumstances and use different ecosystems are put to. This has been dealt with in Chapter 12 in the context of wetlands and agricultural practices followed there. Wetlands contribute toward agricultural practices and livestock maintenance in many parts of the world (Rebelo et al., 2010). However, with changing economic circumstances, ecological components, like wetlands, are being increasingly put under pressure, thus also threatening their ecosystem services and biodiversity. In order to protect such ecological areas of strategic importance, their economic value in terms of their continued existence can be evaluated using both monetary and nonmonetary metrics. In Chapter 12, the use of wetland vegetation in building materials is proposed as a means to enhance their value and promote a resilient and dynamic model of site maintenance.

Thus, there is a need to measure such complex ecosystem functions and explore related economic opportunities. Indicators and modeling approaches are needed to make sound scientific decisions from the perspective of an urban metabolism approach (Verma and Raghubanshi, 2018; Perrotti, 2020). In the next section, a brief account of discussions related to modeling and economic sustainability indicators is given, corresponding to the second theme of the book.

3. Indicators and sustainability
3.1 State of the art

To measure the shift toward a circular and sustainable economy, the environmental, economic, and social dimensions of such a transition need to be measured. Economic institutions are faced with the challenge of achieving sustainable development goals related to the environment, economic production, or social equity. Clarity is needed to develop and select appropriate indicators in this regard, which can be multidimensional and not solely focused on one aspect of sustainability. This struggle is often visible in manufacturing industries (Suhi et al., 2019). In the words of Peter Ducker quoted in Haupt and Hellweg (2019), "What gets measured gets managed" Sustainable and circular economy brings forth the importance of this statement, making it important to have measurable targets and an indicator framework to support such development. However, widely used indicators do not always fully express environmental and economic sustainability in all its facets. This is, for example, reflected in the many critiques of the use of GDP to measure social progress (Haberl et al., 2019), and the growing discourse on the need to craft alterative indicators to assess the contribution of economic growth and material and energy consumption to improve human well-being (Millward-Hopkins et al., 2020). Moreover, apart from measuring, the impact of any approach devised to measure sustainability has to be regularly cross-checked for its suitability over time and in changing socioeconomic circumstances. The primary challenges toward developing such indicator frameworks to measure sustainability lie in lack of empirical studies and data, especially in developing countries (Verma and Raghubanshi, 2018). In a study of certain industries in Bangladesh, a best-worst method was proposed to assess the environmental measures for sustainability. Expert opinion was included in the evaluation of this indicator framework, which helped in selecting the best indicators for environmental sustainability (Suhi et al., 2019). It was found that waste management was the most important indicator of environmental sustainability in industries. Another indicator system proposed for environmental sustainability made use of the fact that reuse, recycling, repair, and remanufacturing hold certain environmental values which can be exploited (Haupt and Hellweg, 2019). This indicator framework accounted for the end-of-life and life cycle of the materials. The integration of socioeconomic aspects with ecological research is essential to creating such indicator frameworks. However, interdisciplinary approaches have several challenges especially when it is hard to identify clear research boundaries among disciplines and the "labelling" of research approaches is difficult or even artificial (Haberl et al., 2019). In some cases, the lack of organic linkages between different approaches in various disciplines can limit scholars' understanding of the nuances of interdisciplinary frameworks. A detailed description of indicator frameworks, the challenges and opportunities toward their development are described in Verma and Raghubanshi (2018), Corredor-Ochoa et al. (2020), and Batalhão and Teixeira (2020). The next section describes the emerging aspects of measuring economic sustainability as discussed in chapters within this theme.

3.2 Highlights from chapters relevant to the indicators and sustainability theme

Production and consumption of goods create waste at different stages. Material recovery is an integral part of creating circular economies (Genovese et al., 2017). Recycling is a direct method of material recovery; however, a rational perspective reveals that it involves recovery at different stages of the product life cycle. The potential for material recovery depends on multiple factors which have been described in Chapter 8. Thus, the economic reasons act as a push-and-pull factor for practical application of such an approach, which involves maturity of recycling technologies, geographical proximity, scalability, and existing marketplaces to minimize the cost and maximize the sustainability of the products. As discussed in Chapter 9, economic indicators can provide "early warning signs for future economic, environmental and social problems."

Further, sustainable economic development needs to move beyond measuring the economic and environmental impacts of economic growth toward a holistic form where the multidimensional aspects of sustainability, namely resilience, social progress, connections between the economic processes, environmental and social well-being, are considered. Thus, a rethinking is required in sustainability indicators of economic development. Such rethinking needs to account for disruptions in economic activities which have an impact on the environment. These can be as simple as disruption in the production line or as complex as road maintenance leading to traffic jams, increased travel times, and more carbon emissions. These disruptions serve as sources of increased inefficiency and are challenging to include in modeling approaches. The subject of Chapter 10 deals with analyzing such modeling approaches where the authors have taken the example of road maintenance sites. Road maintenance leads to disruption in traffic flows.

The next section illustrates examples of alternative, holistic approaches to economic and environmental sustainability, which draw from research into circular economy and urban metabolism, corresponding to the third theme of the book.

4. Circular economy and urban metabolism
4.1 State of the art

Circular economy and urban metabolism are two interdisciplinary research fields that study the functioning of ecological, social, and economic systems as influencing and influenced by a complex series of biophysical and socioeconomic processes, including resource management and decision-making. Circular economy research applied to the valorization of local waste flows is paramount to improve the performance of urban systems in the light of population growth and a variety of other socioeconomic drivers influencing resource demand and waste generation. Recent progress in green economy and industrial ecology has substantially contributed to broadening the methodological spectrum of circular economy research; the number of works on the subject has grown exponentially over the past two decades, showing that the field is undergoing significant development in both science and policy (Schöggl et al., 2020). The growing scientific interest in circular economy frameworks and strategies has also resulted in higher scrutiny of the use of concepts and the narratives deployed, as well as their differences and overlaps with other models in sustainability science and action (Reike et al., 2018; Korhonen et al., 2018). Other critical research questions in the field concentrate on the global impact and contribution of circular economy initiatives toward the UN Sustainable Development Goals (SDG), most notably the Responsible Production and Consumption Goal (SDG 12), and at least four other SDGs on water access, management, and sanitation (SDG 6), affordability of renewable energy sources and energy efficiency (SDG 7), sustainable economic growth (SDG 8), and climate change adaptation and the mitigation of greenhouse gas emissions (SDG 13) (Geng et al., 2019; Schroeder et al., 2018).

The urban metabolism field investigates resource demand and emissions in cities through the quantification of energy, materials, water, and nutrient flows. Cities are studied as socio-ecological systems whose metabolism is the result of the interactions with other—close or remote—anthropogenic systems and the natural environment. The "metabolic" approach to urban sustainability rests on a wide spectrum of analytical tools and methods to assess the resource intensity of urban systems (and the associated waste and pollutant emissions) which bear great potential to inform resource management policy and action at the regional and local scale (Perrotti and Iuorio, 2019). The application of the concept of "metabolism" to societies and cities has a strong interdisciplinary tradition (Haberl et al., 2019). At least three critical moments in the history of the concept as used across social, natural, and engineering sciences are identified in the literature (Newell and Cousins, 2015). Borrowing from Moleschott's (1852) physiological materialism and chemist Justus von Liebig's (1842, 1859) idea of "metabolic rift," Marx and Engels used the concept of "metabolism" to describe the "material exchange" relation between man and nature and the role of the human labor in it. The metaphor of metabolism was implicitly present almost from the outset of urban sociology in the 1920s, with the human ecology of the Chicago School studying the city as an ecosystem in analogy to natural systems (Burgess, 1925). However, it was the rise of environmental awareness and the increase in cultural acceptability of a critical view of economic growth in the late 1960s that brought forward a renewed interest in the idea of metabolism as applied to cities (Wolman, 1965). Finally, the late 1990s saw a rejuvenation of urban metabolic studies, with the "limits to growths" narrative becoming mainstream of the engineering, social, and environmental sciences; this led to the recognition of urban metabolism as a well-established interdisciplinary theme in sustainability science resting on a rich collection of diversified analytical strategies and modeling tools (Kennedy et al., 2011). The field has gained significant momentum over the past years through interdisciplinary contributions spanning from a broad range of disciplines in ecological science (Perrotti 2020) and a growing number of applications of "urban metabolism thinking" in spatial planning and design (Perrotti, 2019; Galan and Perrotti, 2019; Perrotti and Stremke, 2020).

Social ecology and urban ecology have produced the most influential lines of research in metabolic studies. They can be described as "archipelagos" of disciplinary islands and are influenced by a wide range of disciplines in the humanities and social sciences (human geography, political ecology, environmental history, ecological economics), the natural sciences (land-use science, ecosystem, and landscape ecology), and engineering (industrial ecology and civil and environmental engineering) (Fischer-Kowalski and Weisz, 2016; Cadenasso and Pickett, 2013). Both fields have been most influential in bridging multidisciplinary, conceptual approaches to society—nature coevolution pertaining to history, current development, and future sustainability transitions. They have catalyzed efforts to bridge the "great divide" between engineering and the natural sciences on the one hand, and the social sciences and the humanities on the other hand, inevitably engendered by the conceptualization of society and nature as two separate entities in relation. Main directions for future research to advance urban metabolism discussed in the recent literature include promoting continual dialogue between "metabolic" research communities to overcome conceptual and methodological silos and identifying opportunities for epistemological and methodological coordination and synthesis; fostering a deeper understanding of the interdependence between biophysical and socioeconomic aspects of urban material and energy balances; and strengthening the link between metabolic analysis and implementation of real-world solutions to stakeholders' concerns, including advanced design strategies (Perrotti, 2020).

4.2 Highlights from chapters relevant to circular economy and urban metabolism theme

The processes by which "value systems" are generated and studied, as opposed to the more traditional conceptualization of "value chains" are some of the most critical frontiers in circular economy research (Babbitt et al., 2018). Only a few empirical studies to date focus on the generation of "circular" value systems and, in most cases, they lack a comprehensive conceptual foundation. Chapter 11 has embraced the challenge of developing a novel framework through which social-ecological systems can be empirically studied based on their capacity to generate "circular" value. The proposed analytical framework provides a three-dimensional categorization system including agency (actors and their interactions), biophysical components (material and energy flows as well as spaces and artifacts), and framing elements (domains, scale, and rules). Using a comparative lens, the framework is applied to three case studies relative to the sustainable management of three different types of urban material flows (excavated earth, urban wood, and coffee grounds) which are understood as multilayered value systems. Through the discussion of the framework application, the chapter has provided novel insights on value creation in circular economy initiatives, while bridging theoretical approaches and empirical shreds of evidence in metabolic research.

Chapter 12 has opened up novel pathways into the analysis and assessment of the processes underpinning the local cycling of natural resources and the generation of construction-material value chains. It presented a twofold approach to study opportunities for the recovery of four different kinds of plant fibers from a protected wetland area in Normandy, France, and their use as an insulation material for construction purposes (earth-based composites). First, a range of strategies for generating novel value chains are assessed through the identification of territorial resources, agents, and dynamics using "resilience thinking" as an overarching framework and an alternative to conventional approaches to natural resource management (Folke et al., 2010). Consideration of the ecological values of the studied wetland landscape (maintenance of the protected landscape and preservation of the ecosystem services this provides) is combined with the appraisal of socioeconomic factors that can increase the viability of the use of the fibers for the local farms. Second, characterization of the mechanical, physical, and hygrothermal properties of the local wetland plants is conducted through experimental lab testing. The integrated approach allows for studying the circulations and reusing of locally sourced bio-based materials based on their potential to leverage the resilience of the regional system from an environmental, socioeconomic, and cultural perspective.

Finally, Chapter 13 has addressed the limitations that traditional circular economy tools might bear when it comes to supporting decision-making across environmental and economic policy areas. It made the case that, when aiming at replacing a traditional construction material with a more "sustainable" one, recycled options do not automatically lead to better overall sustainability; indeed, indirect effects on the market must also be considered when introducing a new product in a well-established industry sector. Traditional analysis tools to assess direct environmental impacts throughout a product's life, such as Life Cycle Assessment (or "attributional-LCA" ALCA), do not allow to take into consideration the consequences of a change in supply and demand chains (Vázquez-Rowe et al., 2013). The study has illustrated the limits of ALCA by concentrating on the environmental consequences and rebound effects occurring when substituting ordinary Portland Cement (responsible for 10% of global carbon emissions) with stainless steel-slag blocks (an alkaline residue from industrial processes). Consequential Life Cycle Assessment (CLCA) is employed as an alternative tool to ALCA and the Flemish construction industry, in Belgium, is an example. Two different markets affected by the stainless steel-slag blocks production (nonstructural construction blocks and low-quality aggregates) are considered in parallel together with impacts on marginal suppliers. The study has shown that if environmental impacts are reduced through minimizing the use of concrete blocks, the new recycled product requires additional natural aggregates for low-value applications, adding to the complexity of policy support exercises toward "sustainable" decision-making. The next section discusses the several levels of intertwinement that exist between the role of the market and supply/demand chains and the sustainability of technologies, products, and policies (corresponding to the fourth theme of the book).

5. Market and sustainability

The mitigation efforts for the impact of climate have grown to encompass climate finance as its integral part. The inclusion of climate finance in the United Nations Framework Convention on Climate Change (UNFCCC, 1992) stated the importance of Climate Finance arising out of a need for trans-national and multiinstitutional cooperation. Climate finance refers to "local, national or transnational financing—drawn from public, private and alternative sources of financing" (UNFCCC website) to mitigate the impact of climate change and develop adaptation strategies. Rapid development has given rise to megacities, and urban expansion has also resulted in the intensification of economic activities, consumption, and waste production (Verma and Raghubanshi, 2018). This has resulted in a number of challenges for the environment. From the perspective of sustainability, a clear institutional framework is needed, which can collate the biophysical components and the various processes of the earth. Urban areas contribute to about 70% of greenhouse gas emissions while covering only about 2.7% of the global land surface (UN-HABITAT, 2011). The role of climate finance has garnered greater support due to the rapid urban growth, pollution, and environmental destruction. The role of institutions in supporting sustainable development can bridge the gap between consumption and sustainability by creating new markets for sustainable technology, products, and policies. One of the four themes addressed in this book consists in exploring the role of market mechanisms and finance for sustainable development.

5.1 State of the art

The market and sustainability concept has evolved from a number of components into certain broader thematic areas. From the year 2008 to 2013, various thematic areas like "conservation," "trade," "sustainable development," "growth," "city," "products," "management policy," "life-cycle assessment," "innovation," and "reduction" strategies diversified and recombined into much more focused research fields in 2014–17 (Fig. 17.2). The thematic evolution showed diversification of thematic areas of "policy," "management," "conservation," "city," "trade," "sustainable development," and "growth," whereas, "innovation" merged into "management," and "life-cycle assessment" and "reduction" merged into "energy" during the 2008–13 period. Some of the "conservation"-based studies also focused toward "impacts" in 2014–17, whereas, the "conservation" theme focused on adopting a systems approach in 2018–20. "Trade" diversified into "conservation," "sustainable development," and "impacts" in 2014–17. The "sustainable development" and "policy" themes showed the largest diversification during the 2008–17 period. In 2014–17, the "conservation," "policy," and "energy" themes were studied from the perspective of "sustainable development." In 2018–20, "sustainable development" was explored from the perspective of "climate change,"

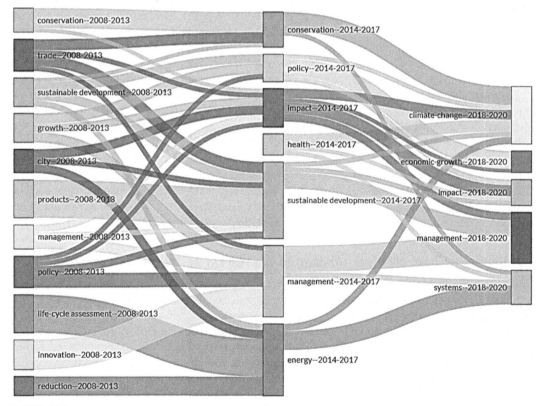

FIGURE 17.2

Thematic evolution of economic and environmental sustainability themes for the 2008–20 time period.

"impacts," and "management." This indicated evolution of the "sustainable development" theme was the focus of "management," "impacts," and "climate change" studies. "Life-cycle assessment" and "reduction" themes were mainly focused on "energy" in the 2014—17 period, which further evolved into "systems" approach and studies on "climate change" 2018—20. Studies encompassing the "policy" theme diversified into "impacts," "sustainable development," and "management" in 2014—17, apart from the "policy" theme. In the 2018—20 period, studies on "policy" aspects of markets and sustainability were mainly approached from the perspectives of "climate change," "impacts," and "economic growth." This was also very significant as it showed the evolution of "policy" studies into one of the most pressing problems in the world today, that is, "climate change." Another significant aspect of the thematic evolution was the emergence of the "health" theme in the 2014—17 period in the market and sustainability field. This showed the direction of economic growth and sustainability toward health. This theme further diversified into "climate change," "impacts," and "systems."

Further analysis of its components showed that the field was concerned with human health, the impact of pollution due to economic activities in different manufacturing industries, healthy environment, and the impact of climate change on such activities. The emergence of "system" in the 2018—20 period also indicated the maturing of this field and inclusion of different components into studies encompassing the economic and environmental sustainability. The thematic evolution showed that the number of themes has reduced and become more focused; there are indications of the development of systems approach in analyzing the economic and environmental sustainability; various aspects of sustainable development are being integrated into economic mechanisms, and vice-versa, which would help in coproduction of knowledge and integrated frameworks.

Coproduction and integrative frameworks are deemed essential to generate knowledge, which can be used to create sustainability practices. Some of the most recent studies focused on different aspects of environmental and economic sustainability are aimed toward creating integrated mechanisms for real estate market analysis (Kauskale and Geipele, 2017) and studying the impact of rural development and sustainability practices (Cole and Ingalls, 2020), various aspects of different industries like energy and natural gas (Yang, 2018), power generation (Dong et al., 2019), multinational industries, corporate sustainability and emerging markets (Park, 2018), emerging green markets, green entrepreneurship (Lofti et al., 2018), tourist market and sustainable travel (Kastenholz et al., 2018), and sustainable food markets like seafood and their impact on the social-ecological systems (Travaille et al., 2019).

5.2 Highlights from chapters relevant to the market and sustainability theme

Mitchell (2010) showed by empirical modeling that the Sustainable Market Orientation could help create corporate benefits when it focuses on the economic, social, and ecological sustainability. The role of sustainable market supply chains is also very important. A shift in environmental economics and sustainability has been observed by Vermeulen and Seuring (2009) in four ways. They have stated that due to development of newer models of environmental politics, environmental policies are increasingly being embedded into sustainable development with producer's responsibility toward the community as well as the distribution of economic benefits arising out of the resource utilization. They have also observed that the producers are increasingly being held responsible for the impact of economic activities on the society, and this responsibility is shared along the supply chain with the driving force in the form of government agencies and the direct consumers.

The role of governance, social awareness, and lack of implementation of environment protection measures are a significant drawback in order to reduce the role financial institutions and the nongovernmental organizations have become very important. The role of financial mechanisms from the aspect of mitigating climate change impacts has been the focus of Chapter [15]. Economic growth, along with sustainability practices, has been the focus of green growth practices. There is a need to create a shift in the conventional economic policies, especially in developing and developed countries, toward inclusive green growth. Thus is the focus of Chapters [14 and 16]. The authors have examined the conventionally used macroeconomic instruments in green economy accounting and discussed the green economic policies practiced in several countries in the European Union, Asia, and the United States of America.

6. Conclusion and common challenges across the four themes

The book has explored multiple aspects of "Environmental Sustainability and Economy" from four complementary perspectives: *Environment and economy, Indicators and sustainability, Circular Economy and urban metabolism,* and *Market and sustainability.* Several common challenges lie across economic, environmental, and social dimensions to sustainability. For example, the markets can help in creating sustainability practices and understanding the impact of such policies on management and governance. However, there is a need for a broader view of economic strategies due to the adverse impact of economic activities on the ecological and socioeconomic systems, such as the forest destruction in Amazon river basin, disappearance of fisheries, the union carbide chemical plant disaster in Bhopal, India, and many others (Mitchell, 2010).

Another challenge lies in the "opportunity cost" which arises due to lack of understanding and integration of the ecosystem and economic systems (Fig. 17.3). A considerable opportunity cost is incurred due to excessive exploitation of resources to fulfill short-term economic or social benefits. This lack of understanding arises due to differing perceptions of the ecosystems and the micro- and macroeconomies. When the macroeconomy is considered as the all-encompassing system, containing the microeconomies and the environmental and ecological components (natural resources and their

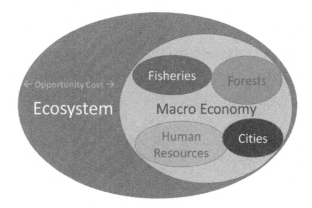

FIGURE 17.3

Challenges of economic sustainability—integration of micro- and macroeconomy within an ecosystem metabolism.

supply chains), the value of the ecosystem is diminished thereby ignoring the ecosystems services and a large number of benefits provided by them. This opportunity cost can be recovered by the inclusion and application of sustainable development practices in economic policies. This integration needs to happen at different scales of supply chains, manufacturing, governance, and social awareness. Future research should be directed toward a more far-reaching integration of the ecological, social, and economic sustainability. This should be based on fine-tuned understanding the metabolic pathways of different human systems, their material and energy fluxes, and impacts on the environment, as well as the cultural embeddedness of the heterogeneous forms of *oiko* (household)-*nomia* (management) embraced by societies across the world.

References

Babbitt, C.W., Gaustad, G., Fisher, A., Chen, W.Q., Liu, G., 2018. Closing the loop on circular economy research: from theory to practice and back again. Resour. Conserv. Recycl. 135, 1–2.

Batalhão, A.C., Teixeira, D., 2020. Cities management and sustainable development: monitoring and assessment approach. In: Urban Ecology. Elsevier, pp. 335–354.

Brand, U., 2012. No lessons learned from failures of implementing sustainable development. Ecol. Perspect. Sci. Soc. 21 (1), 28–32.

Brugnach, M., Ingram, H., 2012. Ambiguity: the challenge of knowing and deciding together. Environ. Sci. Pol. 15 (1), 60–71.

Burgess, E.W., 1925. The growth of the city: an introduction to a research project. Public. Am. Sociol. Soc. XVIII, 85–97.

Cadenasso, M.L., Pickett, S.T.A., 2013. Three tides: the development and state of the art of urban ecological science. In: Pickett, S.T.A., Cadenasso, M.L., McGrath, B. (Eds.), Resilience in Ecology and Urban Design. Springer, Cham, pp. 29–46.

Cole, R., Ingalls, M.L., 2020. Rural revolutions: socialist, market and sustainable development of the countryside in Vietnam and Laos. In: The Socialist Market Economy in Asia. Palgrave Macmillan, Singapore, pp. 167–194.

Corredor-Ochoa, Á., Antuña-Rozado, C., Fariña-Tojo, J., Rajaniemi, J., 2020. Challenges in assessing urban sustainability. In: Urban Ecology. Elsevier, pp. 355–374.

Crépin, A.S., Folke, C., 2015. The economy, the biosphere and planetary boundaries: towards biosphere economics. Int. Rev. Environ. Resour. Econ. 8 (1), 57–100.

Dasgupta, S., Laplante, B., Wang, H., Wheeler, D., 2002. Confronting the environmental Kuznets curve. J. Econ. Perspect. 16 (1), 147–168.

Dong, J., Liu, D., Wang, D., Zhang, Q., 2019. Identification of key influencing factors of sustainable development for traditional power generation groups in a market by applying an extended MCDM model. Sustainability 11 (6), 1754.

Fischer-Kowalski, M., Weisz, H., 2016. The archipelago of social ecology and the island of the Vienna School. In: Haberl, H., Fischer-Kowalski, M., Krausmann, F., Winiwarter, V. (Eds.), Social Ecology. Society-Nature Relations across Time and Space. Springer, Cham, pp. 3–28.

Folke, C., Carpenter, S.R., Walker, B., Scheffer, M., Chapin, T., Rockström, J., 2010. Resilience thinking: integrating resilience, adaptability and transformability. Ecol. Soc. 15 (4), 20.

Galan, J., Perrotti, D., 2019. Incorporating metabolic thinking into regional planning: the case of the Sierra Calderona Strategic Plan. Urban Plann. 4 (1), 152–171.

Geng, Y., Sarkis, J., Bleischwitz, R., 2019. How to globalize the circular economy. Nature 565, 153–155.

Genovese, A., Acquaye, A.A., Figueroa, A., Koh, S.L., 2017. Sustainable supply chain management and the transition towards a circular economy: evidence and some applications. Omega 66, 344–357.

Glazyrina, I.P., Faleichik, L.M., Yakovleva, K.A., 2015. Socioeconomic effectiveness and "green" growth of regional forest use. Geogr. Nat. Resour. 36 (4), 327−334.

Haberl, H., Wiedenhofer, D., Pauliuk, S., Krausmann, F., Müller, D.B., Fischer-Kowalski, M., 2019. Contributions of sociometabolic research to sustainability science. Nat. Sustain. 2, 173−184.

Habitat, U.N., 2011. Hot Cities: The Battle-Ground for Climate Change. UN Sustainable Development Goals.

Haupt, M., Hellweg, S., 2019. Measuring the environmental sustainability of a circular economy. Environ. Sustain. Indicat. 1, 100005.

Kastenholz, E., Eusébio, C., Carneiro, M.J., 2018. Segmenting the rural tourist market by sustainable travel behaviour: insights from village visitors in Portugal. J. Destinat. Market. Manag. 10, 132−142.

Kauškale, L., Geipele, I., 2017. Integrated approach of real estate market analysis in sustainable development context for decision making. Procedia Eng. 172, 505−512.

Kennedy, C., Pincetl, S., Bunje, P., 2011. The study of urban metabolism and its applications to urban planning and design. Environ. Pollut. 159 (8−9), 1965−1973.

Korhonen, J., Honkasalo, A., Seppälä, J., 2018. Circular economy: the concept and its limitations. Ecol. Econ. 143, 37−46.

Lotfi, M., Yousefi, A., Jafari, S., 2018. The effect of emerging green market on green entrepreneurship and sustainable development in knowledge-based companies. Sustainability 10 (7), 2308.

Millward-Hopkins, J., et al., 2020. Providing decent living with minimum energy: a global scenario. Global Environ. Change 65, 102168. https://doi.org/10.1016/j.gloenvcha.2020.102168.

Mitchell, R.W., Wooliscroft, B., Higham, J., 2010. Sustainable market orientation: a new approach to managing marketing strategy. J. Macromarket. 30 (2), 160−170.

Moldan, B., Dahl, A.L., 2007. Challenges to sustainability indicators. Sustain. Indicat. 1.

Najam, A., Halle, M., 2010. Global Environmental Governance: The Challenge of Accountability. Sustainable Development Insights, No. 5. Boston University, Boston. The Frederick S. Pardee Center for the Study of the Longer-Range Future.

Newell, J.P., Cousins, J.J., 2015. The boundaries of urban metabolism: towards a political−industrial ecology. Prog. Hum. Geogr. 9 (6), 702−728.

Moleschott, J., 1852. Der krcislauf des lebens: physiologische antworten auf Liebig's chemische briefe.

Park, S.B., 2018. Multinationals and sustainable development: does internationalization develop corporate sustainability of emerging market multinationals? Bus. Strat. Environ. 27 (8), 1514−1524.

Perrotti, D., 2019. Evaluating urban metabolism assessment methods and knowledge transfer between scientists and practitioners: a combined framework for supporting practice-relevant research. Environ. Plann. B 46 (8), 1458−1479.

Perrotti, D., 2020. Urban metabolism: old challenges, new frontiers, and the research agenda ahead. In: Urban Ecology. Elsevier, pp. 17−32.

Perrotti, D., Iuorio, O., 2019. Green infrastructure in the space of flows: an urban metabolism approach to bridge environmental performance and user's wellbeing. In: Lemes de Oliveira, F., Mell, I. (Eds.), Planning Cities with Nature: Theories, Strategies and Methods. Springer, pp. 265−277.

Perrotti, D., Stremke, S., 2020. Can urban metabolism models advance green infrastructure planning? Insights from ecosystem services research. Environ. Plann. B 47 (4), 678−694.

Rebelo, L.M., McCartney, M.P., Finlayson, C.M., 2010. Wetlands of sub-Saharan Africa: distribution and contribution of agriculture to livelihoods. Wetl. Ecol. Manag. 18 (5), 557−572.

Reike, D., Vermeulen, W.J.V., Witjes, S., 2018. The circular economy: new or Refurbished as CE 3.0? — exploring controversies in the conceptualization of the circular economy through a focus on history and resource value retention options. Resour. Conserv. Recycl. 135, 246−264.

Rockström, J., Steffen, W., Noone, K., Persson, Å., Chapin, F.S., Lambin, E.F., et al., 2009. A safe operating space for humanity. Nature 461 (7263), 472−475.

Rogelj, J., Den Elzen, M., Höhne, N., Fransen, T., Fekete, H., Winkler, H., et al., 2016. Paris Agreement climate proposals need a boost to keep warming well below 2 C. Nature 534 (7609), 631–639.

Schöggl, J.P., Stumpf, L., Baumgartner, R.J., 2020. The narrative of sustainability and circular economy - a longitudinal review of two decades of research. Resour. Conserv. Recycl. 163, 105073.

Schroeder, P., Anggraeni, K., Weber, U., 2018. The relevance of circular economy practices to the sustainable development goals. J. Ind. Ecol. 23 (1), 1–19.

Seto, K.C., Dhakal, S., Bigio, A., Blanco, H., Delgado, G.C., Dewar, D., Ramaswami, A., 2014.). Human settlements, infrastructure and spatial planning.

Steffen, W., Richardson, K., Rockström, J., Cornell, S.E., Fetzer, I., Bennett, E.M., et al., 2015. Planetary boundaries: guiding human development on a changing planet. Science 347 (6223).

Suhi, S.A., Enayet, R., Haque, T., Ali, S.M., Moktadir, M.A., Paul, S.K., 2019. Environmental sustainability assessment in supply chain: an emerging economy context. Environ. Impact Assess. Rev. 79, 106306.

Travaille, K.L.T., Lindley, J., Kendrick, G.A., Crowder, L.B., Clifton, J., 2019. The market for sustainable seafood drives transformative change in fishery social-ecological systems. Global Environ. Change 57, 101919.

Towards a Green Economy: Pathways to Sustainable Development and Poverty Eradication, 2011. UNEP. URL: http://www.unep.org/greeneconomy/Portals/88/documents/ger/ger_final_dec_2011/Green%20EconomyRep ort_Final_Dec2011.pdf. (Accessed 4 June 2015).

UNFCCC, 1992. United Nations Framework Convention on climate change. United Nations. Available at. https:// unfccc.int/.

Vázquez-Rowe, I., Rege, S., Marvuglia, A., Thénie, J., Haurie, A., Benetto, E., 2013. Application of three independent consequential LCA approaches to the agricultural sector in Luxembourg. Int. J. Life Cycle Assess. 18, 1593–1604.

Verma, P., Raghubanshi, A.S., 2018. Urban sustainability indicators: challenges and opportunities. Ecol. Indicat. 93, 282–291.

Verma, P., Singh, R., Bryant, C., Raghubanshi, A.S., 2020a. Green Space Indicators in a Social-Ecological System: A Case Study of Varanasi, India. Sustainable Cities and Society, p. 102261.

Verma, P., Singh, R., Singh, P., Raghubanshi, A.S., 2020b. Urban ecology—current state of research and concepts. In: Urban Ecology. Elsevier, pp. 3–16.

Vermeulen, W.J., Seuring, S., 2009. Sustainability through the market-the impacts of sustainable supply chain management: introduction. Sustain. Dev. 17 (5), 269–273.

von Leibig, J., 1842. Animal chemistry or organic chemistry in its application to physiology and pathology [translated by W. Gregory]. Taylor and Walton, London, England, p. 144.

von Liebig, J.F., 1859. Familiar letters on chemistry: in its relations to physiology, dietetics, agriculture, commerce, and political economy. Walton and Maberly.

Wolman, A., 1965. The metabolism of cities. Sci. Am. 213 (3), 179–190.

Xia, X.H., Hu, Y., Alsaedi, A., Hayat, T., Wu, X.D., et al., 2015. Structure decomposition analysis for energy-related GHG emission in Beijing: urban metabolism and hierarchical structure. Ecol. Inform. 26, 60–69.

Yang, J., 2018. Analysis of sustainable development of natural gas market in China. Nat. Gas. Ind. B 5 (6), 644–651.

Zabortseva, T.I., Kuznetsova, A.N., Violin, S.I., 2017. The potential of a "green" economy in the socioeconomic development of Irkutsk Oblast. Geogr. Nat. Resour. 38 (4), 379–385.

Further reading

Najam, A., Runnals, D., Halle, M., 2007. Environment and Globalization: Five Propositions. International Institute for Sustainable Development (IISD), Winnipeg, MB.

Index

Note: 'Page numbers followed by "f" indicate figures and "t" indicate tables.'

Printed in the United States
by Baker & Taylor Publisher Services